# Condensed Matter Research Using Neutrons

## Today and Tomorrow

# NATO ASI Series

**Advanced Science Institutes Series**

*A series presenting the results of activities sponsored by the NATO Science Committee, which aims at the dissemination of advanced scientific and technological knowledge, with a view to strengthening links between scientific communities.*

The series is published by an international board of publishers in conjunction with the NATO Scientific Affairs Division

| | |
|---|---|
| **A Life Sciences**<br>**B Physics** | Plenum Publishing Corporation<br>New York and London |
| **C Mathematical and Physical Sciences** | D. Reidel Publishing Company<br>Dordrecht, Boston, and Lancaster |
| **D Behavioral and Social Sciences**<br>**E Engineering and Materials Sciences** | Martinus Nijhoff Publishers<br>The Hague, Boston, and Lancaster |
| **F Computer and Systems Sciences**<br>**G Ecological Sciences** | Springer-Verlag<br>Berlin, Heidelberg, New York, and Tokyo |

***Recent Volumes in this Series***

*Volume 108*—Polarons and Excitons in Polar Semiconductors and Ionic Crystals
edited by J. T. Devreese and F. Peeters

*Volume 109*—Percolation, Localization, and Superconductivity
edited by Allen M. Goldman and Stuart A. Wolf

*Volume 110*—Quantum Electrodynamics and Quantum Optics
edited by A. O. Barut

*Volume 111*—Monopole '83
edited by James L. Stone

*Volume 112*—Condensed Matter Research Using Neutrons: Today and Tomorrow
edited by Stephen W. Lovesey and Reinhard Scherm

*Volume 113*—The Electronic Structure of Complex Systems
edited by P. Phariseau and W. M. Temmerman

*Volume 114*—Energy Transfer Processes in Condensed Matter
edited by Baldassare Di Bartolo

***Series B: Physics***

# Condensed Matter Research Using Neutrons

## Today and Tomorrow

Edited by

Stephen W. Lovesey

Rutherford Appleton Laboratory
Chilton, Oxfordshire, United Kingdom

and

Reinhard Scherm

PTB
Braunschweig, Federal Republic of Germany

Plenum Press
New York and London
Published in cooperation with NATO Scientific Affairs Division

Proceedings of a NATO Advanced Research Workshop on
Condensed Matter Research Using Neutrons: Today and Tomorrow,
held March 26-29, 1984, in Abingdon, Oxfordshire, United Kingdom



Library of Congress Cataloging in Publication Data

NATO Advanced Research Workshop on Condensed Matter Research Using Neutrons: Today and Tomorrow (1984: Abingdon, Oxfordshire)
Condensed matter research using neutrons—today and tomorrow.

(NATO ASI series. Series B, Physics; v. 112)
"Proceedings of a NATO Advanced Research Workshop on Condensed Matter Research Using Neutrons: Today and Tomorrow, held March 26-29, 1984, in Abingdon, Oxfordshire, United Kingdom"—T.p. verso.
"Published in cooperation with NATO Scientific Affairs Division."
Includes bibliographical references and index.
1. Condensed matter—Congresses. 2. Neutrons—Scattering—Congresses. I. Lovesey, S. W. (Stephen W.) II. Scherm, Reinhard. III. Title. IV. Series.
QC173.4.C65N37 1984 530.4'1 84-17807
ISBN 0-306-41821-5


A Division of Plenum Publishing Corporation
233 Spring Street, New York, N.Y. 10013



Printed in the United States of America

# PREFACE

The Advanced Research Workshop (ARW) on Condensed Matter Research Using Neutrons, Today and Tomorrow was held in Abingdon, Oxfordshire for four days beginning 26 March 1984. The Workshop was sponsored by NATO and the Rutherford Appleton Laboratory. A total of 32 lecturers and participants attended.

An objective of the Workshop was to review some dynamic properties of condensed matter that can be studied using neutron spectroscopy. A second objective, no less important, was to identify new topics that might be investigated with advanced spallation neutron sources. The twelve lectures reproduced in this volume bear witness, largely by themselves, to the success of the Workshop in meeting these objectives. The many discussions generated by lecturers and participants meant that, in the event, the objectives were indeed amply satisfied. I should like to thank all those who attended the Workshop for their part in making it so beneficial and rewarding.

I am most grateful to Reinhard Scherm, who acted as my advisor in the organisation of the Workshop. The efforts of Mrs. M. Sherwen and Miss J. Warren made light my burden of administrative duties. The preparation of the manuscript for publication was simplified by the assistance of Miss C. Monypenny.

Stephen W. Lovesey

Rutherford Appelton Laboratory
Oxfordshire
Director of the Advanced
Research Workshop

CONTENTS

# MECHANISMS FOR THE DYNAMICS OF PHASE TRANSFORMATIONS

Kurt Binder
Institut für Physik
Johannes-Gutenberg Universität
Postfach 3980
D-6500 Mainz, W-Germany

## ABSTRACT

An introductory review of the dynamics of (first-order) phase transitions is given. Concepts describing the initial stages of the transition, such as nucleation, spinodal decomposition (in the case of unmixing) are introduced, and their validity is critically discussed. The theoretical results are compared to recent computer simulations and pertinent experiments. Then the scaling concepts describing the late stages of domain growth are discussed, and open problems are outlined.

## I. INTRODUCTION

We consider the dynamics of phase transformations, induced by a sudden change of external parameters of the system.[1-8] For instance, we consider a system which is suddenly cooled from the disordered phase to a state below the critical temperature of an order-disorder transition (Fig.1a). Then the initially disordered system is unstable, and immediately small ordered domains will form. As the time t after the quench passes, the order parameter in these domains must reach its equilibrium value, and the domain size L(t) must grow ultimately to macroscopic size. In binary alloys or in chemisorbed layers, for instance, this domain growth is diffusion limited and hence slow enough to be observed experimentally by scattering techniques.

Alternatively, we treat a binary system AB with a miscibility gap (Fig.1b). "Quenching" the system at time t=0 from an equilibrium state in the one-phase region to a state underneath the coexistence curve leads to phase separation: in thermal equilibrium, macroscopic regions of both phases with concentrations $c^{(1)}_{coex}$, $c^{(2)}_{coex}$ coexist. Of course, there occur also phase changes combining an ordering process (such as in Fig.1a) with an unmixing process (such as in Fig.1b); being interested mainly in the basic concepts of these processes we shall not consider such more complex phenomena explicitly.

In all these cases the system develops towards equilibrium through states far from equilibrium. These states are inhomogeneous at certain length scales. The final equilibrium is inhomogeneous at macroscopic scales and is reached only for $t \to \infty$ in the thermodynamic limit $N \to \infty$. This problem is of interest in the context of irreversible

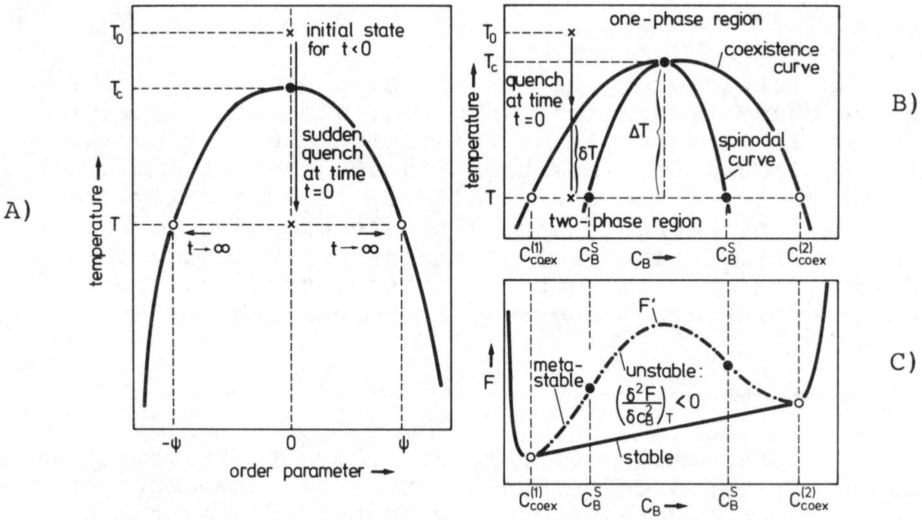


Fig.1: A) Order parameter of a second-order phase transition plotted vs. temperature.
B) Phase diagram of a binary system with a critical temperature $T_c$ of unmixing, in the temperature-concentration plane. The quenching distances from the coexistence curve and from $T_c$ are denoted as $\delta T$ and $\Delta T$, respectively. C) Free energy versus composition at temperature T (schematic). From Ref.7.

thermodynamics, and has practical applications in metallurgy, surface science, etc.

The basic ideas for the kinetic mechanisms of phase separation stem from a discussion of the free energy F (Fig.1c). F decreases first with concentration $c_B$ in the one-phase region, due to the entropy of mixing, while it varies linearly with $c_B$ in the two-phase region. There the amounts of the two coexisting phases change. Theories of the mean-field type suggest now that one can also introduce a free energy $F' \geqq F$ describing one-phase states within the two-phase region. The regime where $F' > F$ but where still $(\delta^2 F'/\delta c_B^2)_T > 0$ is called "metastable" (Fig.1c). The regime where $(\delta^2 F'/\delta c_B^2)_T < 0$ is called "unstable". The locus of inflection points $(\delta^2 F'/\delta c_B^2)_T = 0$ is the "spinodal curve" $c_B^s$ (T).

As we shall see later, in general a unique F' cannot be defined, and hence such a sharp distinction between metastable and unstable states does not exist. At this moment, however, we disregard this problem. Then it is natural to postulate two different transformation mechanisms, Fig.2: in the unstable regime,

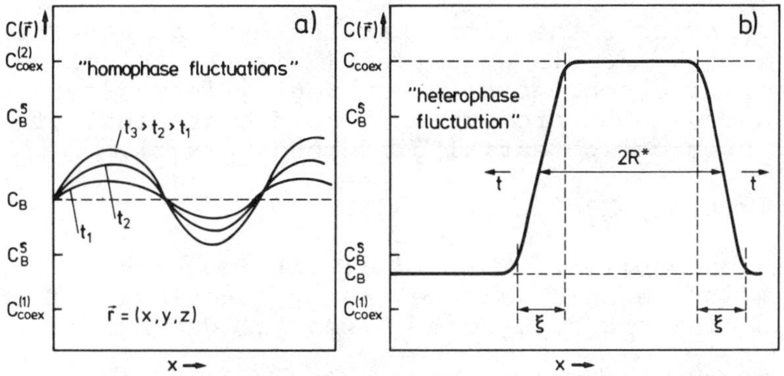


Fig.2: Unstable fluctuations in the two-phase regime: the local concentration $c(\vec{x})$ where $\vec{x} = (x,y,z)$ is plotted vs. the coordinate x in the unstable regime (a) and in the metastable regime (b), at some time(s) after the quench. The radius of a critical droplet is denoted as $R^*$ and the correlation length of concentration fluctuation as $\xi$. From Ref.7.

long-wavelength small-amplitude delocalized concentration fluctuations ("homophase fluctuations") will grow with time after the quench, and thus achieve an unmixing on a local scale. As a second step, the inhomogeneous concentration distribution coarsens. This mechanism is called "spinodal decomposition". In the metastable regime, however, localized large-amplitude fluctuations ("heterophase fluctuations" or "droplets (clusters) of the new phase") must form in order to start the transformation. This mechanism is called nucleation".

In these lectures, we shall first consider the theory of spinodal decomposition in its linear version (Sec.II) and beyond (Sec.III). Nucleation theory (Sec.-IV) and a discussion of the significance of spinodal lines (Sec.V) are given next. We conclude with a discussion of the late stages of domain growth (Sec.VI) and characterize open problems (Sec.VII).

## II. CONTINUUM THEORY OF SPINODAL DECOMPOSITION (CAHN-HILLIARD THEORY[9])

The local concentration $c(\vec{x},t)$ of the binary system satisfies the continuity equation

$$\delta c(\vec{x},t)/\delta t + \nabla \cdot \vec{j}(\vec{x},t) = 0, \tag{1}$$

which expresses the fact that the average concentration $\int d\vec{x} c(\vec{x},t)/V = c_B$ is conserved (V is the volume of the system). The concentration current $\vec{j}$ introduced in Eq.(1) is assumed to be proportional to the gradient of the local chemical potential difference $\mu(\vec{x},t)$,

$$\vec{j}(\vec{x},t) = -M\nabla\mu\ (\vec{x},t), \tag{2}$$

M being a mobility. To find $\mu(\vec{x},t)$ in terms of the $c(\vec{x},t)$, one assumes a free energy functional $\Delta\mathfrak{F}$ of the Landau-Ginzburg type form [d=spatial dimensionality]

$$\Delta\mathfrak{F}/T = \int d^d x\ \{f[c(\vec{x})]/T + \frac{1}{2}\ r^2 [\nabla c(\vec{x})]^2\}, \tag{3}$$

where r is the range of the interaction and f is the density of F' (Fig.1c). According to mean-field theory of an Ising model of a mixture, f(c) would be given by

$$\frac{1}{T}f(c) = c\ln c + (1-c)\ \ln\ (1-c) + 2(T_c/T)\ c(1-c), \tag{4}$$

which near $T_c$ is of the Ginzburg-Landau form

$$f(c)=f_o+A(c-c_B^{crit})^2+B(c-c_B^{crit})^4+\ldots, A<0,\ B>0. \qquad (5)$$

From Eq.(3) $\mu(\vec{x})$ follows from the functional derivative,

$$\mu(\vec{x}) \equiv \delta(\Delta\mathcal{F})/\delta c(\vec{x})=(\partial f/\partial c)_T - r^2T\nabla^2c(\vec{x}). \qquad (6)$$

Eqs.(1,2,6) then yield the Cahn-Hilliard equation[9]

$$\partial c(\vec{x},t)/\partial t=M\nabla^2\left[(\partial f(c(\vec{x},t))/\partial c)_T - r^2T\nabla^2c(\vec{x},t)\right]. \qquad (7)$$

Due to its nonlinearity it is not analytically soluble. Thus one assumes that in the initial stages of unmixing $c(\vec{x},t)$ <u>everywhere</u> is not much different from $c_B$ (Fig.2a); the validity of this assumption will be studied in Sec.V. With this assumption, Eq.(7) reduces by linearization around $c_B$ to

$$\partial(c-c_B)/\partial t = M\nabla^2\left[(\partial^2f/\partial c^2)_T\Big|_{c_B} - r^2T\nabla^2\ (c-c_B)\right]. \qquad (8)$$

By fourier transformation

$$\delta c(k,t)\equiv \int d^dx\ \exp(i\vec{k}\cdot\vec{x})\left[c(\vec{x},t) - c_B\right] \qquad (9)$$

one readily finds from Eq. (8)

$$\delta c(\vec{k},t) = \delta c(\vec{k},0)\ \exp\left[R\ (\vec{k})t\right], \qquad (10)$$

with the "amplification factor" $R(\vec{k})$ given as

$$R(\vec{k}) \equiv -M\vec{k}^2\left[(\partial^2f/\partial c^2)_T\Big|_{c_B} + r^2Tk^2\right]. \qquad (11)$$

The equal-time structure factor at time t after the quench then is

$$S(\vec{k},t) \equiv\langle\delta c(-\vec{k},t)\,\delta c(\vec{k},t)\rangle_T =\langle\delta c(-\vec{k})\,\delta c(\vec{k})\rangle_{T_o}\cdot$$

$$\cdot\exp\left[2R(\vec{k})t\right]. \qquad (12)$$

Eq.(12) shows that the only thermal fluctuations included so far are those of the initial state, $\langle\delta c(-\vec{k},0)\,\delta c(\vec{k},0)\rangle =\langle\delta c(-\vec{k})\,\delta c(\vec{k})\rangle_{T_o}$. For metastable states where $(\partial^2f/\partial c^2)_T\Big|_{c_B} >0$, see Fig.1c, $R(\vec{k})<0$ for all k: hence all fluctuations die out. Of course only homophase fluctuations (Fig.2a) are included by the linearization approximation, Eq.(7),- heterophase fluctuations (Fig.2b) require a nonlinear treatment.

In the unstable regime, where $(\partial^2 f/\partial c^2)_T \big|_{c_B} < 0$, the amplification factor is positive for long wavelengths $\lambda = 2\pi/k$, namely for

$$0 < k < k_c \equiv 2\pi/\lambda_c = \left[ -(\partial^2 f/\partial c^2)_T \big|_{c_B} / (r^2 T) \right]^{1/2}; \quad (13)$$

fluctuations with wavelengths $\lambda > \lambda_c$ increase with time (i.e. they get amplified), and their growth should be <u>exponential</u> [Eq.(12)] until the neglected nonlinear terms become important. Maximal growth would occur for $k = k_{max} = k_c/\sqrt{2}$, while at $k_c$ $S(\vec{k},t)$ would be time-independent. For the description of the kinetics of the formation of ordered structures out of disordered ones by long wavelength unstable fluctuations (Fig.1a) the treatment is analoguous. Assuming a (non-conserved) one-component (scalar) order parameter $\Psi$, the equation analoguous to Eq.(7) is

$$\partial\Psi(\vec{x},t)/\partial t = -\Gamma_o \left[ (\partial f(\Psi(\vec{x},t))/\partial\Psi)_T - r^2 T \nabla^2 \Psi(\vec{x},t) \right], \quad (14)$$

i.e. the term $(-M\nabla^2)$ is replaced by a rate factor $\Gamma_o$. The linearized treatment is fully analoguous to Eqs. (8)-(13), but the maximum growth rate - and hence the peak of the structure factor $S(\vec{k},t)$ - occurs at $\vec{k}=0$ rather than at non zero finite $k_{max}$. Note also that in general $\vec{k}$ has to be interpreted as a wavevector measured relative to the wavevector $K_{Bragg}$ where Bragg-scattering in the ordered phase occurs, $\vec{k} = \vec{K} - \vec{K}_{Bragg}$ {for unmixing, $\vec{K}_{Bragg}=0$, of course}.

Some experimental evidence for Eqs.(12),(13) has been seen in quenched amorphous alloys such as Al-22 at % Zn[10] {e.g., curves for $S(\vec{k},t)$ intersect at a time-independent value $k_c$ for a range of times}, but even there the growth of $S(k_{max},t)$ is much slower than exponential. For most other experiments on solid(e.g. Ref.11) or fluid mixtures (e.g.Ref.12), the predictions of the linearized Cahn-Hilliard theory are not observed - the notable exception being polymer mixtures.[13] We shall comment on this point in Sec.V.

The linearized theory presented so far contains two neglects: (i) nonlinear effects,- (ii) fluctuations in the final state to which one quenches. The nonlinear terms in Eq.(7) limit the exponential growth of the modes $\delta c(\vec{k},t)$, which are no longer independent of each other, but rather strongly coupled. Due to this coupling, there is neither a time-independent $k_c$ where $dS(\vec{k},t)/dt=0$ nor a time-independent $k_{max}$: rather one has $k_{max}(t)\to 0$ as $t\to\infty$, as the formed inhomogeneous concentration distribution coarsens to form macroscopic domains. This behavior will be analyzed in the next section, while here we return to (ii), fluctuations in the final state. To include these, a random force $\eta(\vec{x},t)$ is added[14] on the right-hand side of Eqs.(7),(8), which describes gaussian noise but satisfies the conservation law for the concentration,

$$\langle\eta(\vec{x},t)\eta(\vec{x}',t')\rangle_T = \langle\eta^2\rangle_T\nabla^2\delta(\vec{x}-\vec{x}')\delta(t-t'),\langle\eta^2\rangle_T=TM. \tag{15}$$

As is usual for fluctuations near thermal equilibrium, their strength $\langle\eta^2\rangle_T$ is linked to a kinetic coefficient (the mobility M) via an "Einstein relation" (fluctuation dissipation theorem). The linear Eq.(8) can be solved in the presence of these fluctuations, one finds

$$dS(\vec{k},t)/dt=-2Mk^2\{\left[(\partial^2 f/\partial c^2)_T\Big|_{c_B} + r^2Tk^2\right]S(\vec{k},t)-T\}. \tag{16}$$

The stationary solution of Eq.(16) has Ornstein-Zernike form,

$$S(\vec{k},\infty)= S_T(\vec{k})= T/\left[(\partial^2 f/\partial c^2)_T\Big|_{c_B} + r^2Tk^2\right] = r^{-2}(\xi^{-2}+k^2)^{-1}, \tag{17}$$

with $\xi=\left[r^2T/(\partial^2 f/\partial c^2)_T\Big|_{c_B}\right]^{1/2}$ being the correlation length of concentration fluctuations.

The time evolution of the structure factor in a quenching experiment (Fig.1b) is then found integrating Eq.(16) as

$$S(\vec{k},t)=S_{T_0}(\vec{k})\exp\left[2R(\vec{k})t\right]+ S_T(\vec{k})\{1-\exp\left[2R(\vec{k})t\right]\}. \tag{18}$$

Eq.(18), which can also be justified by rather different arguments[15] based on master equations, is a reasonable description for quenches which do not cross the coexistence curve, both T and $T_o$ at $c_B$ being in the one-phase region. Recent experiments[16] agree with Eq. (18) reasonably well.

The linear theory including fluctuations still does not distinguish stable and metastable states {which still would have infinite lifetime}. Both $\xi$ and the critical wavelength $\lambda_c$ diverge (according to this theory) at the spinodal, Fig. 3a; in the mean-field theory, this curve is just a line of critical points, where also "critical slowing down" of fluctuations occur. Also the radius R* of the "critical cluster" (Fig. 2b) having the size of just overcome the nucleation barrier $\Delta F^*$ diverges there, although $\Delta F^*$ itself vanishes there for d < 6 (see Sec. IV), Fig. 3b. While the spinodal is well-defined for infinitely weak and infinitely long-ranged forces, where nucleation is suppressed and metastable states have infinite life-

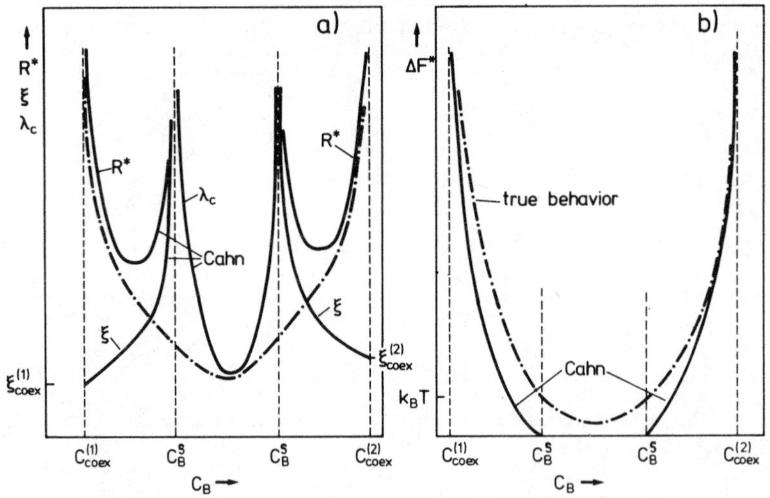


Fig.3: Characteristic lengths R*, $\xi$, $\lambda_c$ (a) and nucleation barrier $\Delta F^*$ (b) plotted vs. concentration (schematic). Full curves are the predictions of the Cahn-Hilliard mean field theory of nucleation and spinodal decomposition,- dash-dotted curve shows (on a different scale) the conjectured smooth behavior of an extremely short-range system, for which the spinodal singularity is completely washed out. From Ref.7.

time,[17] the spinodal singularity (Fig.3a) is rounded off for short range interactions,[18-20] see Sec.V. At the moment, we only note the following inconsistency[6,7] : inside the spinodal region near $c_B$ the Cahn-Hilliard unstable modes ($\lambda > \lambda_c$) produce a very slow rate of transformation only, due to the above critical slowing down. But, for a short-range system, also fluctuations of finite amplitude and smaller linear dimension (cf. Fig.2b) can form there, involving non-zero but small energy barriers $\Delta F \cong T$ (Fig.3b, dash-dotted curve). These fluctuations also are unstable and transform the system by a much quicker rate: hence the slow Cahn-Hilliard modes have no chance to come into play. The situation differs from equilibrium, where all fluctuations decay, and hence the slowest ones dominate the late stages of relaxation: in the nonequilibrium relaxation from one state to another the quickest mechanism dominates. In addition, also the position of the spinodal is ill-defined[19,21,22,23], see Sec.V. Thus, going beyond mean-field theory one no longer has a well-defined spinodal $c_B$, but rather a concentration region $\delta c_S$ over which a gradual transition from nucleation to spinodal decomposition occurs.[18,21] Similarly, in other first-order transitions there is also no sharply defined limit of metastability.

## III. TOWARDS A NONLINEAR THEORY OF PHASE SEPARATION

The nonlinear Cahn-Hilliard equation, Eq.(7), supplemented by the random force $\eta(\vec{x},t)$ Eq.(15) , should be a valid continuum theory describing adequately both spinodal decomposition, nucleation, and late-stage coarsening. Unfortunately, however, so far one has been able to treat the nonlinear terms on this basis only in an approximation valid for the initial stages of spinodal decomposition.[24-28,21] Eqs.(7,15) also contain still full information on the static and dynamic critical behavior (in the one-phase region and at the coexistence curve), which is itself a formidable problem requiring the renormalization group approach.[29,30] Thus, to get rid of the latter problem, one assumes that a coarse-graining is performed over a volume $V_\alpha = (\xi/\alpha)^d$, where $\alpha$ is a constant of order unity, and one assumes the critical behavior to be known. This approach restricts the validity of the theory to the regime[25] $k\xi < 1$, of course, as shorter wavelengths are averaged out.

The crucial approximation, however, consists in the following steps[25]: one starts by deriving exact equations of motion for the probability distributions $\rho_1[c(\vec{x})]$, $\rho_2[c(\vec{x}_1), c(\vec{x}_2)]$. Here $\rho_1$ is the probability density that at point $\vec{x}$ the concentration $c(\vec{x})$ occurs, and $\rho_2$ is the corresponding two-point function. As expected, the equation of motion for $\rho_1$ involves $\rho_2$, and the equation of motion for $\rho_2$ would involve the three-point function $\rho_3$, etc., so that an infinite hierarchy of equations of motion would be generated. This hierarchy is decoupled by the following approximation $[\delta c(\vec{x}) = c(\vec{x}) - c_B]$

$$\rho_2\,[c(\vec{x}_1), c(\vec{x}_2)] = \rho_1[c(\vec{x}_1)]\,\rho_1[c(\vec{x}_2)]\cdot$$

$$\cdot\{1 + \frac{\langle\delta c(\vec{x}_1)\,\delta c(\vec{x}_2)\rangle_T}{\langle(\delta c)^2\rangle_T}\;\frac{\delta c(\vec{x}_1)\;\delta c(\vec{x}_2)}{\langle(\delta c)^2\rangle_T}\} \tag{19}$$

The motivation for Eq.(19) is the following: if there were no correlation between concentrations at points $\vec{x}_1$, $\vec{x}_2$, the probability $\rho_2$ would just be the product of the one-point probabilities. Therefore the correction to this factorization approximation is put proportional to the two-point correlation function $\langle\delta c(\vec{x}_1)\,\delta c(\vec{x}_2)\rangle_T$. In this way, Eq.(19) yields a closed equation of motion for the probability $\rho_1$. This equation is then solved only approximately, assuming that the coarse-grained free energy still has the Ginzburg--Landau form, Eq.(5), where terms of higher order than $(c-c_B^{crit})^4$ are omitted ; the coefficients A,B and r in Eq.(3) are adjusted self-consistently such that the resulting static equilibrium is still correctly described in the critical region. Note that continuum theories introduced in this way are restricted to the critical region, as the coarse-graining makes only sense if the coarse-graining volume $V_\alpha$ is much larger than the volume of the elementary cell.

This approach has been worked out both for the case of conserved order parameter(by Langer, Baron and Miller[24]) and for the case of a non-conserved order parameter (by Billotet and Binder[25]). Denoting the scaled order parameter by m, the scaled conjugate field by h, the scaled wave vector by q, the scaled time by $\tau$, and the scaled structure factor by $\tilde{S}(q,\tau)$, one obtains the following hierachy of equations for the non-conserved case

$$\frac{\partial \langle m\rangle}{\partial \tau} = \langle m\rangle - \langle m^3\rangle + h, \tag{20a}$$

$$\frac{\partial \langle m^2\rangle}{\partial \tau} = 2\{\langle m^2\rangle - \langle m^4\rangle + h\langle m\rangle + \frac{1}{f_o}[1 - \frac{d}{q_{max}^d}\int_0^{q_{max}} dq\, q^{d+1}\, \tilde{S}(q,\tau)]\}, \tag{20b}$$

$$\frac{\partial \langle m^3\rangle}{\partial \tau} = \ldots, \quad \frac{\partial \tilde{S}(\vec{q},\tau)}{\partial \tau} = -2\{[q^2 + A(m)]\tilde{S}(\vec{q},\tau) - 1\}; \tag{20c}$$

in these equations $q_{max}$ is a cutoff of order unity (proportional to the constant $\alpha$ introduced via the coarse-graining volume $V_\alpha = (\xi/\alpha)^d$), $f_o$ is the scale factor for the coarse-grained free energy density, and the function A(m) is

$$A(m) = [\langle m^4\rangle - \langle m\rangle\langle m^3\rangle]/[\langle m^2\rangle - \langle m\rangle^2] - 1. \tag{21}$$

Eqs.(20),(21) exhibit a scaled universal form:the temperature dependence has been absorbed in the scale factors $\{m \propto \psi(1-T/T_c)^{-\beta}$, $h \propto H(1-T/T_c)^{-(\gamma+\beta)}$, $\tilde{S} \propto (1-T/T_c)^{-\gamma} S$, $q \propto (1-T/T_c)^{-\nu} k$ etc., where $\beta, \gamma, \nu$ are the standard critical exponents}; thus the concept that scaling also applies to relaxation far from equilibrium[31] is explicitly borne out.

Neglecting correlations between the local order parameter at different sites, one would have $\langle m^3\rangle \cong \langle m\rangle^3$ etc. and Eq. (20a) would decouple from the other equations; then also A(m) would turn to a constant, and Eq. (20c) would be equivalent to the linear theory discussed previously [Eq.(16)]. Since in the prescence of correlations A(m) becomes itself time-dependent, and depends on $S(q,\tau)$ itself [cf.Eq.(20b)], the solution of Eqs.(20),(21), which is still a nonlinear problem, requires numerical work.[24,25]

A closer inspection of these equations reveals that the importance of the nonlinear effects is governed by the parameter $f_o$ [if $f_o \to \infty$ the moments $<m^l>$ would decouple from $\tilde{S}(q,\tau)$,] and one also would recover the linearized theory . This parameter $f_o$ can be expressed in terms of the critical amplitudes $\hat{\xi}_o, \hat{B}, \hat{C}$ of correlation length, order parameter and susceptibility as follows,

$$f_o \propto \alpha^{-d}(1-T/T_c)^{-d\nu+(2\beta+\gamma)} \hat{\xi}_o^d \hat{B}^2/\hat{C}. \qquad (22)$$

If the critical exponents satisfy the "hyperscaling relation"[29] $d\nu = 2\beta+\gamma$, which they do in the nontrivial critical region, the critical amplitude combination $\hat{\xi}_o^d \hat{B}^2/\hat{C}$ becomes a universal constant of order unity[25] ("two-scale-factor universality"[32]); since $\alpha$ should not exceed unity, $f_o$ then universally is of order unity, which fact indicates that nonlinear effects are very important. The numerical calculations[24] in fact show that $\tilde{S}(\vec{q},t)$ increases with time only linearly or weaker: thus there is no regime of times where the linearized Cahn-Hilliard theory holds! The calculated structure factor $S(q,t)$ Fig.4A is in good qualitative agreement with simulations[33] Fig.4B and experiment[11] Fig.4C.

Fig. 4: A) Scaled structure factor $\tilde{S}(q,\tau)$ plotted vs. q at various $\tau$ for an alloy at the critical concentration according to Langer-Baron-Miller theory[24].
B) Time evolution of the structure factor according to a Monte Carlo simulation[33] of a three-dimensional nearest-neighbor Ising model of an alloy at critical cencentration and temperature $T=0.6T_c$. Due to the periodic boundary condition for the 30x30x30 lattice k is defined only for discrete multiples of $(2\pi)/30$ [in between these discrete values of $S(k,t)$ are connected by straight lines].
C) Neutron small angle scattering intensity observed by Singhal et al.[11] for a quenched Au-60 at % Pt alloy.

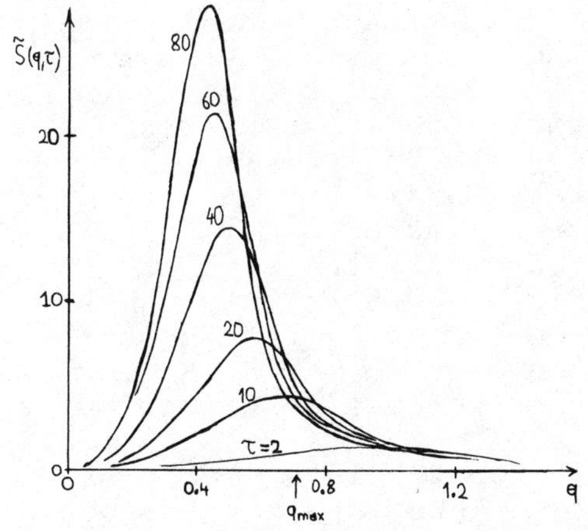

$\tilde{S}(q,\tau)$
80
60
40
20
10
τ=2
0
10
20
0.4
0.8
1.2
q
$q_{max}$


A

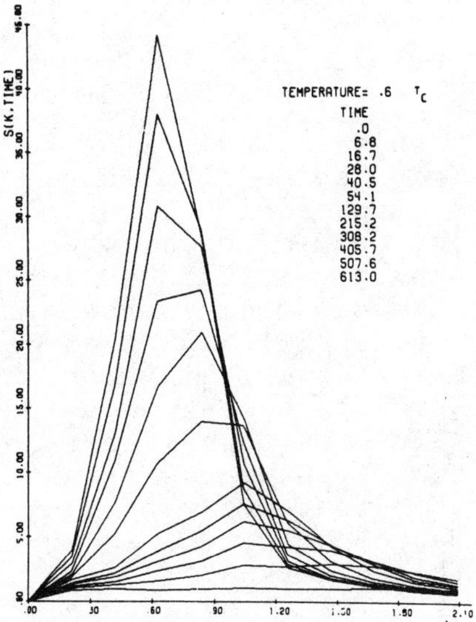

S(K,TIME)
TEMPERATURE= .6 $T_C$
TIME
.0
6.8
16.7
28.0
40.5
54.1
129.7
215.2
308.2
405.7
507.6
613.0


B

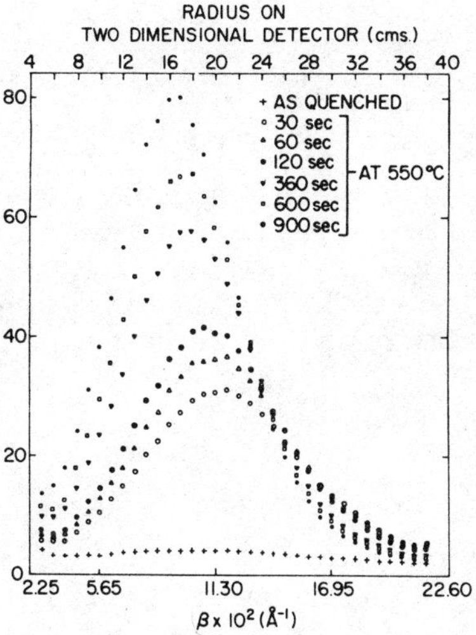

RADIUS ON
TWO DIMENSIONAL DETECTOR (cms.)
4 8 12 16 20 24 28 32 36 40
+ AS QUENCHED
30 sec
60 sec
120 sec
360 sec
600 sec
900 sec
AT 550 °C
dΣ/dΩ x 10² (cm⁻¹ Sr⁻¹)
0 20 40 60 80
2.25 5.65 11.30 16.95 22.60
β x 10² (Å⁻¹)


C

This theory hence is quite successful for the initial stages of spinodal decomposition, and it is also easily extended to systems near tricritical points[28] and to the time-displaced structure factor[26] $<\delta c(\vec{k},t)\,\delta c(-\vec{k},t')>$. On the other hand, this theory also has numerous shortcomings which we list below:

(i) The theory still exhibits metastable states of infinite lifetime.[21,25] The "spinodal" (where no longer $\xi$ but only $d\xi/dc_B$ diverges) is located in between Cahn's spinodal and the coexistence curve (Fig.5). This "spinodal" clearly is an artefact of the theory, as its precise location depends on the parameter $\alpha$ introduced above, which to some extent is arbitrary and only introduced for computational convenience. If the "spinodal" were physically meaningful, it could not depend on such an arbitrary parameter. In this metastable regime, the structure factor $\tilde{S}(q,\tau)$ also develops a peak whose position shifts the smaller values of q as $\tau$ increses (Fig.5), but at late times $\tilde{S}(q,\tau)$ saturates at an envelope which is nothing but the structure factor of the metastable state (Fig.5). Hence nucleation and growth are not yet described by this theory.

(ii) The theory is unreliable for the late stages of spinodal decomposition, where it predicts for the position $q_{max}(\tau)$ where $\tilde{S}(q,\tau)$ is maximal[15,25]

$$q_{max}(\tau)\propto\tau^{-1/4}, \quad \tilde{S}(q,\tau\to\infty)\propto q^{-2}. \tag{23}$$

The first of these relations can be interpreted as saying that the characteristic domain size L(t) increases as $L(t)\propto q_{max}^{-1}\propto t^{1/4}$, which is believed to be incorrect: one rather expects $L(t)\propto t^{1/3}$, following Lifshitz and Slyozov[34]. The second of these relations says that the system does not even decompose into macroscopic domains, but rather stays in a sort of "critical state" - one reaches only the "spinodal" in this theory of unmixing, rather than the coexistence curve!

While the latter difficulty is avoided by the theory of Horner and Jüngling,[27] the behavior $L(t)\propto t^{1/3}$ so far has not yet been reproduced. The theory hence fails to give a correct description of the late stage coarsening behavior. At late stages, we expect an Ornstein-Zernike background $[S(\vec{k},t)\propto(k^2+\xi^{-2})^{-1}]$,

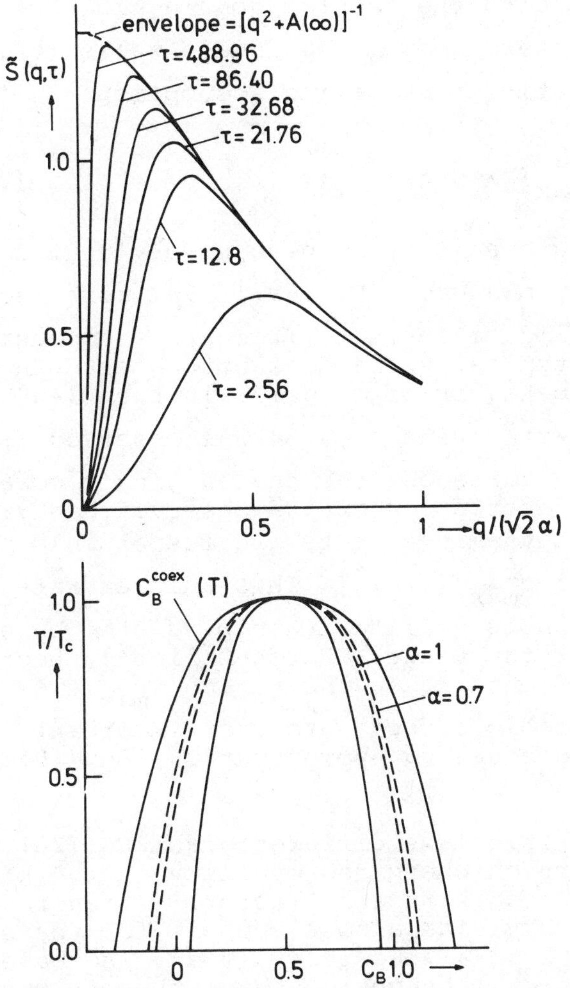


Fig.5: Phase diagram exhibiting the coexistence curve and classical spinodal (full) and the "spinodal" of the Langer-Baron-Miller decoupling (broken curves)[25], lower part. Upper part shows the scaled structure factor[25] at $(c_B-c_B^{crit})/(c_{coex}^{(1)}-c_B^{crit})=0.88$; here $A(\infty)$ is the limiting value to which $A(m)$ {Eq.(21)} tends for $\tau\to\infty$.

describing scattering from fluctuations within domains, superimposed by a high and sharp peak [its position being related to the inverse domain size $L^{-1}(t)$], describing scattering from the domain walls (cf.Fig.6b,c). This peak satifies a scaling assumption[15,21,25,35,36],

$$\tilde{S}(q,\tau) = q_{max}^{-3}(\tau)\ \tilde{\tilde{S}}\{q/q_{max}(\tau)\}. \qquad (24)$$

This scaling behavior has been nicely confirmed for fluid binary mixtures (Fig.6a)[12] and more recently also for solid mixtures.[37] It must be emphasized, however, that typically one does not observe the correct asymptotic behavior of $q_{max}(\tau)$: in the Al-6.8% Zn alloy, $q_{max} \propto \tau^{-a'}$ with $a' \cong 0.24$ was observed[37]; in fluids it is quite obvious that a single power law is not a good description of the observed behavior[12]. In fluids hydrodynamic effects are responsible for the prediction[38] $q_{max}(\tau) \propto \tau^{-1}$. This quicker rate of decomposition is qualitatively predicted by an extension of the Langer- Baron-Miller[24] theory due to Kawasaki and Ohta[39]. In the regime $k_{max}(t)\xi \ll 1$, where the hydrodynamic effects are most important, the underlying decoupling approximation, Eq.(19), clearly is unreliable.

Asymptotic laws different from Eq.(23) are found from an approach where the mobility in the kinetic equation for the structure factor is "renormalized"[40]. This approach yields also a support for the scaling assumption, Eq.(24). However, it is also based on some phenomenological assumptions, and thus it must be

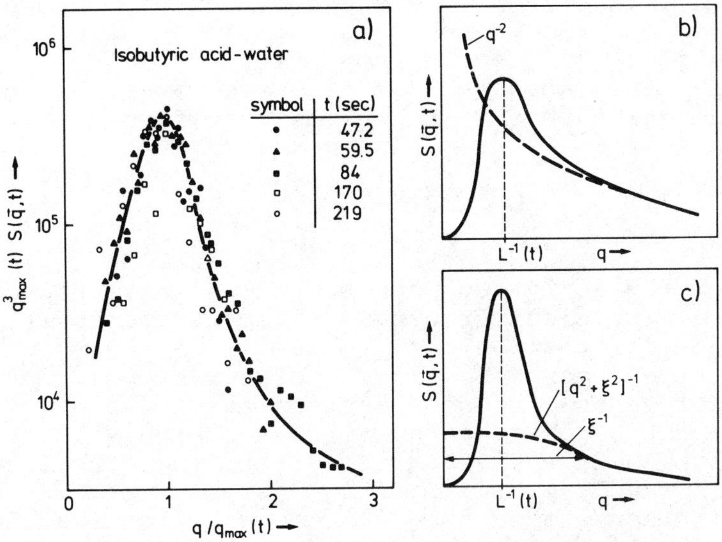


Fig.6: a) Scaling plot for the scattering from a fluid mixture studied via light scattering by Chou and Goldburg[12] (1981).
b) Schematic variation of the structure factor at late stages resulting from the Langer-Baron-Miller theory[24,25]
c) Correct qualitative behavior of the structure factor at late stages.

admitted that a fully satisfactory nonlinear theory of phase separation does not yet exist.

A rather complementary and highly phenomenological description of phase separation is a picture of an "assembly of growing droplets",[3,15,21,35,41,42]. Clearly, this approach can be viewed as an extension of nucleation theory, to which we turn next.

## IV. NUCLEATION AND CLUSTER GROWTH

As mentioned in the introduction, nucleation is a thermally activated process where clusters exceeding a certain critical size have to be formed to start the phase transformation. In the "classical" version of nucleation theory[3,4,8,43] one describes all clusters by a certain set of coordinates: their "size" l [i.e., the amount of order parameter associated with a cluster], their surface area s, etc. Then the formation free energy F (l,s,...) of a cluster in the metastable state is considered as a function of these variables $\vec{l}$=(l,s,..). F will exhibit a saddle point geometry: for small clusters the surface area is relatively large, and although their bulk energy is negative, F is still positive (and increasing with l) due to the unfavorable surface energy. For large clusters, the negative bulk energy dominates, and hence F is steadily decreasing there. In between there must occur a saddle point $\vec{l}^*$ of height $F(\vec{l}^*)$.

If this free energy barrier $F(\vec{l}^*)$ is large, the decay of the metastable state is so slow that nucleation can be approximated as a steady-state process[3]. The theory then becomes equivalent to Maxwell's equations for steady state currents in the "cluster size space" $\{\vec{l}\}$:

$$\nabla \times \vec{E}(\vec{l})=0, \quad \nabla \cdot \vec{j}(\vec{l})=0. \tag{25}$$

Here the "field" is

$$\vec{E}(\vec{l})=-\nabla\phi, \quad \phi=-\bar{n}(\vec{l})/n(\vec{l}), \tag{26}$$

where $\bar{n}(\vec{l})$ is the steady-state cluster concentration while $n(\vec{l})$ is the cluster concentration which would be in equilibrium in the metastable state, $n(\vec{l}) \propto \exp[-F(\vec{l})/k_BT]$. The current density $\vec{j}(\vec{l})$ is related to $\vec{E}(\vec{l})$ via the "conductivity tensor" $\underline{\sigma}(\vec{l})$,

$$\vec{j}(\vec{l}) = \underline{\sigma}(\vec{l})\vec{E}(\vec{l}) \equiv n(\vec{l})\ \underline{R}(\vec{l})\vec{E}(\vec{l}), \tag{27}$$

$\underline{R}(\vec{l})$ being tensor of cluster reation rates describing purely kinetic factors. This set of equations is completed by boundary conditions for the "potential" $\phi$: since small clusters are in thermal equilibrium, $\bar{n}(o) \approx n(0)$ and thus $\phi(0)=-1$; since there are not yet infinite domains of the new phase present, $\phi(\vec{l}\to\infty)=0$. The nucleation rate, i.e. the number of large clusters formed per unit volume and unit time, then is the total current in large distances from the "source" at the origin, $J=\oint \vec{j}(\vec{l})d\vec{f}$. Just as in the electric analogy a battery supplies electrons to maintain an electric current, thermal fluctuations supply small clusters (in this continuum description at $\vec{l}=0$) which are fed into the nucleation process and maintain the steady-state nucleation current.

The problem with this description is threefold: (i) if the barrier $F(\vec{l}^*)$ is not very much larger than T, a fully dynamic theory is needed; (ii) the functions $n(\vec{l})$, $R(\vec{l})$ are not known explicitly very well; (iii) even if they were known, the problem in non-trivial, Eqs. (25)-(27) have no known analytic solution. One usually resorts to an expansion around the saddle point,

$$F(\vec{l})=F(\vec{l}^*) + (\vec{l}-\vec{l}^*)\ \underline{G}\ (\vec{l}-\vec{l}^*)+\ldots.., \tag{28}$$

where the matrix $\underline{G}$ has one negative eigenvalue (-g) and otherwise positive ones, since $\vec{l}^*$ is a saddle point. Introducing the effective cross section area $\vec{A}^*$ of the saddle point region , one finds[3]

$$J \propto \exp\{-F(\vec{l}^*)/T\}\sqrt{g}\ (\vec{A}^*\ \underline{R}(\vec{l}^*)\vec{A}^*)/|\vec{A}^*|. \tag{29}$$

In the conventional Becker-Döring nucleation theory[43] one uses a single cluster coordinate l only, i.e. all cluster properties are fixed uniquely by their size. Then Eqs.(25)-(27) can be generally solved as

$$\bar{n}(l)/n(l) = J \int_{l}^{\infty} \frac{dl'}{R(l')n(l')}, \quad J = \left[\int_{0}^{\infty} \frac{dl}{R(l)n(l)}\right]^{-1} \approx$$

$$\approx \quad R(l^*)\sqrt{g}\,\exp\{-F(l^*)/T\} \qquad (30)$$

It is seen that in the framework of the saddle point approximation, due to the neglect of fluctuations in the cluster properties (described by other cluster coordinates) one gets a different (and simpler) pre-exponential factor.

The crucial question then is to estimate $F(l^*)$. The simplest possibility is the "capillarity approximation",

$$F(l) = -\Delta\mu l + \bar{s}(l) f_s, \qquad (31)$$

where $\bar{s}(l)$ is the surface area of the droplet (usually assumed spherical), $f_s$ is the interface tension for an infinitely large planar interface, and $\Delta\mu$ is the chemical potential difference between the metastable state and the state at the coexistence curve. The term $(-\Delta\mu l)$ just represents the volume energy of the cluster.

While Eq.(31) is expected to be valid for $l\to\infty$, for the finite $l^*$ of practical interest one expects correction terms. These correction terms have been discussed extensively[2-4,43], but a fully satisfactory solution to this problem does not seem to exist within this framework.

An alternative theoretical approach[5,8,44-46] is also based on Eq.(3), noting that the saddle point is a situation of unstable equilibrium between the critical droplet and the surrounding metastable phase. It hence can also be found by extremizing this free energy functional: one looks for a spherically symmetric solution of the order parameter profile $\psi(\rho)$ which is non-uniform and reduces to the order parameter $\psi_{ms}$ of the supersaturated metastable state for large distances $\rho\to\infty$ from the center of the droplet (Fig.7B,C). Putting this solution for the order parameter profile of the critical droplet coexisting with the surrounding metastable phase into Eq.(3), one can obtain the free energy excess $F(l^*)$ relative to the spatially uniform solution.

If the metastable state is close to the coexistence curve, $\psi_{ms}$ near $\psi_{coex}$, one finds an order parameter profile (Fig.7B) which closely resembles that of a planar interface (Fig.7A). The result for $F(\vec{l}^*)$ in this limit hence is consistent with the result based on the conventional nucleation theory, Eq.(31). On the other hand, the present mean-field theory of

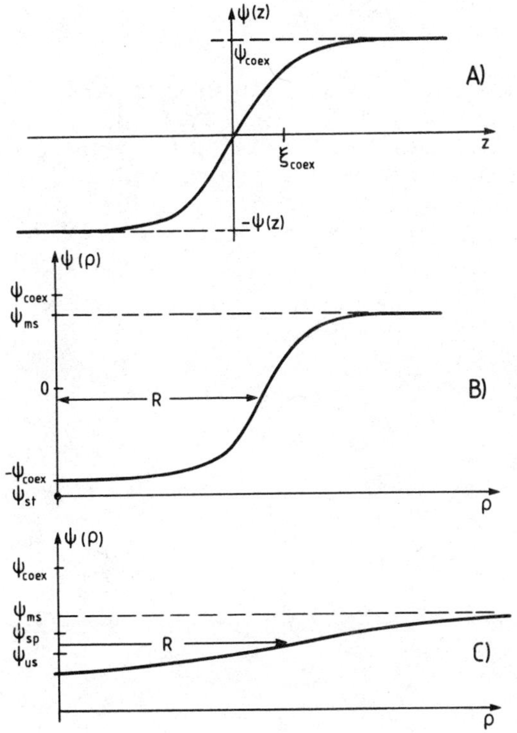


Fig.7A) Order parameter profile $\psi(z)$ across an interface between two coexisting bulk phases. B) Radial order parameter profile for a critical droplet existing in a metastable state $\psi_{ms}$ near $\psi_{coex}$. C) Same as B) for $\psi_{ms}$ near $\psi_{sp}$. R is the droplet radius, $\xi_{coex}$ the correlation length at the coexistence curve. From Ref.20.

nucleation can also be worked out[45] for metastable states up to the mean-field spinodal curve $\psi_{sp}$. Since the correlation length $\xi$ diverges there, the radius R of the critical droplet resulting from this calculation diverges again (Fig.3a).The order parameter profile (Fig.7C) then only yields a small order parameter variation from the interior to the exterior of the droplet, of the order of $\psi_{ms}-\psi_{sp}$.The critical droplet there is hence rather ramified[45,46].

As always with mean field theory, one has to worry whether the neglected fluctuations might change the results substantially. For the compact droplets near coexistence (Fig.7B), the most important fluctuations are capillary wave excitations of the droplet surface[44]. An essentially rigorous analysis[44] then confirms that Eq.(31) yields the leading terms of $F(\vec{l}^*)$ correctly, and in addition yields a correction term proportional to ln(l). In the nucleation rate J, this correction then shows up as a pre-exponential factor, different from those obtained in Eqs.(29), (30).

Away from the coexistence curve, such a field-theoretic analysis of nucleation barriers including fluctuations has not yet been given; thus this problem has been studied by numerical Monte Carlo methods.[47, 48] In one approach[47], the coexistence between the critical droplet and the surrounding metastable phase in a finite volume (where the equilibrium is stable rather

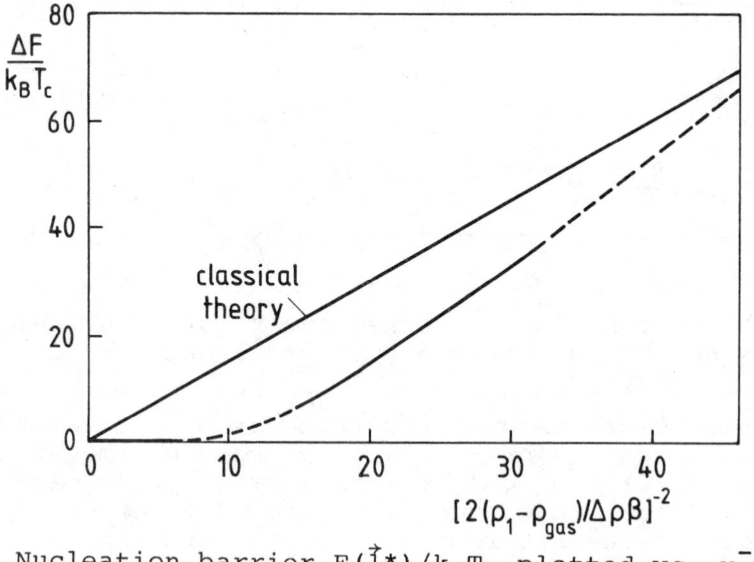


Fig.8: Nucleation barrier $F(\vec{l}^*)/k_BT_c$ plotted vs. $x^{-2}$ in the critical region. Full curve is based on Monte Carlo results[47].

than unstable, if the total order parameter is fixed) was simulated directly for an Ising model in the critical region. There Eq.(31) would predict that[3]

$$F(l^*)/T_c=(x_o/x)^2, \quad x=(\psi_{ms}/\psi_{coex}-1)/\beta \qquad (32)$$

where $\beta$ is the order parameter critical exponent, and $x_o$ is a universal constant.[47] The numerical results (Fig.8) seem to support that $F(l^*)/T_c$ is distinctly smaller than the classical prediction, Eq.(32), for barrier heights of experimental interest (i.e.,$20<F(\vec{l}^*)/T_c<60$). On the other hand, an independent calculation[48] where cluster numbers were counted and $F(\vec{l})$ is estimated from $n(\vec{l})\propto\exp\{-F(\vec{l})/T\}$ does not seem to confirm these results. All these calculations are also rather difficult because it is not entirely clear how to precisely define a "cluster".[3,4,8,31,47,48]

Also the experimental situation is rather unclear, since the nucleation rate J can hardly be measured directly. Rather one quenches the system to a metastable state a distance $\delta T$ away from the coexistence curve and asks whether an appreciable fraction of the metastable phase has been transformed during a given time $\tau$.[49] A theoretical understanding of these experiments needs to consider the combined effects of nucleation and cluster growth.[3,15,21,41] One is still far from a fully satisfactory solution to this problem.

Approximate theories for these processes are extensions of nucleation theory, Eqs.(25)-(27), where the conservation of order parameter is taken into account[15,21,34,41] (and sometimes[15,21] also nonlinear effects such as cluster coagulation events are considered). Assuming then that the radial order parameter distribution function reflects a sort of excluded-volume interaction among the growing clusters, when many clusters grow at the same time (Fig.9a), one can calculate the time-dependent structure factor $S(\vec{k},t)$[21,42], Fig.9b. The structure factor resulting from this generalized nucleation theory strikingly resembles the structure factor calculated from nonlinear theories of spinodal decomposition[24] or simulations[33], Fig.4. This result has re-emphasized the conclusion[3,18,19,31] that there is no sharp boundary between metastable and unstable states, for systems with short range forces, and that the transition between nucleation and spinodal decomposition is not sharp but rather gradual: there is no essential distinction between a wave packet of (strongly coupled!) growing concentration waves, and an assembly of growing clusters (which are in a quasi-periodic spatial arrangement due to their excluded volume interaction). Thus nucleation and spinodal decomposition turn out to be extreme limits of an unique phase separation mechanism.

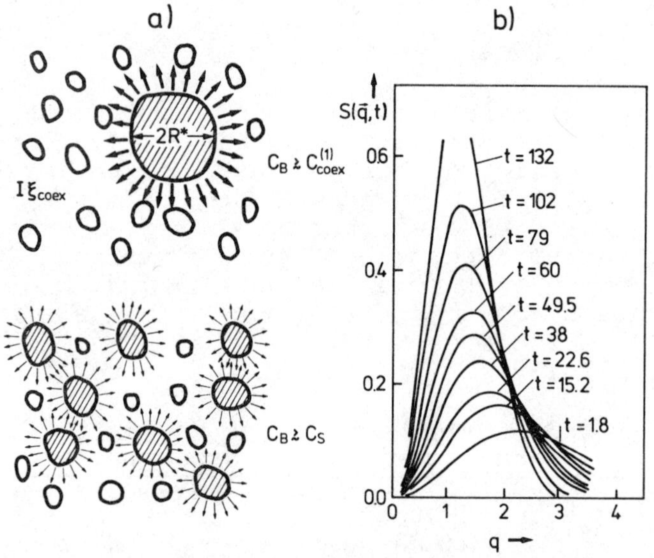


Fig.9a) Schematic "Snapshot picture" of fluctuations of an unmixing system in the metastable regime (upper part) and unstable regime (lower part; unstable fluctuations are shaded).

b) Structure factor $S(\vec{k},t)$ plotted vs. k for various times t after the quench from infinite temperature to $T=0.6T_c$ for a three-dimensional Ising model at $c_B=0.1$, as calculated from a generalized nucleation theory[21] (all units contain scale factors).

## V. SIGNIFICANCE OF SPINODAL LINES

The question then arises: in which sense do spinodal curves have a physical meaning? One answer to this question starts from considering more explicitly the coarse-grained free energy density $f_L(c)$ to be used in Eq.(3): we may define it in terms of "constrained partition functions", averaging over the microscopic degrees of freedom but keeping the concentration in cells of size $L^d$ fixed at c. For L finite $f_L(c)$ has a double-peak structure.[19] One can now define a spinodal $c_s(L)$ from the condition $\delta^2 f_L/\delta c^2=0$, but it will depend on the linear dimension L of the coarse-graining volume.[22,23]

An attempt to estimate $c_s(L)$ (or the associated "magnetization" $M_s(L)$ in an Ising model) has been

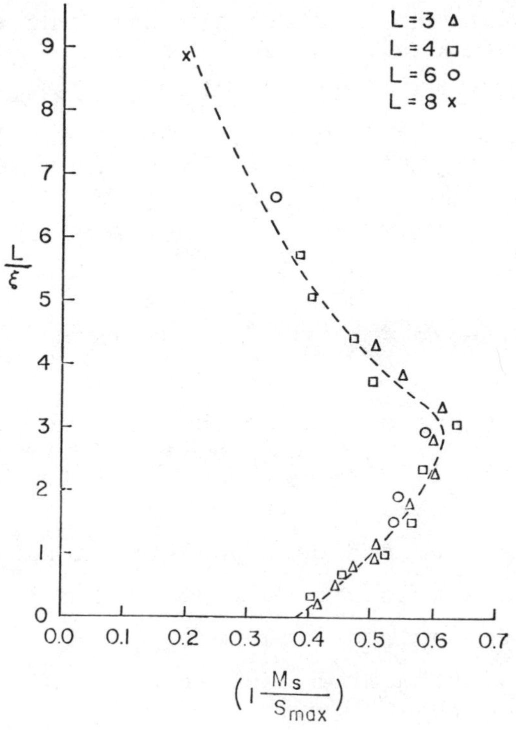


Fig.10 Monte Carlo results for the cell-size dependence of the "spinodal line" defined from cell-distribution functions[22], for the Ising model in the critical region.

based on a Monte-Carlo study of the distribution function $P_L(s) \propto \exp[-L^d f_L/T]$ of the magnetization s in finite cells.[22] In the critical region $M_s(L)$ can again be represented in a scaled universal form, normalizing it by the value $S_{max}(L)$ where $P_L(s)$ has its maximum, and plotting it vs. $L/\xi$ (Fig.10). While for $L/\xi$ small $1-M_s(L)/S_{max}(L)$ agrees with the mean-field value, $1-1/\sqrt{3} \approx 0.41$, for $L/\xi$ large $M_s(L)$ comes closer and closer towards $S_{max}(L)$; i.e., $c_s(L)$ moves towards $c_{coex}$, and a unique spinodal does not exist.

On the other hand, a spinodal is well-defined in the limit of infinitely weak infinitely long-range forces.[17] In this limit, the nucleation barriers are infinitely high, nucleation hence is suppressed and metastable states have infinite lifetime.[18]

Thus it is interesting to consider the case where the range r of the interaction is large but finite.[20] Then in the regime $r^d(1-T/T_c)^{2-d/2} \gg 1$ one has critical behavior of mean-field type. It is interesting to note that the scale for the energy barrier $\Delta F^*=F(l^*)$ is also set by a factor $r^d(1-T/T_c)^{2-d/2}$: near the coexistence curve, we have[20]

$$\Delta F^*/T_c \propto r^d(1-T/T_c)^{2-d/2}(1-\psi/\psi_{coex})^{-(d-1)}. \tag{33}$$

Thus even for a large distance away from the coexistence curve {where the term $(1-\psi/\psi_{coex})^{-(d-1)}$ is no longer large} $\Delta F^*/T_c$ may be still large. The extreme limit of metastability is obtained putting $\Delta F^*/T_c \approx 1$, however; hence metastable states are well-defined up to a narrow region near the spinodal curve $\psi_{sp}$, where then still a gradual transition from nucleation to spinodal decomposition occurs. The width of this transition region is given by

$$[(\psi-\psi_{sp})/\psi_{coex}]^{(3-d/2)} \propto [r^d(1-T/T_c)^{2-d/2}]^{-1}. \tag{34}$$

For d=3 and in the mean-field critical region $[r^d(1-T/T_c)^{2-d/2} \gg 1]$ one hence can come rather close to the spinodal, Fig.11A, while for $r^d(1-T/T_c)^{2-d/2} \approx 1$ one enters the regime discussed previously, where the gradual transition from nucleation to spinodal decomposition occurs over a wide range of compositions, and a spinodal line is no longer meaningful(Fig.11B).

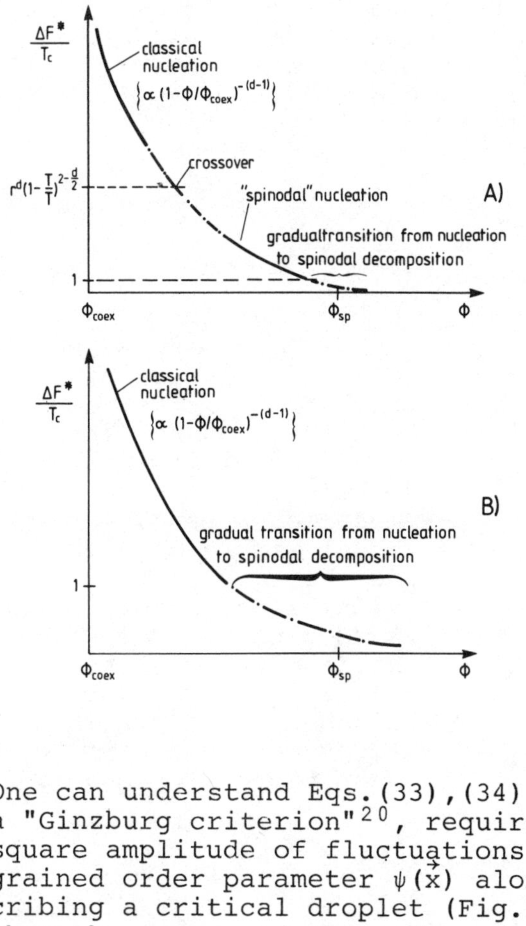


Fig.11:A) Schematic plots of the free energy barriers for the mean-field critical region (A) and the non-mean-field critical region (B). From Ref.20.

One can understand Eqs.(33),(34) simply in terms of a "Ginzburg criterion"[20], requiring that the mean-square amplitude of fluctuations of the coarse-grained order parameter $\psi(\vec{x})$ along the profile describing a critical droplet (Fig.7) must be smaller than the square of the order parameter difference between $\rho=0$ and $\rho\to\infty$ as described by the profile itself,

$$\langle\{\delta\psi(\vec{x})\}^2\rangle_T \ll \{\psi(\rho\to\infty)-\psi(\rho=0)\}^2. \tag{35}$$

Similarly one recognizes that the parameter $f_o$ in Eq.(22), setting the scale for the coarse-grained free energy density, in the mean-field critical region also is large, $f_o \propto r^d(1-T/T_c)^{d/2-2} \gg 1$. As a consequence, one does recover an initial regime regime of times, where nonlinear effects in the unstable regime are not yet important, and one has exponential growth of fluctuations in accord with the linear theory of spinodal decomposition, Sec.II. This prediction[20] has been verified by recent computer simulations[50], see Fig.12. As expected, the initial linear regime becomes less pronounced when one approaches the spinodal curve. Near $\psi=0$ instead of Eq.(35) the Ginzburg criterion ruling the validity of the linear theory of spinodal decomposition simply is $\langle\{\delta\psi(\vec{x})\}^2\rangle_T \ll \psi_{sp}^2$.

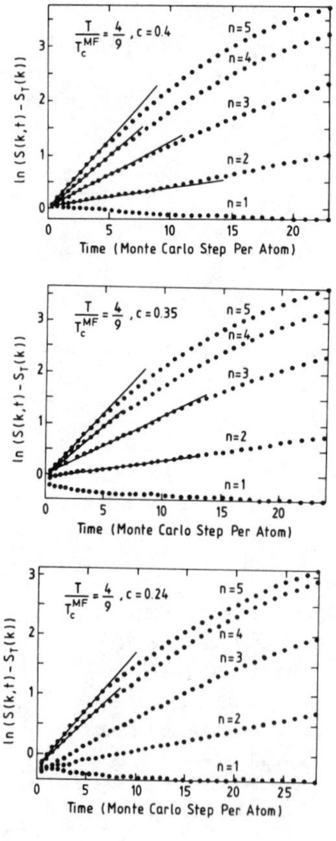


Fig.12: Evolution of the structure factor after the quench into the miscibility gap of a binary system, modelled by an Ising model with equivalent neighbor interaction with 124 neighbors (d=3). Wave-vectors shown are $k=(2\pi/60)n$, with n= 1,2,3,4,5. Straight lines are the predictions of the linear Cahn theory. From Ref.50.

These results are of interest particularly for polymer mixtures,[51] where the chainlengths N of the polymers correspond to the interaction range r of the Ising problem via[20] $r^d \propto N^{(d/2-1)}$. This mapping results from using the Flory Huggins expression[52] for the polymer problem, instead of Eq.(4) [v is a temperature-dependent effective interaction parameter]

$$f_{Polymer}/T= \frac{1}{N} \{c\ln c + (1-c)\ln(1-c)\} + vc(1-c). \qquad (36)$$

Since one can reduce the polymer problem to Eqs.(3), (4) by writing $Nf_{Polymer}=f(c)$, $\Delta F_{Polymer}=\Delta F/N$ with $r^2=r^2_{Polymer}N$, the scale for the energy barriers is set by $r^d_{Polymer}N^{d/2}/N$. Of course, the physical interaction range for the polymer problem $r_{Polymer}$ is of short range. Thus, it is gratifying that agreement with the predictions of linear theory has recently indeed been observed,[13] Fig.13.

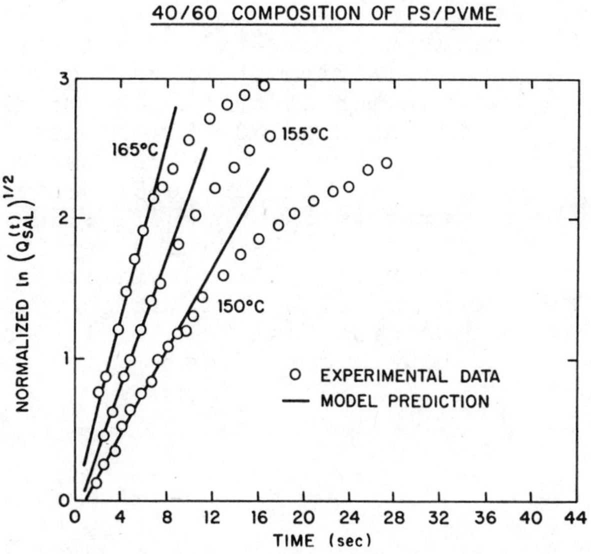


Fig.13: Time evolution of the structure factor at maximum wavevector for a polymer mixture quenched into the unstable region, as observed by light scattering.[13]

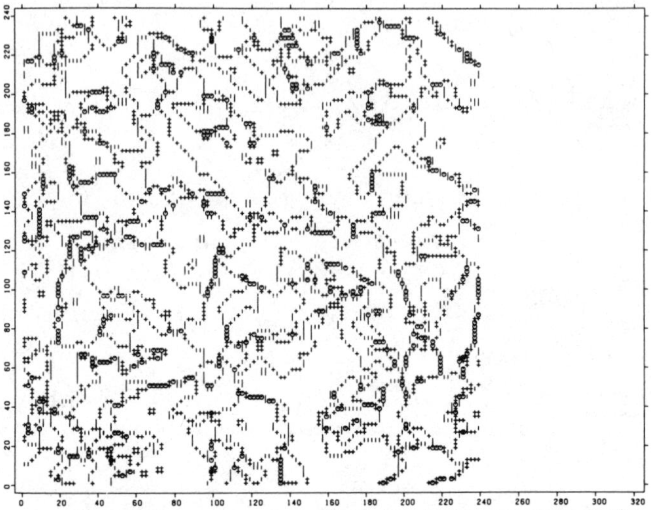


Fig. 14: Snapshot pictures[54] of a 120 x 120 square lattice at times t = 20 Mcs/ site (upper part), t = 40 (middle part) and t ≈ 140 (lower part) after the quench to a temperature T/J = 1.33 at density c = 0.5. (critical temperature $T_c/J \approx 2.07$). Sites belonging to walls are indicated by + if the unit cell contains positive excess mass, by o if it contains no excess mass, and by | if it contains negative excess mass.

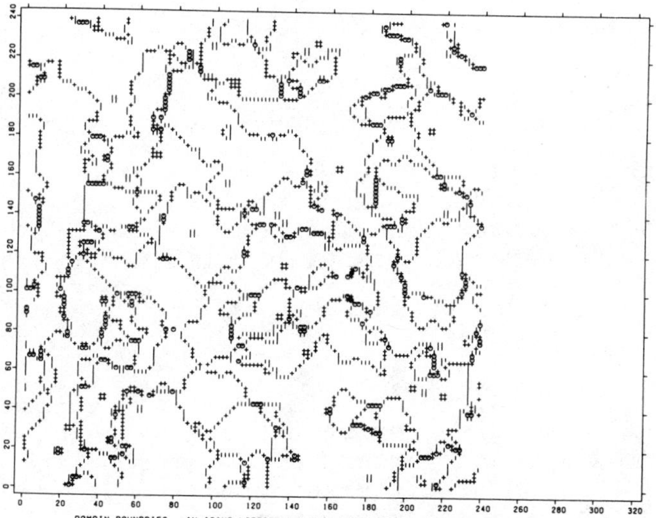

DOMAIN BOUNDRIES IN ISING LATTICE GAS WITH EQUAL NEAREST AND NEXT-NEARST REPULSIONS USING GLAUBER DYNMICS. LATTICE SIZE=(120,120). T=1.33,C=0.5,AFTER 40 MCS+ EXCESSMASS,- DEFICIT MASS,O NO EXCESS MASS

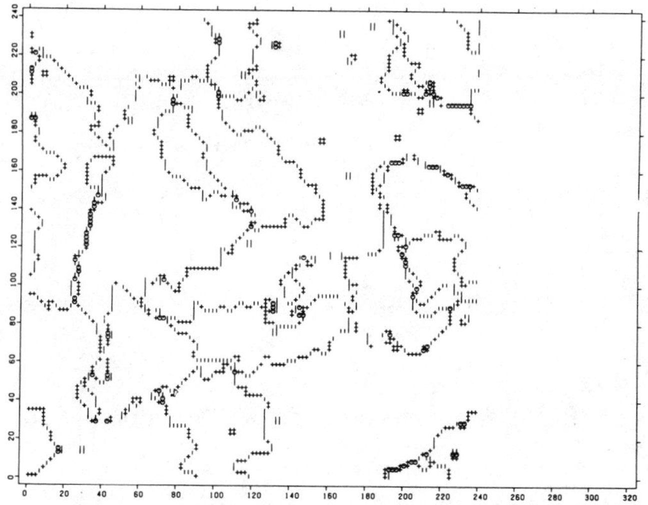

DOMAIN BOUNDRIES IN ISING LATTICE GAS WITH EQUAL NEAREST AND NEXT-NEARST REPULSIONS USING GLAUBER DYNMICS. LATTICE SIZE=(120,120). T=1.33,C=0.5,AFTER 140 MCS+ EXCESSMASS,- DEFICIT MASS,O NO EXCESS MASS

We also conclude that spinodal curves have much more significance for polymer mixtures, which have a pronounced mean-field critical regime, than for ordinary mixtures, which have not. For polymer mixtures and d=3, the width over which the spinodal is smeared out is[51] $(\psi-\psi_{sp})/\psi_{coex} \propto [N(1-T/T_c)]^{-1/3}$.

## VI. LATE STAGES OF DOMAIN GROWTH

In this section we return to the problem of the scaling of the structure factor during the late stages of a phase transformation, cf. Eq.(24): we are now interested in the power law with which the typical domain size L(t) grows,

$$L(t) \propto q_{max}^{-1}(t) \propto t^x. \qquad (37)$$

While for phase separation Lifshitz-Slyozov theory[34] yields x=1/3, different exponents are expected for the related problem of the growth of domains of an ordered phase out of an initially disordered state. For one-component order parameter one finds x=1/2 (see e.g.Ref.53). For more-component order parameter, however, this problem is theoretically not yet well understood, and much information on this problem so far is solely due to Monte Carlo simulations. We here only describe a typical example, rather than giving a full account of the problem.

We consider a square lattice where ordering of (2x1) superstructures occurs. This ordering is two-component {described by Bragg peaks $\vec{k}_1 \equiv (\pi/a)(1,0)$ and $\vec{k}_2 \equiv (\pi/a)(0,1)$ where a is the lattice spacing of the square lattice}. This ordering is obtained e.g. for an Ising lattice gas with equal nearest and next-nearest neighbor repulsive interaction.

The ordering of this lattice was simulated both for the case of conserved density and for the case without any conservation laws.[54] In the latter case x=1/2 also was obtained. This result is nontrivial, since there are several types of walls competing with each other, and the domain shapes are very irregular (Fig.14). In the case of conserved density, however, one finds $x \cong 1/3$. The interpretation suggested in Ref. 54 is that this domain growth again is ruled by the Lifshitz-Slyozov[34] diffusion mechanism: atoms have to diffuse over typical distances of order $L(t)$ from walls with positive excess density to walls with negative one, for promoting further growth.

While for spinodal decomposition the peak of the structure factor occurs at $q_{max}(t)$, off the Bragg positions, for ordering kinetics the peaks occur precisely at the positions where Bragg peaks occur in the fully ordered structure (Fig.15a). A scaling of the form of Eq.(24) still holds, if $q$ is measured relative to the Bragg position, and $q_{max}(t)$ is replaced by the halfwidth $\sigma(t)$.

At lower temperatures in this model[54], as well as for other models with larger order parameter degeneracy[55], exponents $x$ different from $x=1/2$ or $x=1/3$ are found. It is as yet unclear whether this reflects true non-universal behavior[56] or just crossover effects.

## VII. DISCUSSION

These lectures have summarized the state of the art with respect to the dynamics of phase transformations, with an emphasis on phase separation. Obviously, the theory still is quite incomplete: while one can easily work out the linear theory of spinodal decomposition, it is valid only for systems with rather long range of the forces, or polymer mixtures; classical nucleation theory, on the other hand, is expected to be valid very close to the coexistence curve only - where nucleation hardly is observable on realistic timescales. Theories describing the combined effects of nucleation and growth, incorporating also the gradual transition from this mechanism to nonlinear spinodal decomposition, still need to be developed. And theoretical approaches describing how ordered domains form and grow are also rather incomplete.

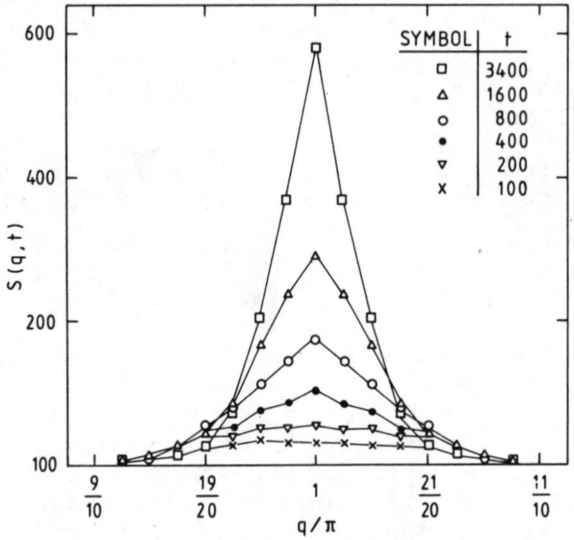


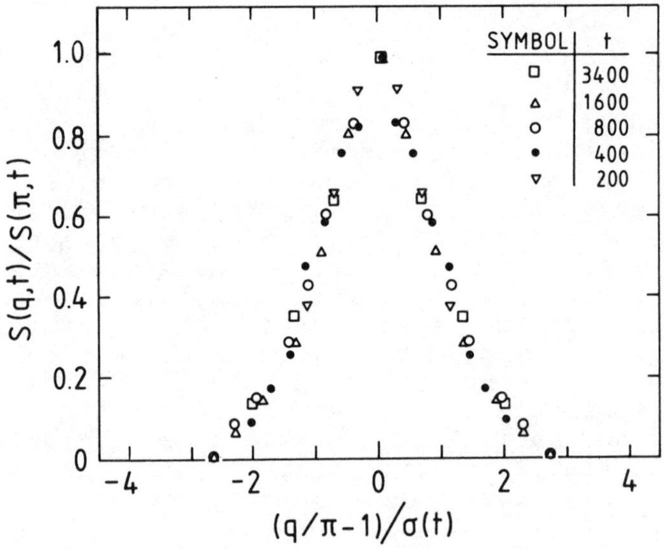


Fig.15: Structure factor S(q,t) plotted vs. wavenumber q[lattice spacing chosen to be unity] for various times after a quench from infinite temperature to T/J=1.33, for conserved density c=1/2(upper part). Scaling representation of the structure factor where the distance from the Bragg position is scaled with the halfwidth of the peak[54] (lower part).

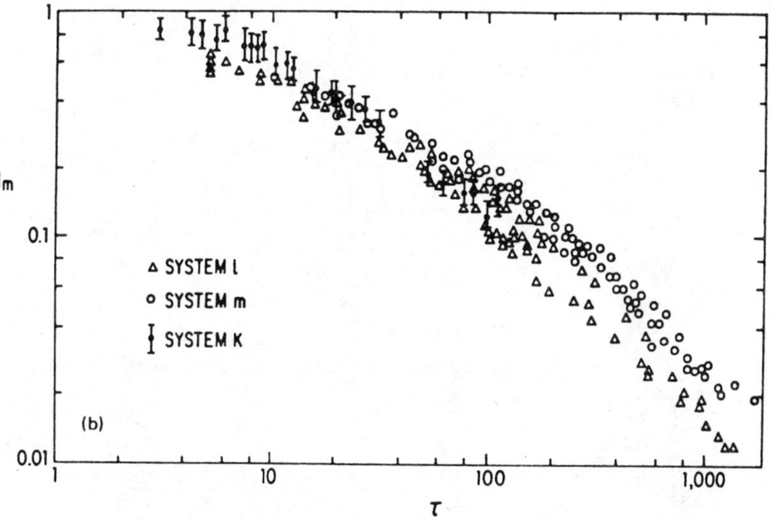


Fig.16: Comparison of $q_m$ vs. $\tau$ behavior found in a PS-PVME mixture (k) with isobutyric acid-water (l) and 2,6 lutidine-water mixtures (m).[57]

On the other hand, the experimental situation also is quite unsatisfactory. Snyder and Meakin[57] have recently pointed out that one finds hardly any distinction between ordinary mixtures and polymer mixtures, if one rescales time $\tau$ and wavevectors $q_m$ properly (Fig.16). While our analysis implies that polymer mixtures and mixtures of small molecules at least for small times $\tau$ should behave differently, the accuracy of the data (Fig.16) hardly is sufficient to discuss this issue. Similarly, the scatter of data points in Fig.6 precludes any final conclusion as to whether the scaling of Eq.(24) actually holds. Clearly, more powerful scattering techniques for a study of the structure functions in quenching experiments would be very desirable.

Recent experimental studies indicate two modifications of the scaling picture presented so far: (i) the scaling function $\tilde{\tilde{S}}$ depends on the direction of $\vec{q}$ in solids, reflecting the anisotropy of the precipitated clusters[58]. (ii) In Fe-Cr alloys, there is evidence for two regimes with exponents $a' \approx 0.17$ and $a' \approx 0.33$, respectively, which have different scaling functions[59], and seem to confirm conjectures of Ref.15.

## REFERENCES


1. Previous reviews of this subject include Refs. 2-8.
2. K. Binder, in Fluctuations,Instabilities and Phase Transitions (T.Riste,ed.)p.53,Plenum, New York 1975.
3. K. Binder and D. Stauffer, Adv.Phys. 25,343 (1976).
4. K. Binder, J.phys.(Paris) 41,C4-51(1980).
5. J.S. Langer, in Systems Far from Equilibrium (L.Garrido, ed.)p.1, Springer, Berlin 1980.
6. K. Binder, in Ref.5, p.63.
7. K. Binder, in Stochastic Nonlinear Systems in Physics, Chemistry and Biology (L.Arnold and R. Lefever, eds.)p.62, Springer, Berlin 1981.
8. J.D. Gunton, M. San Miguel and P.S. Sahni, in Phase Transitions and Critical Phenomena, Vol.8 (C.Domb and J.L. Lebowitz, eds.) Academic Press, New York 1983 p. 267.
9. J.W. Cahn and J.E. Hilliard, J.Chem.Phys. 28, 258 (1958); J.W. Cahn, Trans.Metall.Soc. AIME 242, 166 (1968).
10. A. Agarwal and H. Herman, Scripta metall. 7, 503 (1973).
11. G. Laslaz and P. Guyot, Acta Met. 25, 277 (1977); V. Gerold and J. Kostorz, Appl.Cryst. 11, 376 (1978); G. Laslaz, P. Guyot and G. Kostorz, J.phys.(Paris) 38,406 (1977); M. Hennion, D. Ronzaud, and P. Guyot, Acta Met. 30, 599 (1982); S.P. Singhal, H. Herman and G. Kostorz, J.Appl. Cryst. 11, 572 (1978); S. Katano and M. Iizumi, J.Phys.Soc. Japan 51,347 (1982); M.Furusaka et al., Physica 120B, 383 (1983).
12. Y.C. Chou and W.I. Goldburg, Phys.Rev. A20,2105 (1979); A23,858 (1981); N.C. Wong and C.M. Knobler, J.Chem.Phys. 69,727 (1978); Phys.Rev. A24, 3205 (1981).
13. H.L. Snyder, P. Meakin and S. Reich, Macromolecules 16, 757 (1983).
14. H.E. Cook, Acta Met. 18, 297 (1970).
15. K. Binder, Phys.Rev. B15, 4425 (1977).
16. N.C. Wong and C.M. Knobler, Phys.Rev.Lett. 43, 1733 (1979); 45, 498 (1980).
17. O. Penrose and J.L. Lebowitz, J. Statist.Phys. 3, 211 (1971).
18. K. Binder, Phys.Rev. B8, 3423 (1973).
19. J.S. Langer, Physica 73, 61 (1974).
20. K. Binder, Phys.Rev. A29, 341 (1984).

21. K. Binder, C. Billotet, and P. Mirold, Z.Phys. B30, 183 (1978).
22. K. Kaski, K. Binder, and J.D. Gunton, J.Phys. A16, L623 (1983); Phys.Rev. in press.
23. K. Kawasaki, T. Imaeda, and J.D. Gunton, in Perspectives in Statistical Physics, p. 203 (H.J. Raveché, ed.) North-Holland, Amsterdam 1981.
24. J.S. Langer, M. Baron, and H.D. Miller, Phys.Rev. A11, 1417 (1975).
25. C. Billotet and K. Binder, Z.Phys. B32, 195 (1979).
26. C. Billotet and K. Binder, Physica 103A, 99 (1980).
27. H. Horner and K. Jüngling, Z.Phys. B36, 97 (1979).
28. G. Dee, J.D. Gunton, and K. Kawasaki, Progr.theor. Phys. 65, 365 (1981).
29. M.E. Fisher, Rev.Mod.Phys. 46, 597 (1974).
30. P.C. Hohenberg and B.I. Halperin, Rev.Mod.Phys. 49, 435 (1977).
31. K. Binder and E. Stoll, Phys.Rev.Lett. 31, 47 (1973); K. Binder and H. Müller-Krumbhaar, Phys.Rev. B9, 2328 (1974); K. Binder, D. Stauffer, and H. Müller-Krumbhaar, Phys.Rev. B12, 5261 (1975).
32. D. Stauffer, M. Ferer and M. Wortis, Phys.Rev. Lett. 29, 245 (1972).
33. J. Marro, A.B. Bortz, M.H. Kalos, and J.L. Lebowitz, Phys.Rev. B12, 2000 (1975).
34. I.M. Lifshitz and V.V. Slyozov, J.Phys.Chem. Solids 19, 35 (1961).
35. K. Binder and D. Stauffer, Phys.Rev.Lett. 33, 1006 (1974).
36. J. Marro, J.L. Lebowitz, and M.H. Kalos, Phys. Rev.Lett. 43, 282 (1979).
37. S. Komura, K. Osamura, H. Fujii, and T. Takeda, Hiroshima university preprint (1983).
38. E. Siggia, Phys.Rev. A20, 595 (1979).
39. K. Kawasaki and T. Ohta, Progr.Theor.Phys. 59, 362 (1978).
40. H. Furukawa, Phys.Rev.Lett. 43, 136 (1979); Phys. Rev. A23, 1535 (1981).
41. J.S. Langer and A.J. Schwartz, Phys.Rev. A21, 948 (1980).
42. P.A. Rikvold and J.D. Gunton, Phys.Rev.Lett. 49, 286 (1982).
43. A.C. Zettlemoyer (ed.) Nucleation: M.Dekker, New York 1969.
44. J.S. Langer, Ann.Phys. (N.Y.) 41, 108 (1967); N.J. Günther, D.A. Nicole, and D.J. Wallace, J.Phys. A13, 1755 (1980).

45. J.W. Cahn and J.E. Hilliard, J.Chem.Phys. 31, 688 (1959).
46. W. Klein and C. Unger, Phys.Rev. B28, 445 (1983), and to be published.
47. H. Furukawa and K. Binder, Phys.Rev. A26, 556 (1982); see also K. Binder and M.H. Kalos, J. Stat.Phys. 22, 363 (1980).
48. D. Stauffer, A. Coniglio, and D.W. Heermann, Phys.Rev.Lett. 49, 1299 (1982); and to be published.
49. A.J. Schwartz, S. Krishnamurthy, and W.I. Goldburg, Phys.Rev. 22, 2147 (1980).
50. D.W. Heermann, preprint.
51. K. Binder, J.Chem.Phys. 79, 6387 (1983).
52. P.J. Flory, Principles of polymer chemistry Cornell University Press, Ithaca, N.Y. 1976.
53. T. Ohta, D. Jasnaw and K. Kawasaki, Phys.Rev. Lett. 49, 1223 (1982).
54 A. Sadiq and K. Binder, Phys.Rev.Lett. 51, 674 (1983), and to be published.
55. P.S. Sahni, G.S. Grest, M.P. Anderson, and D.J: Srolowitz, Phys.Rev.Lett. 50, 263 (1983).
56. H. Furukawa, preprints.
57. H.L. Snyder and P.A. Meakin, J.Chem.Phys. 79, 5588 (1983).
58. J.P. Simon, P. Guyot and A. Ghilarducci de Salva, Phil.Mag. 49, 1 (1984).
59. S. Katano and M. Iizumi, Phys.Rev.Lett. 52, 835 (1984).

# NEUTRON SCATTERING AT HIGH PRESSURE

Daniel Bloch and Jean Voiron

Laboratoire Louis Néel, C.N.R.S.
166X
38042 Grenoble-cédex, France

## INTRODUCTION

Progress in high pressure technology nowadays allows to characterize the properties of matter submitted to high pressure with approximately the same degree of sophistication and accuracy as obtained in experiments performed under usual atmospheric conditions.

In the first chapter of this lecture we discuss the compressibility of solids and we give some examples of pressure-induced "crystallographic" or "electronic" phase transitions.

Neutron scattering experiments are now commonly used to characterize condensed matter at high pressure. We introduce the technology of such measurements in chapter II.

In chapter III, we give some experimental results obtained using this technique. They allow for instance advances in the physics of vibrational spectra, crystallographic structural changes crystal field effects and magnetic order-disorder phase transitions. Electro-elastic or magneto-elastic coupling will be introduced using a ld-compound as an example.

Partial or total chemical substitutions, in a crystal lattice, may simulate lattice compression or dilatation, and may thus appear similar to high pressure experiments. We will however emphasize the absence of validity of this "chemical pressure" concept through the comparison between real pressure experiments and chemically simulated pressure experiments. Furthermore we will demonstrate, using a variety of examples, how high pressure experiments make it possible to distinguish between volume and thermal effects, and to induce T = 0 original crystallographic or electronic phase transition.

## I - SOLID STATE PHYSICS AT HIGH PRESSURE

### 1 - Solids are compressible

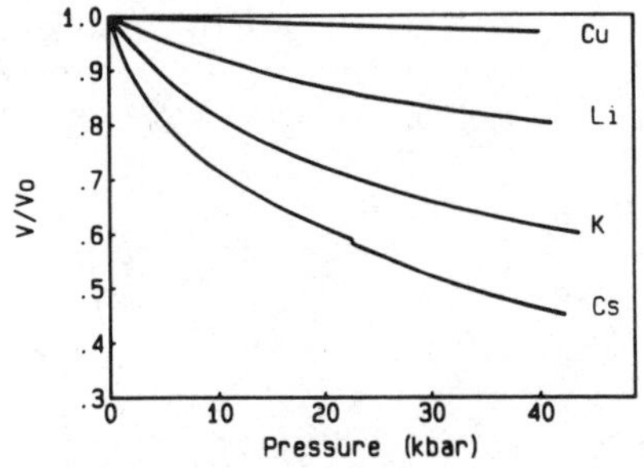


Fig. 1 : Pressure volume relation of selected materials (room temperature)

When a hydrostatic pressure is applied to a single crystal, it first produces continuous modifications of the lattice parameters and angles as well as in some cases changes of the atomic positions within the unit cell. These continuous variations may be disturbed with discontinuous crystallographic phase transitions, associated, or not, with changes in the crystal symmetry.

The first step to describe the effect of high pressure is to give the variation of the volume of the unit cell as a function of the applied pressure (Fig. 1). The softness of the lattice may be expressed from its compressibility $K = -(1/V)(dV/dp)$, which depends both on temperature and pressure. In the following, we will restrict ourselves to isothermal compressibility, which results from standard high pressure laboratory experiments. One should notice however that pressures above 2000 kbar can only be obtained from shock-wave experiments ; in that case along with volume reduction there occurs a temperature increase. The magnitude of the applied pressure will be described using as a unit the kilobar (kbar) with 1 kbar = $10^8$ Pa ; the Pascal (Pa) is the S.I. unit. To give orders of magnitude, 1 kbar is about the pressure to which materials are submitted at a depth of 10 km in water, whereas the pressure at the center of the earth core is $\sim$ 3600 kbar.

Most experiments we will refer to during this lecture have been undertaken between 0 and 45 kbar. In that range the compressibility of soft solids can vary considerably with pressure (Fig. 1) ; its temperature variation should also be taken into account. Inner electrons or the crystal structure have little influence on the compressibility, which depends mainly on the outer electron configuration and on chemical bonding.

Table 1

| | Cs | K | Na | Li | U | Cu |
|---|---|---|---|---|---|---|
| Initial compressibility $K_o$ ($10^{-3}$ kbar$^{-1}$) | 32 | 23 | 13 | 6.3 | 1.0 | 0.73 |
| Compression energy for $\Delta V/V_o = 1$ % (meV/at.) | 0.11 | 0.10 | 0.10 | 0.11 | 0.64 | 0.51 |
| Equivalent pressure to $\Delta T = 0 \rightarrow 300$ K (kbar) | 1.7 | 2.3 | 3.6 | 3.8 | 7.6 | 13 |

To produce a volume reduction $\Delta V/V_o = 1/100$ requires a compression energy $\simeq (V_o/2K)(\Delta V/V_o)^2$ of typically a few tenth of a meV. The thermal contraction of solids between room temperature and 0 Kelvin is equivalent to that which results at room temperature, from an applied pressure of $\sim$5 kbar (Table 1).

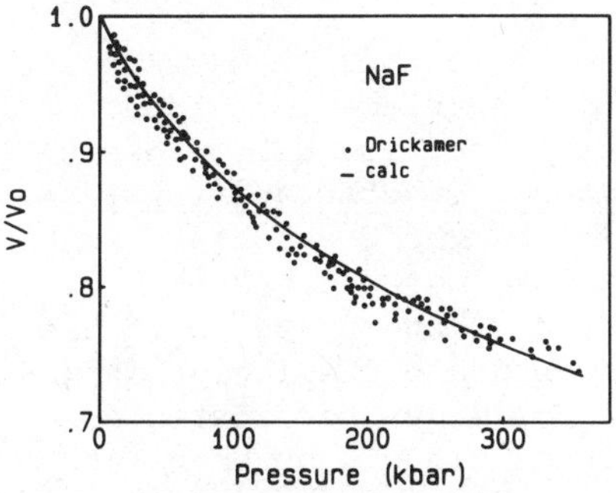


Fig. 2 . Pressure volume relation in NaF (from Ref. 2).

Numerous works have been devoted to the evaluation of the cohesive energy of noble gases, ionic crystals, covalent or metallic crystals, and to their associated compressibility and equilibrium lattice constants. We give, as an example, figure 2, the P-V relationship for an ionic crystal (NaF) as determined experimentally[1] from X-ray experiments, and evaluated theoretically[2] from a self-consistent KKR band calculation. Such a model leads for instance to a lattice constant "a" for the p = 0 NaCl rocksalt structure of 4.58 Å, in good agreement with the 4.62 Å experimental value, but is uncapable for the time being of predicting whether the p = 0 ground state will be of the NaCl or CsCl type.

## 2 - Structural phase transitions

The cohesive energies of the various possible pressure induced crystallographic structures are usually very close to each other. It is therefore very difficult to predict with accuracy the occurrence of pressure induced crystallographic phase transitions. Of course the density of the high pressure phase is larger than the density of the low pressure phase.

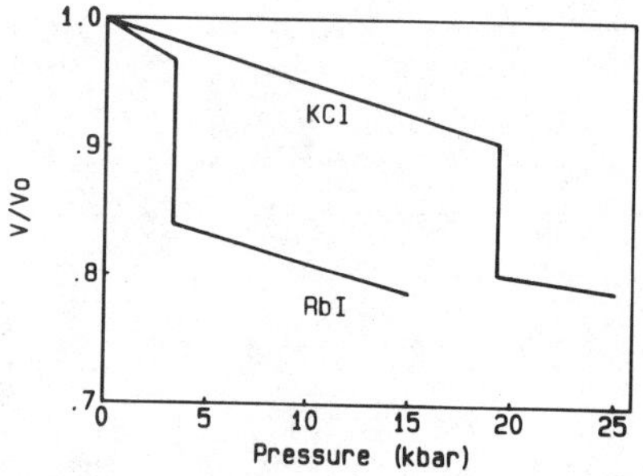


Fig. 3. The NaCl-CsCl pressure induced transition in ionic compounds (from Ref. 3).

As an example we give the NaCl-CsCl transition which has been studied quite extensively in many ionic compounds, both theoretically and experimentally. In such a transition there is a change not only in the volume but also in the coordination number. Many alkali halides feature a transition from a B1, NaCl-type at low pressure to a B2, CsCl-type structure at high pressure (Fig. 3). In the B1 structure (face centered cubic) the cation in surrounded by six nearest neighbour anions and in the B2 structure (primitive cubic) the cation is surrounded by eight nearest neighbour anions.

Time resolved studies of the NaCl-CsCl transition have recently been performed, using X-rays from synchrotron radiation[4]. Most crystallographic phase transitions occur discontinuously. However, they may also occur continuously. This has been observed for instance in many ferroelectrics[5], where the soft mode associated with the occurrence of ferroelectricity is a long-wave length ($k = 0$) optical phonon. This soft mode may also occur at any point in the Brillouin zone, giving rise to an incommensurate displacive structural transition. In SbSI (antimony-sulfo-iodide) for instance (Fig. 4) the structural ferroelectric transition is observed at 290 K at atmospheric pressure, where it is discontinuous, but appears to be continuous above 1.4 kbar and below 235 K[5]. Note the $T = 0$ pressure induced ferro-paraelectric transition at $\sim$ 8 kbar. Continuous crystallographic phase transitions have been evidenced in $NiF_2$[6] or CsI[7] (Fig. 5) for example, using high pressure X-ray scattering.

## 3 - Electronic transitions

Pressure can induce drastic electronic transitions, such as

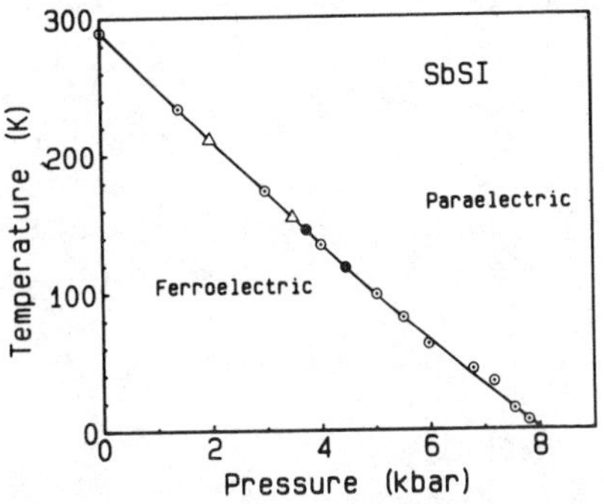


Fig. 4. Pressure dependence of the ferroelectric transition temperature of SbSI (from Ref. 5).

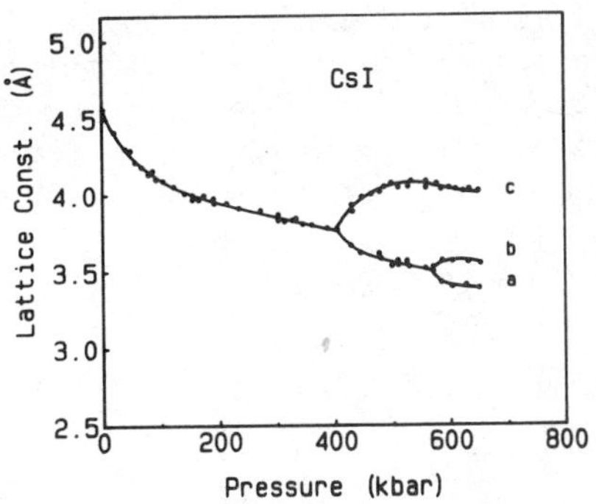


Fig. 5. Pressure induced structural transitions in CsI (from Ref. 7) from cubic to tetragonal and from tetragonal to orthorhombic.

those which give rise to lattice collapse observed for cerium, samarium sulfide or cesium respectively at 7.6, 6.5 and 42.5 kbar (Fig. 6). These transitions present a common feature the invariance of the crystal structure on either side of the electronic transition. The volume collapse is $\sim$18 % for cesium ; the low pressure phase allows a Curie-Weiss ionic-like paramagnetism whereas the high pressure phase is a strongly Pauli enhanced metallic-like paramagnet. The 6.5 kbar transition in SmS is a semi-conductor to metallic transition, accompanied by $\sim$ 10 % volume decrease.

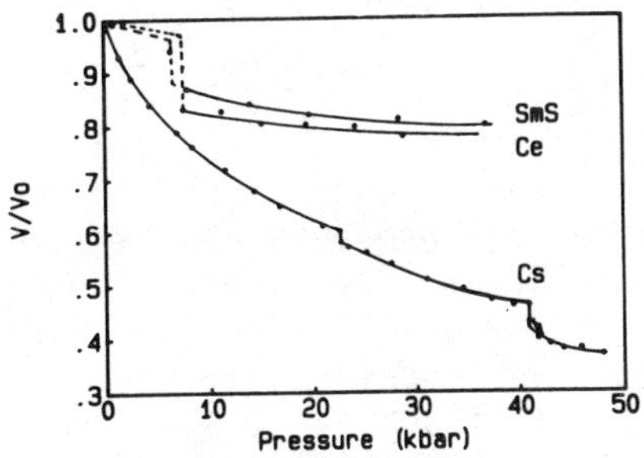


Fig. 6. Pressure volume relationship for Ce (from Ref. 8), SmS[9] and Cs[10,11] (increasing pressure).

In fact continuous transitions of similar nature have been observed in SmSe, SmTe or $SmB_6$. In $SmB_6$ for instance, the gap between the conduction and the valence band decreases linearly with pressure, to a zero value at $\sim$ 30 kbar. The 42.5 kbar fcc-fcc transition in cesium is associated with a volume collapse of $\sim$8 % ; it corresponds to an increasing 6s-5d character. In fact the electronic transitions in Ce, SmS or Cs occur partly discontinuously and partly continuously; note that superconducting states are

obtained respectively above 50 kbar in cerium and above 120 kbar in cesium. Many other elements as Y, Ba, Si, P, Ge, As, Sb, Se, Te, Bi and U have been shown to become superconductors at high pressure, whereas they are normal metals at atmospheric pressure. Futhermore some compounds are insulators at zero pressure and superconductors at high pressure. A typical example is the 1-dimensional organic compound $(TMTSF)_2PF_6$[12] with a phase diagram given in Fig. 7[13].
diagram given in Fig. 7[13].

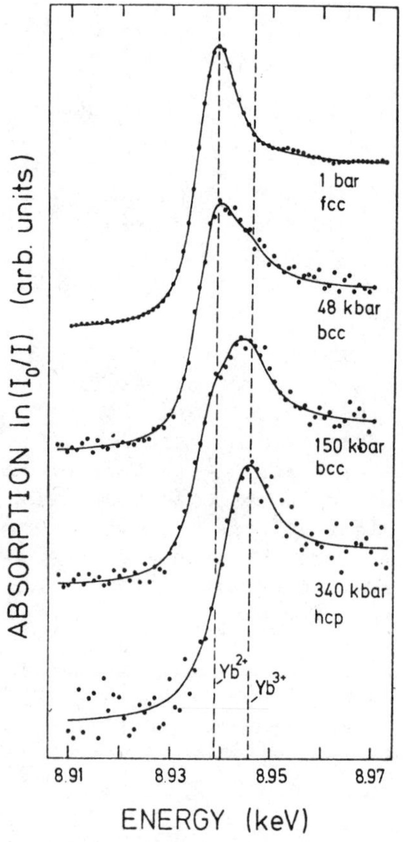


Fig. 8. The $Yb^{2+} \rightarrow Yb^{3+}$ transition. Edge absroption spectroscopy from Ref. 14.

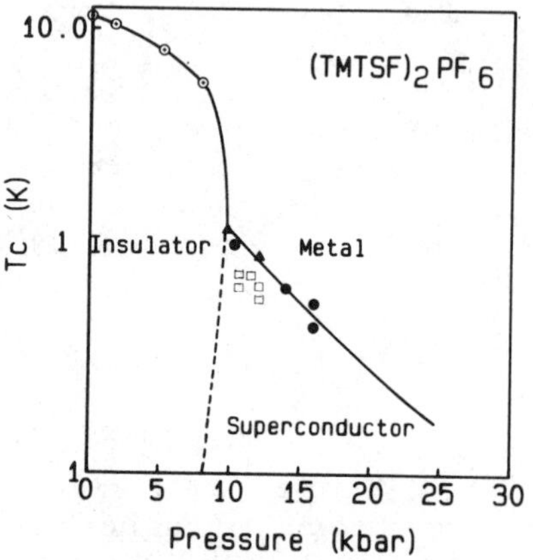


Fig. 7. The phase diagram of $(TMTSF)_2PF_6$ (from Ref. 13).

A large variety of techniques have been used to characterize those high pressure electronic transitions. Edge absorption spectroscopy (Fig. 8) was used to establish for instance the $Yb^{2+} \rightarrow Yb^{3+}$ transition in pure ytterbium[14]. Pressure thus favours the magnetic $4f^{13}$ state instead of the p = 0 non-magnetic $4f^{14}$ configuration. In fact the susceptibility of many ytterbium compounds (YbCuAl, $YbInAu_2$,...) increases with pressure.

These electronic transitions may be associated with strong modifications of the optical, magnetic, transport,... properties. One of the most famous examples is that of the molecular crystal iodine (Fig. 9), where a continuous metallization takes place. The broadening of both valence and conduction bands results in the

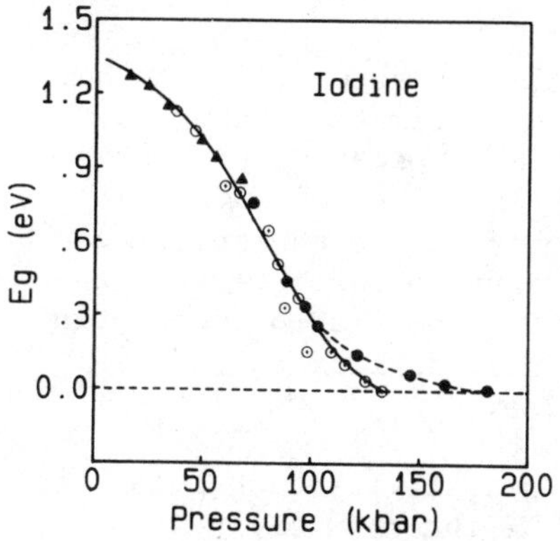


Fig. 9. Gap Eg versus pressure in iodine (from Ref. 15).

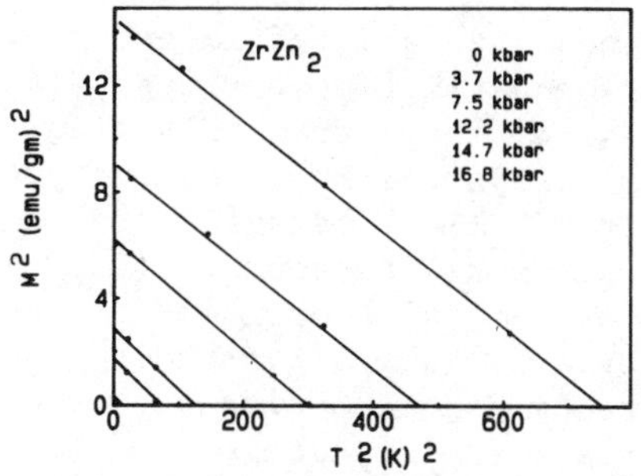


Fig. 10. Square of the magnetic moment versus square of the temperature for $ZrZn_2$ at various pressures (from Ref. 17)

disappearance of the energy gap at ∿ 180 kbar. In fact a new phase transition has been obtained at ∿ 210 kbar[16] through X-ray diffraction techniques, indicating the onset of the dissociation of molecular iodine.

The electronic band broadening may also drive continuous magnetic to non-magnetic transitions as observed in $ZrZn_2$[17]. In fact the condition for the occurrence of ferromagnetism in a collective electron picture is a large density of states at the Fermi level. Associated with the broadening of the electron energy bands, there is a general tendency towards a decrease of the density at the Fermi level, which therefore induces magnetic to non-magnetic transition (Fig. 10) as observed in many elements and compounds.

In an ionic state, electronic transitions may also be experienced through a transition between states with different magnetic moment. For instance a high spin-low spin transition for Fe in $MnS_2$ has been detected above ∿ 60 kbar[18].

## II - TECHNOLOGY OF HIGH PRESSURE NEUTRON SCATTERING[19,20]

Neutron scattering experiments often require large samples, which leads to severe limitations on the maximum applied pressure. In fact massive vessels would lead to unacceptable absorption and coherent or incoherent extra scattering. The upper limit of the hydrostatic pressure does not depend on the sample, but is limited by its environment. This is in contrast with the uniaxial stress which is severely restricted by the mechanical properties of the sample itself. We will describe two kinds of hydrostatic pressure cells and one kind of uniaxial stress cell developed in C.N.R.S., Laboratory in Grenoble and which have been frequently operated in particular at the Laue-Langevin Institute Reactor in Grenoble.

We will note that :

a - the hydrostatic nature of the applied pressure, or the homogeneity of the applied stress, can be estimated through an analysis of the diffraction lines at increasing pressures.

b - the value of the applied pressure or stress themselves can be determined from the value of the lattice parameters of the sample under consideration as given from neutron scattering experiment, if they are already known or from an auxiliary well-characterized sample otherwise.

More standard techniques such as the resistivity technique, when pressure is given from the variation of the resistivity of a calibrated manometer sample located in the high pressure cell, may also be used.

1 - Materials for high pressure cells

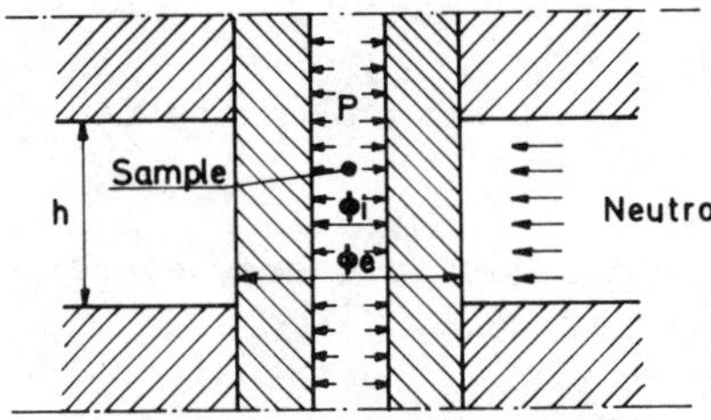


Fig. 11. Neutron window.

These materials should have sufficiently good transparency for neutrons and sufficiently high mechanical properties for pressure. They are mainly aluminium alloys (Al-Zn-Mg-Cu), copper alloys (Cu-Be), maraging steels (Fe-Ni-Co-Mo) or alumina ($Al_2O_3$). The mechanical properties of these materials, as well as their associated maximum working pressures, decrease rapidly above a certain temperature equal to about 100°C for Al-alloys, 200°C for Be-Cu alloys, 300°C for maraging steels and 700°C for $Al_2O_3$.

Table 2. Materials for high pressure cells. Bursting pressure for $h = 5$ mm, $\emptyset_i = 5$ mm and $\emptyset_e = 20$ mm.

| Materials | Al-alloy | Ti-Zr | Cu-alloy | maraging steel | $Al_2O_3$ |
|---|---|---|---|---|---|
| Bursting pressure (kbar) | 27 | 40 | 64 | 85 | 43 |
| Absorption (%) | 16 | 40 | 67 | 82 | 48 |

Table 2 gives the bursting pressure and absorption for these alloys in similar conditions as defined figure 11.

The pressure and temperature range and the specific neutron properties are of prime importance to determine the pressure transmitting media : for instance hydrogen compounds are not used, due to their large incoherent scattering. The most common pressure media are helium with freezing pressure estimated to 115 kbar at room temperature, $C_6F_{12}$, "fluorinert" with a freezing pressure 20 kbar at room temperature as well as deuterated ethanol-methanol mixture, which remains hydrostatic to almost 100 kbar. In fact, experiments performed on a single crystal at the highest pressure necessitate, in order to keep the orientation of the crystal and to leave the mosaic spreads at acceptable level, a transmitting media which is fluid or sufficiently soft, as solid helium.

## 2 - Large volume-moderate pressures aluminium cell

The unsupported aluminium cell given figure 12 has been used from 4 to 300 K and working pressures up to 6 kbar. The sample has a volume up to 5 $cm^3$. A shielding allows for safety in a reactor context. These cells have been used for 2-axes elastic scattering, 3-axes inelastic scattering as well as 4-circles elastic scattering experiments. Maraging belts allow pressures in the 20 kbar range. Such vessels are suitable for any conventional neutron scattering experiments, except small angle scattering.

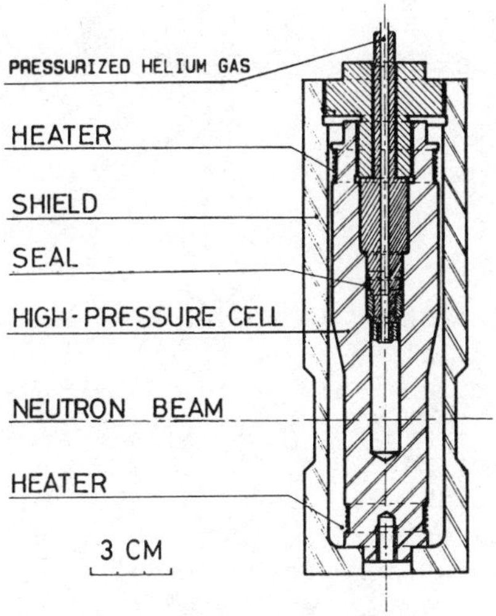


Fig. 12. Aluminium cell.

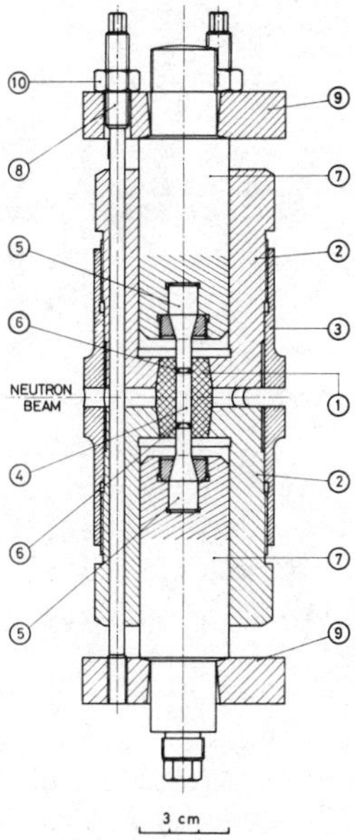


Fig. 13. Alumina cell
1 Alumina part,
5 tungsten carbide piston.

3 - Small volume-high pressures alumina cell[21]

This portable (4.5 kg) clamped cell[21] (fig. 13) has a small outside diameter (68 mm). The maximum working pressure is 45 kbar, for an internal diameter $\emptyset_i$ = 5 mm and a window h = 6 mm. Typical sample may be only 0.1 $cm^3$. The sintered alumina cell which has partial external support is similar to that introduced by McWhan et al[22] but has a clamped external support as well as a clamped internal pressure. This type of cell has been used down to 1.5 K.

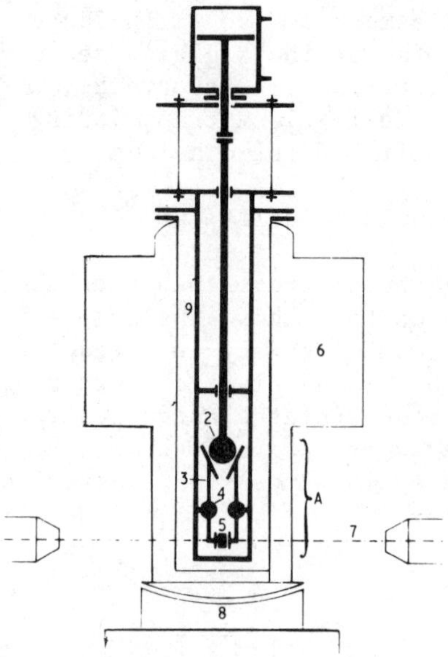


Fig. 14. 2 - piston, 3 - lever, 4 - pivot, 5 - sample, 6- crystal, 7 - neutron beam.

4 - Uniaxial stress experiments[14]

Since the material under investigation is in contact with external pistons on two opposite faces only[23,24], with the direction in the scattering plance or orthogonal to its direction, the major problem results from the difficulty to get strain of constant direction and amplitude within the sample.

III - APPLICATIONS

In this chapter we shall introduce examples where original informations have been extracted from high pressure neutron scattering experiments.

1 - Lattice

As a first approximation, thermal effects in solids are analysed assuming the interatomic potential to be as to give rise to harmonic atomic motions. Detailed informations on the vibrational spectra may be obtained from inelastic neutron scattering ; the density of states and therefore, for instance, the total vibrational energy and its thermal derivatives, that is the "spectroscopic" specific heat of the material under consideration can then be deduced. Deviations from this harmonic potential behaviour have important consequences. They are evidenced in insulating solids through the thermal expansion of the crystal, through the high temperature linear term of the specific heat, from the thermal dependence of the elastic coefficients, or from the thermal conductivity itself. These anharmonic effects may be estimated from the inelastic scattering spectra as determined at various temperatures, since they lead to a thermal dependence of the effective frequences of the oscillations. Alternatively, we may assume that they correspond to quasiharmonic mode frequencies $\omega_i(\vec{q})$ which depend linearly on the volume

$$\frac{\Delta\omega_i(\vec{q})}{\omega_i(\vec{q})} = \gamma_{i,\vec{q}} \frac{\Delta V}{V}$$

where $\gamma_{i\vec{q}}$ is the Grüneisen constant for mode i at wave vector $\vec{q}$. Differentiating the free energy with respect to volume leads to a thermal expansion linear in T at high temperature, and proportional to $T^4$ at low temperature. The Grüneisen modes may be obtained directly through inelastic scattering experiments at high pressures. Furthermore, experiments performed at various pressures and temperatures make it possible to separate between the intrinsic dependence of a particular phonon as a function of temperature and volume

$$d\omega = \left(\frac{\partial\omega}{\partial T}\right)_V dT + \left(\frac{\partial\omega}{\partial V}\right)_T dV$$

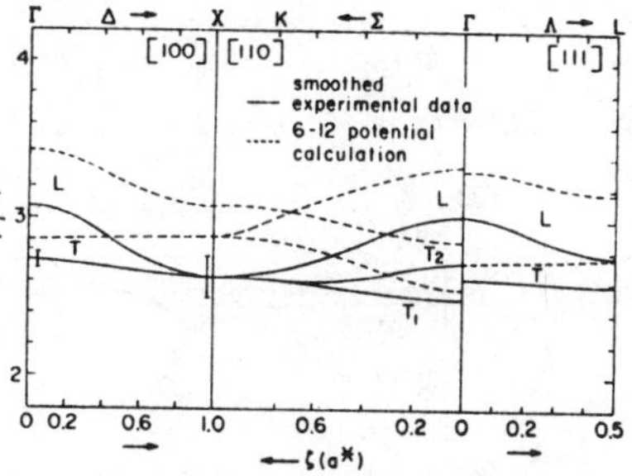


Fig. 15. Grüneisen mode parameters of neon along the three symmetry directions (from Ref. 25).

which allows for isothermal T to determine the true Grüneisen mode. The separation into thermal and volume dependent contributions is of conceptual and practical value.

A series of beaufiful experiments have been performed for instance on neon[25] using crystals prepared at various volumes through cooling of the gas under various applied pressures. Typical results are given on Fig. 15. These results have been compared with those given from various theoretical interatomic potential models as well as with the bulk macroscopic Grüneisen constants as deduced from overall elastic properties, thermal expandion and specific heat measurements.

b - Continuous structural phase transitions

The examples we will introduce have been chosen to demonstrate that crystallographic structures may, using neutrons, be determined at high pressure with a degree of accuracy and sophistication similar to that deduced from zero pressure experiments and better than those deduced from high pressure X-ray experiments. The use of a collimator permits, in time-of-flight experiments, to observe almost only neutrons scattered from the sample, eliminating contamination from the high pressure cell materials. The crystallographic phase transitions observed in both $NiF_2$ (at 18 kbar)[6] and $TeO_2$ (at 9 kbar)[26] lead from the tetragonal to the orthorhombic structure. In both cases it is accompanied by a softening of an acoustic shear mode which propagates along the [110] direction, and corresponds to the elastic constant $C' = 1/2\ (C_{11} - C_{12})$. Arrows on the F atoms (Fig. 16) indicate the atomic motion associated with the transition. No volume or unit cell parameter discontinuity occurs at the crystallographic transition. The main feature is the splitting of the tetragonal "a" lattice parameter into the unequal "a" and "b" orthorhombic lattice parameters (Fig. 17). The primary order parameter is the strain (b-a). The mean field character of the transition is indicated from the linear dependence of $(b-a)^2$ versus pressure.

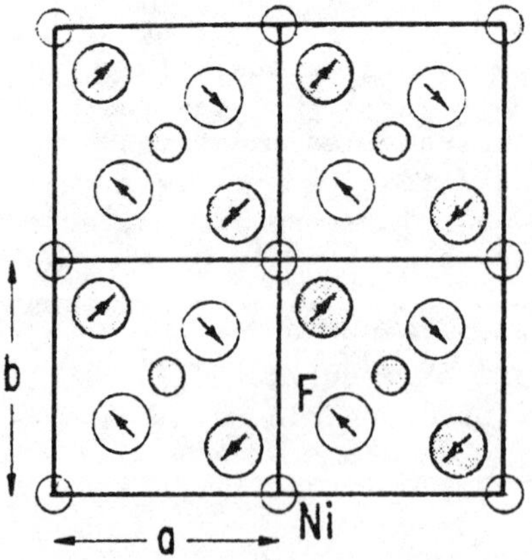

Fig. 16. Projection of $NiF_2$ structure onto the x-y plane (from Ref. 6).

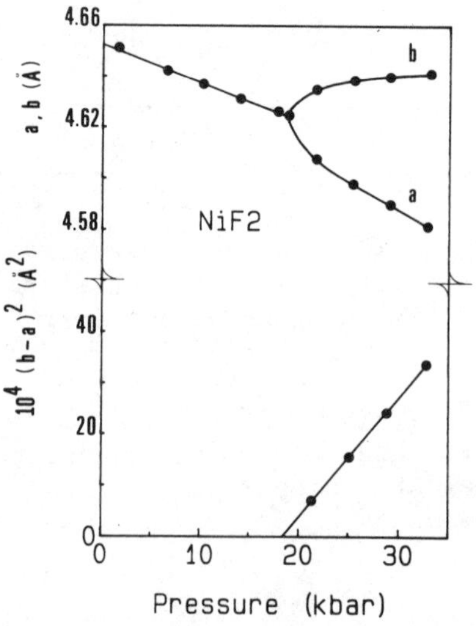

Fig. 17. Lattice parameters of $NiF_2$ versus pressure (from Ref. 6).

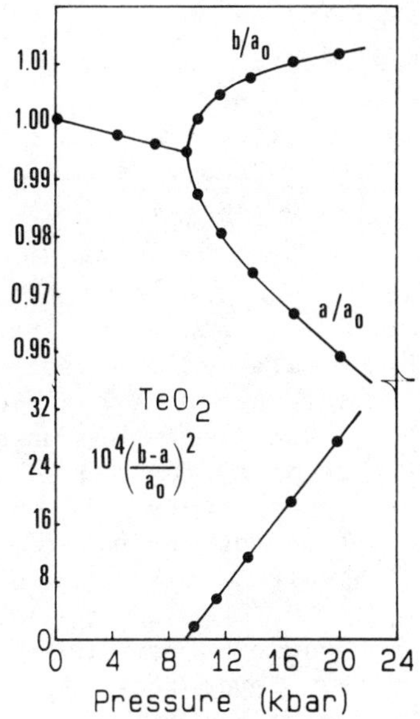


Fig. 18. Lattice parameters of $TeO_2$ versus pressure (from Ref. 26).

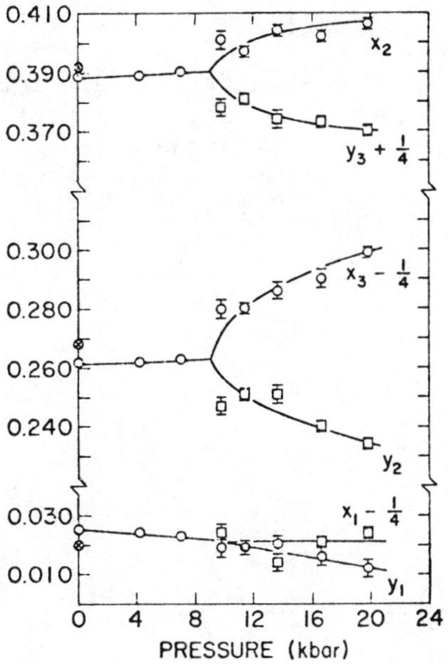


Fig. 19. x, y atomic positions versus pressure in $TeO_2$ (from Ref. 26).

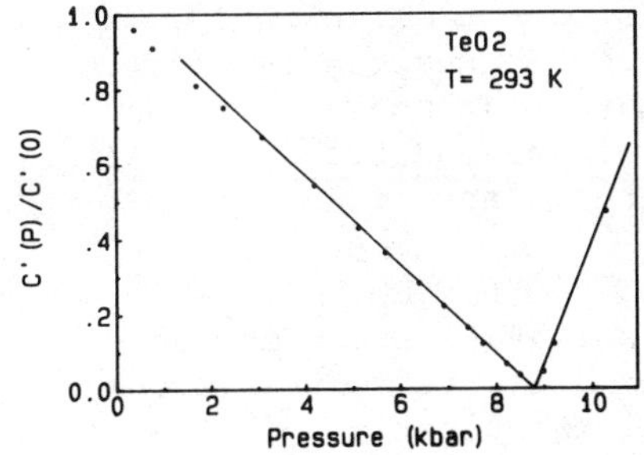


Fig. 20. $C' = 1/2\ (C_{11}-C_{12})$ versus pressure (from Ref. 27).

The same behaviour is observed in paratelluride $TeO_2$ (Fig. 18)[26] where accurate atomic position determinations (Fig. 19) demonstrate that the motion of oxygen atoms is the driving force behind the transition. The elastic constant C' (Fig. 20) goes smoothly and nearly linearly to zero as the transition is approached from either side[27]. Single crystal inelastic neutron scattering[28] shows that complete softening of the acoustic phonon mode is restricted to a narrow region at the Brillouin zone center.

2 - Crystal field[29]

Ions submitted to electrical fields present in solids may have properties largely different from that of the free ion. Therefore these properties may be severely modified through the variation of hydrostatic pressure or uniaxial stress. We have chosen as an illustration the $Pr^{3+}$ ion which possesses a moment

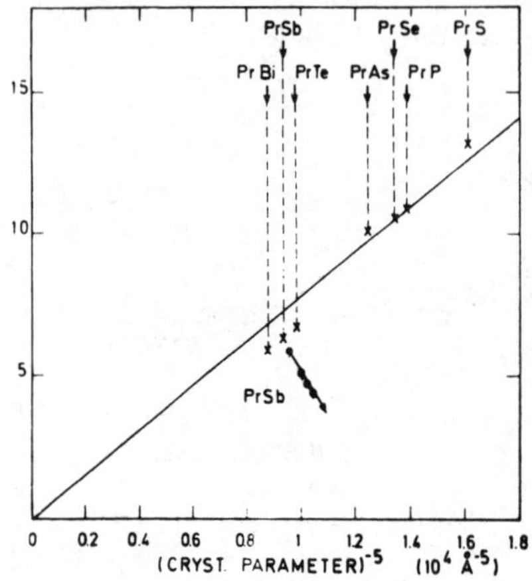


Fig. 21. Ground state to first excited state energy separation versus cubic crystallographic parameter a from neutron spectroscopy. The full line corresponds to a $a^{-5}$ punctual charge model, x to "chemical pressure" (powdered samples)[30] and • to hydrostatic pressure (single crystal, q = 0)[31,32] (from Ref.29).

$g_J J = 4/5 \times 4\ \mu_B$. When embedded in Pr or PrSb the 2J + 1 = 9 fold degeneracy of the free ion is lifted which gives rise to a singlet ground state. The crystal field levels can be determined using various physical measurements. Neutron spectroscopy is of particular interest for metals. High pressure neutron spectroscopy permits to follow the relevant energy level modifications and therefore to test the models or hypothesis needed for their quantitative analysis.

When pressure experiments were not available a common procedure to test the effect of interatomic distance modifications consisted in partial or total chemical substitution. The overall distance modifications were assigned to this "chemical" pressure. For instance (Fig. 21) the variation of the energy separation with the substituted pnictide or chalcogenide in PrSb has been attributed[30] to the variation of the interatomic distances alone, since it follows a simple $a^{-5}$ law as anticipated from a point charge model. This, in fact, has not been corroborated through recent direct or indirect experiments[31,33] which indicate for a few of these compounds (PrSb, PrP, PrN, PrAs, PrBi) a pressure variation opposite to that anticipated from the point charge model. This underlines the importance of intratomic and chemical properties on crystal field energy separation.

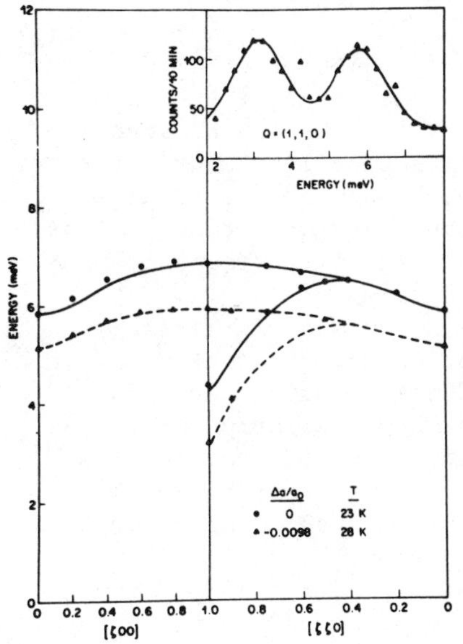


Fig. 22. The effect of pressure on the crystal-field levels. The results suggest that a soft-mode magnetic transition will occur at higher pressures. The insert shows typical data obtained at P = 1.6 GPa (16 kbar) and T = 28 K (from Ref. 31).

As a typical example for the determination of crystal field excitations, results obtained on PrSb at 0 and 16 kbar[32] are given on Fig. 22.

A great deal of efforts has been devoted in the last 20 years to those compounds where inter-atomic exchange J and crystal field splitting Δ are comparable in magnitude, and especially to those with a non magnetic singlet crystal field ground state. For a two singlets level with energy separation Δ:

$$\chi = \chi_c/(1 - J\chi_c).$$

where $\chi_c$, the Van Vleck susceptibility, is inversely proportional to Δ. A spontaneous polarization can occur when the product $J\chi_c$ passes through 1.

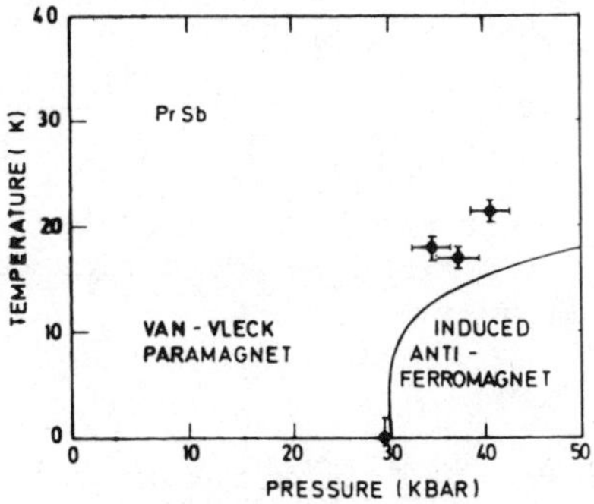


Fig. 23. Phase diagram of PrSb. The full line is calculated within a molecular field approximation[32].

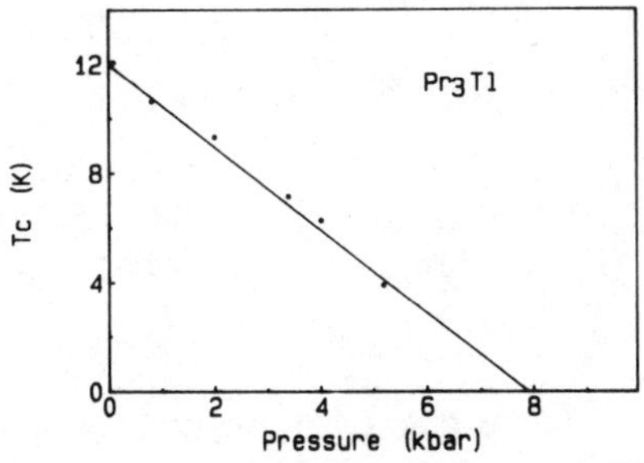


Fig. 24. $T_c$ versus pressure in $Pr_3Tl$ (from Ref. 34).

A substitution of a small amount of magnetic $Nd^{3+}$ to non-magnetic $Pr^{3+}$ in $Pr_{1-x}Nd_x$ alloys leads to an overall $Nd^{3+}$ magnetic moment ; conversely the substitution of non-magnetic $La^{3+}$ for $Pr^{3+}$ in ferromagnetic $Pr_3Tl$ leads to the disappearance of the magnetic moment of all $Pr^{3+}$ ions. The specific features of such magnetic-non-magnetic phase transition are however usually obscured through the inhomogeneous spatial situation which occurs from partial chemical substitution. High pressure can be used to tune the value of $J\chi_c$ through 1, and therefore to study the T = 0 magnetic-non-magnetic transition in homogeneous materials.

Such T = 0 non-magnetic to magnetic phase transitions have been observed via the effect of hydrostatic pressure in PrSb[31,32] (Fig. 23) and $Pr_3Tl$[34] (Fig. 24) and through the action of stress in Pr[35] (Fig. 25), where an uniaxial stress of 800 bar creates a modulated antiferromagnetic structure with $T_N$ = 7.5 K. In this last case hydrostatic pressures does not lead to a formation of a magnetic moment.

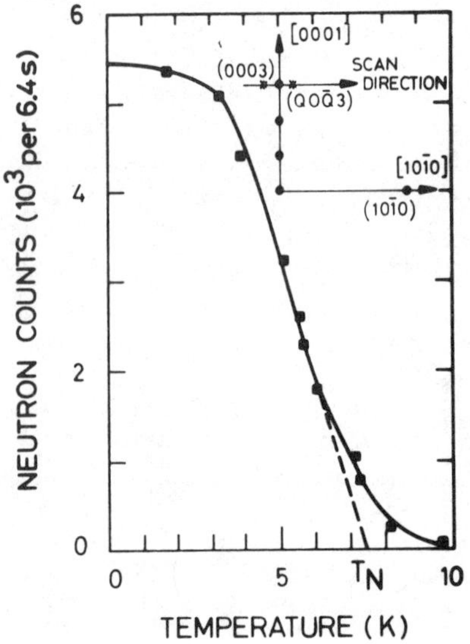


Fig. 25. Temperature dependence of the peak intensity of the (Q0Q̄3) satellite reflection observed in Pr when an uniaxial stress of 800 bar is applied along [12̄10]. (from Ref. 35).

Thus hydrostatic pressure or uniaxial stress may be used to study the T = 0 phase transitions between polarized and paramagnetic states in exchange induced antiferro- or ferromagnets.

3 - Magnetic phase transitions : MnO

Hydrostatic pressure or uniaxial stress may also be used to study the thermally induced order-disorder transition. The sub-lattice magnetization can be determined through neutron scattering experiments. From its thermal dependence in the critical order-disorder transition region, important characteristics of the phase transition may be deduced. Pressure or stress are then added as an ingredient to study the effect of modifications of interatomic distances and/or crystal symmetry on the critical parameters.

As demonstrated by Alben[36], the dimensionality of the order parameter can be larger than 3 for antiferromagnets or helimagnets. Let us consider manganese oxide (MnO) which possesses a fcc structure. The magnetic structure consists of ferromagnetic (111) planes, that is planes where the magnetic moments are all parallel together. The successive (111) planes are stacked antiferromagnetically. As four equivalent body diagonals exists, there are four different propagation vectors for this magnetic structure. For a chosen propagation direction, the magnetic moment is defined from its three components and then 12 components are to be specified in order to describe the actual situation. When the magnetic moments lie in a (111) plane, which is the case for MnO, due to dipolar forces, the dimensionality is reduced to 8.

We will describe neutron scattering experiments on a single crystal of MnO[37]. The intensity of the magnetic peak, at low temperature, is proportional to the square of the magnetic moments, and thus from diffraction experiments performed at various temperatures, one can obtain the temperature dependence of the magnetic moments, and therefore the temperature dependence of the order parameter.

At zero (or small) stress, one notes a discontinuous jump of the sublattice magnetization at $T_1 \sim 120$ K associated with the discontinuous character of the antiferro-paramagnetic transition in MnO[38]. The crystallographic structure of MnO is cubic above 120 K ; it undergoes a rhombohedral distortion below 120 K, the (111) antiferromagnetic planes move closer together and the ferromagnetically aligned atoms within (111) plane move further apart.

If we apply a stress along the [111] diagonal, this rhombohedral distortion is increased by an amount readily calculated from the values of the elastic coefficients. If we look at the temperature dependence of the order parameter at various applied stresses[38], one notices that the discontinuity of the order parameter decreases when the stress increases. The antiferro-paramagnetic transition temperature itself increases with stress at a rate of approximately 3 K/kbar. At 5.5 kbar, the stress remains uniaxial, as observed from the width of the lattice reflection lines. However, the paramagnetic transition is then clearly continuous.

The effect of the dimensionality of the spin system on the character of the transition has been noticed in the last years independently by Brazovskii and Dzyaloshinskii[39], and Bak, Krinsky, and Mukamel[40], which have demonstrated that in the space of parameters, no stable fixed point exists for $n > 4$, which is indicative of a discontinuous transition, as observed at zero stress in manganese oxide. This argument has been used by Bak et al[41], to give an interpretation of the tricritical properties of MnO, where

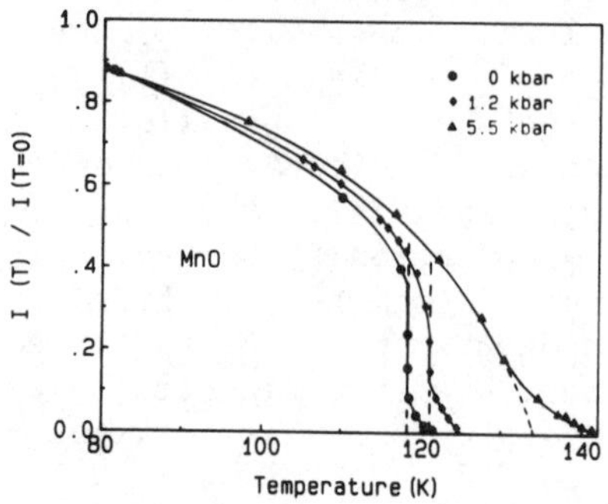


Fig. 26. Variation of the integrated intensity of the (333) lines with temperature for several applied stresses (from Ref. 37).

the applied [111] compressive stress reduces from 8 to 2 the dimensionality of the order parameter[36] and therefore changes the character of the antiferro-paramagnetic transition from discontinuous (n = 8) to continuous (n = 2). The treatment however does not give an account of the tricritical behaviour observed in MnO. The effects of fluctuations for propagation vectors out of the stress direction are in fact not entirely suppressed for small stresses and a more accurate description of the critical region should be used in order to explain the observed phase diagram[42]. These experiments demonstrate the interest of stress experiment for a new kind of studies related to phase transition.

## 4 - Magnetoelastic interactions : MEM(TCNQ)$_2$

In this paragraph, we discuss a category of materials which are characterized by their very large compressibility, typically one or two orders of magnitude larger than measured in usual magnetic materials. Pressure thus produces drastic physical modifications. Furthermore, the P = 0 magnetoelastic ground state itself may be completely at variance from that of an uncompressible material. As an example, for the material under consideration, the T = 0 ground state will be non-magnetic, whereas the T = 0 ground state of the infinitely rigid lattice would be magnetic.

MEM(n-methyl-n-ethylmorpholinium) is a donor molecule which combines with TCNQ (ditetracyanoquinodimethanide) to give $MEM^+(TCNQ)_2^-$. Above 335 K planar TCNQ forms columns along the $\vec{c}$-axis of the triclinic structure, with a perpendicular to their planes lying approximately along the $\vec{c}$-axis, and almost uniform (U) stacking. At 335 K, a dimerization (D) occurs[43,44] producing distances between planes of 3.22 and 3.36 Å and considerable atomic rearangement. The one-dimensional electrical conductivity along the $\vec{c}$-axis in the U-phase is associated with the TCNQ columns, with 1/2 electron per TCNQ molecule. The (D) phase has semi-conducting properties. The U-D transition characterized by both the occurrence of dimerization and a metal-insulator transition is therefore associated with an Electronic-

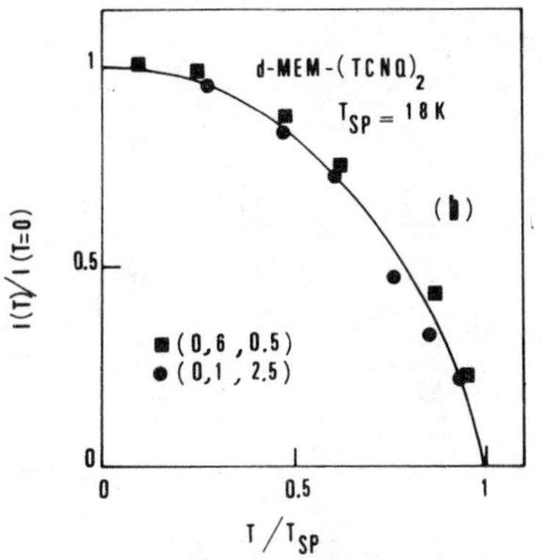


Fig. 27. Deuterated MEM-$(TCNQ)_2$. The dimerization of pairs of $(TCNQ)_2$ dimers (quadramerization) below 18 K from neutron scattering experiments.

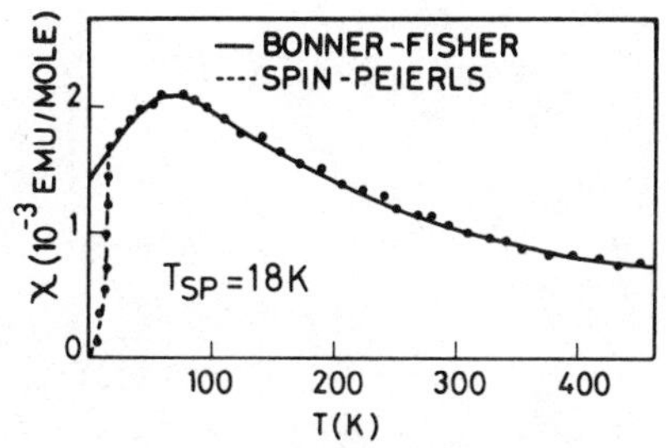


Fig. 28. Above $T_{SP}$ = 18 K the susceptibility of MEM-$(TCNQ)_2$ follows the Bonner-Fisher[46] law for 1d-antiferromagnet. A continuous magnetic-non-magnetic transition occurs at 18 K.

Peierls (EP) transition. In the D-semiconducting phase, a magnetic moment of 1 $\mu_B$ is attached to the S = 1/2 unpaired electron of the $(TCNQ)_2^-$ dimer. $MEM^+$ is non magnetic. This dimerization is followed at 18 K by a quadramerization associated with a transverse displacement of the TCNQ with respect to each other (Fig. 27). Simultaneously (Fig.28) one notes the occurrence of a continuous magnetic to non-magnetic transition which is the spin-Peierls transition.

Let us first consider a 1-d antiferromagnetic (J < 0, S = 1/2) Heisenberg chain with exchange interactions restricted to first neighbours, at a distance a

$$H = - J \sum_{i=1}^{N} \vec{S}_i . \vec{S}_{i+1}$$

The linear chain exhibits a non-magnetic $S_t$ = 0 ground state and an excited magnetic state with $S_t$ = 1. The energy spectrum has been calculated by Descloiseaux and Pearson[45] ; it is characterized by the absence of an energy gap at k = 0 and k = $\Pi$/a (Fig. 29a). Quantum zero point fluctuations therefore populate the magnetic states, leading to a non-vanishing T = 0 magnetic susceptibility (see Fig. 28, Bonner and Fisher[46]). In alternating chains, with distances a(1-$\varepsilon$) and a(1+$\varepsilon$) and therefore inequivalent exchange interactions J, a gap occurs at k = 0 and k = $\Pi$/a (Fig. 29b), which implies that the susceptibility vanishes exponentially at T = 0. Various models have been developed[47] which demonstrate that a T = 0

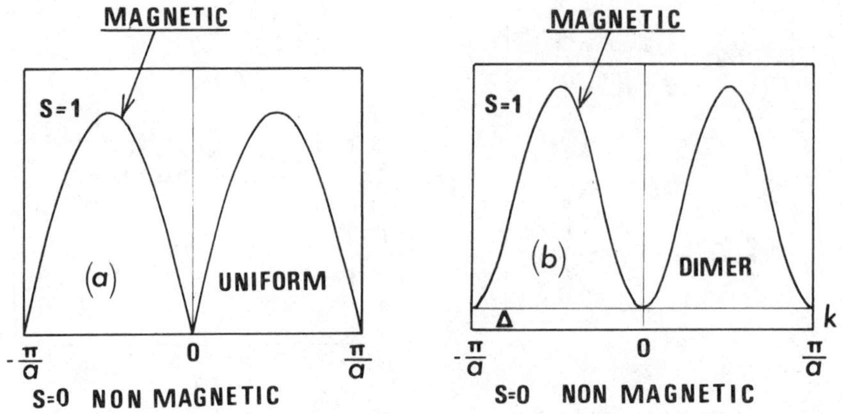


Fig. 29. For an uniform magnetic S = 1/2 antiferromagnetic Heisenberg chain (a) there is no gap between the non-magnetic $S_t = 0$ ground state and the excited magnetic $S_t = 1$ state. Therefore (Fig. 28) the T = 0 susceptibility is not zero. For a dimerized chain (b) a gap takes place between the $S_t = 0$ and $S_t = 1$ state.

dimerization in an uniform chain at T = 0, costs an elastic energy which is overcome through the decrease of the magnetic energy as associated with the opening of the gap in the magnetic excitation spectrum. The transition to the distorted dimerized state can occur at finite temperature, due to the 3d character of the phonon field.

This purely quantum mechanism, associated with a linear dependence of exchange interaction on distances, is sufficient to explain the unstable magnetic state of a linear uniform Heisenberg chain versus dimerization ; it is the magnetic analogous of the electronic Peierls transition. It has been called the spin-Peierls transition. Note[48] that a non linear dependence of exchange interaction distances may give rise to such a magnetic to non-magnetic spin-phonon T = 0 instability.

Neutron scattering experiments have been performed[49] using an aluminium cell for pressure up to 5 kbar and an alumina cell for pressures up to 13 kbar. Data have been taken at 4.2 K (from an initial Q state) and at 300 K (from an initial D state). There exists a crystallographic pressure phase transition at ∿ 1.5 kbar (300 K) and ∿ 3.5 kbar (4.2 K) (Fig. 30). The high pressure crystallographic parameters are, at any of those two temperatures, very similar to those obtained at atmospheric pressure and T = 348 K, in the uniform U phase. A pressure of 7 - 8 kbar is sufficient, as noticed previously to reduce the volume from ∿ 10 %.

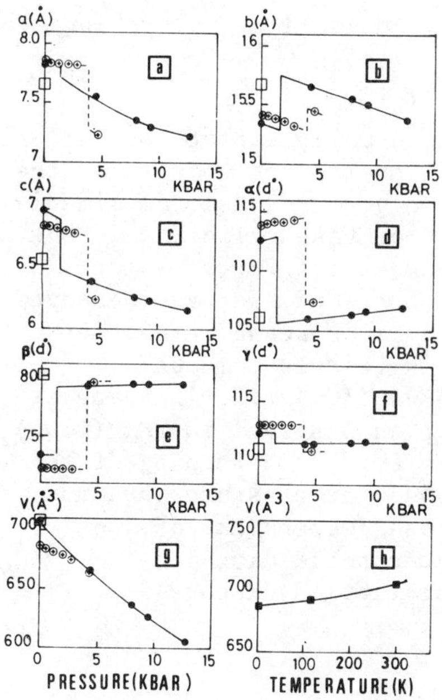


Fig. 30. Pressure dependence of the crystallographic parameters of the deutereted MEM-$(TCNQ)_2$ at T = 4.2 K (⊕) and T = 300 K (●). The data (□) correspond to the U-phase at P = 0, T = 348 K

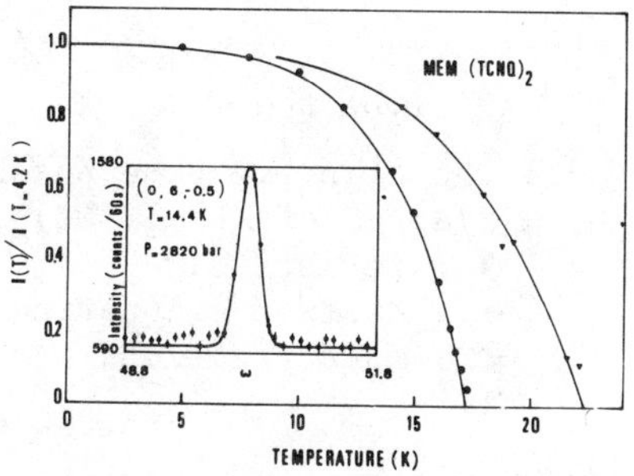


Fig. 31. Neutron scattering intensity of the (0, 6, -0.5) reflection measured as a function of temperature at P = 2800 (▼) and P = 1 bar (●). Insert : ω-scan at P = 2800 bar, T = 14.4 K, counting time : 60 s/point.

At low temperature the tetramerization of the TCNQ molecules gives rise to superlattice peaks along the c-direction. The experiments (Fig. 31) indicate an increase of the spin-Peierls temperature from 17.4 at atmospheric pressure to 22.6 K at 2.6 kbar. No tetramerization has been detected between 300 and 1.5 K at 4.6 kbar.

The high pressure phase diagram, as obtained from neutron scattering, magnetization, resistivity experiments is shown in Fig. 32. We thus note a total disappearance of both electronic-Peierls transition and spin-Peierls transition above 4 kbar. The pressure induces a cross-over from a 1d-behaviour at low pressure to a 3d-one at high pressure. For a linear dependence of exchange interactions on interatomic distances

$$T_{SP} \alpha (dJ/dr)^2/\omega_s^2$$

and

$$\frac{d \log T_{SP}}{dp} = -2 \frac{d \log \omega_s}{dp}$$

One usually expects an hardening of the lattice mode $\omega_s$ at the wave vector where the dimerization takes place, and thus a decrease of $T_{SP}$ as a function of pressure. We have observed such a decrease in another spin-Peierls compounds TTF-BDT(Cu). The increase of $T_{SP}$ as observed in MEM-$(TCNQ)_2$ may imply a softening of the $\omega_s$ mode with pressure or a non linear dependence of J on interatomic distances (or lattice distortion).

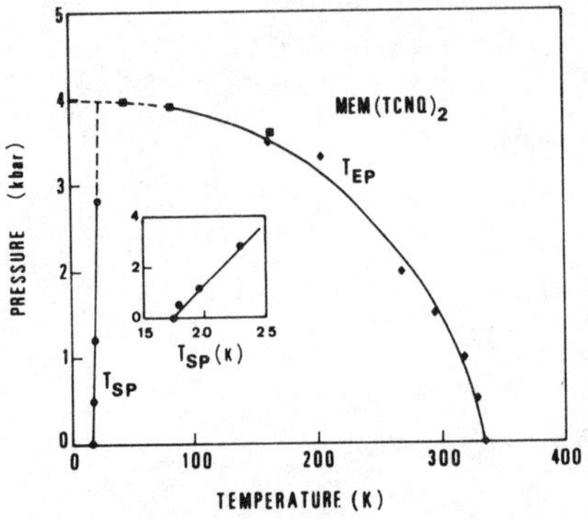


Fig. 32. The high pressure phase diagram of MEM-$(TCNQ)_2$.
◆ : resistivity ;
● : neutron scattering ;
■ : magnetization.

## ACKNOWLEDGEMENTS


The authors acknowledge D.B. McWhan and C. Vettier for their large participation to the development of high pressure neutron scattering at Grenoble, and therefore, for their contribution to a large number of the experiments described in this lecture.


## REFERENCES


1. H.G. Drickamer, R.W. Lynch, R.L. Cledenen and F.A. Perez-Albuerne, in "Solid State Physics", 17, p. 135, F. Seitz and D. Turnbull Edit., Academic press, New York (1966).
2. J. Yamashita and S. Asano, J. of the Phys. Soc. Japan, 52:3506 (1983).
3. S.N. Vaidya and G.C. Kennedy, J. Phys. Chem. Solids, 32:951 (1971).
4. a - E.F. Skelton, W.T. Elam, S.B. Quadri and A.W. Webb,
   b - N. Hamaya, T. Yagi, S. Yamaoko, O. Shimomura and S.A. Kimoto
   c - K. Tsuji
   To be published in the Proc. of the International Symposium on Solid State Physics unde Pressure at IZU-Nagoya, Japan, (January 1984).
5. G.A. Samara, in High pressure and low temperature Physics, p. 255, C.W. Chu and J.A. Wollam Edit., Plenum (1977).
6. J.D. Jorgensen and T.G. Worlton, Phys. Rev. B, 17:2212 (1977).
7. K. Asaumi, Phys. Rev. B, 29:1118 (1984).
8. E. Franceschi and G.L. Olcese, Phys. Rev. Lett., 22:1299 (1969).

9. A. Jayaraman, P.D. Dernier and L.D. Longinotti, Phys, Rev. B, 11:2783 (1975).
10. D.B. McWhan, G. Parisot and D. Bloch, J. Phys. F : Metal Phys. 4:L69 (1974).
11. K. Takemura and S. Minomura in Physics of solid under high pressure, p. 131, J.S. Schilling and R.N. Shelton (Edits), North Holland Publishing Co,(1981).
12. D. Jerome, A. Mazaud, M. Ribault and K. Bechgaard, J. Phys. Lett., 41:L95 (1980).
13. D. Jerome, Mol. Cryst. Liq. Cryst., 79:155 (1982).
14. G. Wortmann, K. Syassen, K.H. Frank, J. Feldhaus and G. Kaindl, in Valence instabilities, p. 159, P. Wachter and H. Boppart (Edits), North-Holland Publishing Co, (1982).
15. a - A.S. Balchan and H.G. Drickamer, J. Chem. Phys., 34:1948 (1961).
b - B.M. Riggleman and H.G. Drickamer, J. Chem. Phys., 37:446 (1962) ; 38:2721 (1963).
16. K. Takemura, S. Minomura, O. Shimomura and Y. Fujii, Phys. Rev. Lett., 45:1881 (1980).
17. J.G. Huber, M.B. Maple and D. Wohlleben, Sol. Stat. Commun., 16:211 (1975).
18. C.B. Bargeron, M. Avinor and H.G. Drickamer, Inorganic Chem., 10:1338 (1971).
19 - D. Bloch and J. Paureau, in High pressure chemistry, p. 111, H. Kelm (Edit.) (1978).
20. C.J. Carlile, D.C. Salter, High temperatures, high pressures, 10:1 (1978).
21. D. Bloch, J. Paureau, J. Voiron and G. Parisot, Rev. Scient. Instr., 47:296 (1976).
22. D.B. McWhan, D. Bloch and G. Parisot, Rev. Scient. Instr., 45:643 (1974).
23. A. Draperi, D. Hermann-Ronzaud,and J. Paureau, J. Phys. E, 9:174 (1976).
24. J. Pretchel, E. Lüscher and J. Kalus, J. Phys. E, 10:432 (1977).
25. J. Eckert, W.B. Daniels, J.D. Axe, Phys. Rev. B, 14:3649 (1976).
26. G.T. Worlton and R.A. Beyerlein, Phys. Rev. B, 12:1899 (1975).
27. P.S. Peercy I.J. Fritz and G.A. Samara, J. Phys. Chem. Solids, 36:1105 (1975).
28. D.B. McWhan, R.J. Birgeneau, W.A. Bonner, H. Taub and J.D. A J.D. Axe, J. Phys. C : Sol. Stat., 8:L81 (1975).
29. D. Bloch and C. Vettier, J. Magn. Magn. Mat., 15:589 (1980).
30. K.C. Tuberfield, L. Passell, R.J. Birgeneau and E. Bucher, J. Appl. Phys., 42:1746 (1971).
31. C. Vettier, D.B. McWhan and E.I. Blount, Phys. Rev. Lett., 39:1028 (1977).
32. D.B. McWhan, C. Vettier, R. Youngblood and G. Shirane, Phys. Rev. B, 20:4612 (1979).

33. R.P. Guertin, J.E. Crow, L.D. Longinotti, E. Bucher, L. Kupferberg and S. Foner, Phys. Rev., 12:1005 (1975).
34. R.P. Guertin, J.E. Crow, F.P. Missell and S. Foner, Phys. Rev. B, 17:2183 (1978).
35. K.A. McEwen, W.G. Stirling and C. Vettier, Phys. Rev. Lett., 41:343 (1978).
36. R. Alben; C.R. Acad. Sc. Paris, 279:111 (1974).
37. D. Bloch, D. Herrmann-Ronzaud, C. Vettier, W.B. Yelon and R. Alben, Phys; Rev. Lett., 35:963 (1975).
38. D. Bloch, R. Maury, C. Vettier and W.B. Yelon, Phys. Lett., 49A:354 (1974).
39. S.A. Brazorskii and I.E. Dzyaloshinskii, J.E.T.P. Lett., 21:164 (1975).
40. P. Bak, S. Krinsky and D. Mukamel, Phys. Rev. Lett., 36:52 (1976).
41. P. Bak, S. Krinsky and D. Mukamel, Phys. Rev. Lett., 36:829 (1976).
42. E. Domany, D. Mukamel and E. Fisher, Phys. Rev. B, 15:5432 (1977).
43. S. Huzinga, J. Kommandeur, G.A. Sawatsky, B.T. Thole, K. Kopinga, W.J.W. De Jonge and J. Ross, Phys. Rev. B, 119:4723 (1979).
B. Van Bodegom and A. Bosch, Acta Cryst., B37:863 (1981).
A. Bosch and B. Van Bodegom, Acta Cryst., B33:3013 (1977).
V. Van Bodegom, Acta Cryst., B37:857 (1981).
B. Van Bodegom, B.C. Larson and H.A. Mook, Phys. Rev. B, 24:1520 (1981).
44. R.J.J. Visser, S. Oostra, C. Vettier and J. Voiron, Phys. Rev. B, 28:2074 (1983).
45. J. Descloiseaux and J.J. Pearson, Phys. Rev., 128:2131 (1962).
46. J.C. Bonner and M.E. Fisher, Phys. Rev., 135:A640 (1964).
47. J.W. Bray et al, in "Extended linear chain materials", III, p. 353, J.S. Miller (Edit.), Plenum (1982).
48. D. Bloch, J. Voiron and L.J. de Jongh, in "High field magnetism", p. 19, M. Date (Edit.), North-Holland Publishing Co (1983).
49. D. Bloch, J. Voiron, C. Vettier, J.W. Bray and S. Oostra, Physica B+C, 119:43 (1983) ; J. de Physique C3-44:1317 (1983).

# BAND STRUCTURE EFFECTS IN THE ELECTRONIC PROPERTIES OF SOLIDS ACCESSIBLE IN NEUTRON SCATTERING STUDIES*

J. F. Cooke

Solid State Division
Oak Ridge National Laboratory
Oak Ridge, TN 37849

J. A. Blackman, and T. Morgan

Physics Department
University of Reading
Reading, England

ABSTRACT

The introduction of high-flux high-energy pulsed neutron sources may allow the study of electronic transitions and elementary excitation phenomena not previously accessible by inelastic neutron scattering techniques. Magnetic scattering from semiconductors and ferromagnetic transition metals are two important areas of interest from the point of view of this new technique. In order to determine what new information could be obtained from such experiments, it is necessary to carry out rather detailed theoretical calculations of the magnetic cross-section based on realistic band structures. Results of total magnetic cross-section calculations for silicon and germanium based on an empirical pseudopotential description of the electronic band structure are presented and analyzed in terms of proposed experiments in the eV range. In addition, some preliminary results from magnetic scattering calculations for ferromagnetic nickel are given. These calculations are based on energy bands which were previously found

*Research sponsored by the Division of Materials Sciences, U.S. Department of Energy under contract W-7405-eng-26 with Union Carbide Corporation.

to yield an excellent description of the complicated spin dynamics of nickel for relatively low energy and wave-vector transfer. Areas of interest such as the determination of the spin-splitting energy, Stoner modes, and high-energy spin waves are discussed.

## I. INTRODUCTION

Inelastic neutron scattering experiments using thermal neutrons have traditionally been limited to neutron energy transfers of ~ 100 meV or less. This, of course, is ideal for the study of a wide variety of the elementary excitation spectra in solids. However, most electronic excitations which can occur in solids lie well outside this range of energy. The development of high-flux high-energy spallation neutron sources could, in principle, raise this limit significantly, perhaps into the few-eV range. This type of source, therefore, has the potential of becoming an important tool for determining band structural information as well as dispersion curves for high-energy elementary excitations.

Because of the potential importance of this neutron source we have initiated a study of the inelastic magnetic neutron scattering for metallic and semiconductor systems. There are two important categories of questions that must be addressed. One is associated with the type of information which might be obtained. What does the structure in the cross section look like? Can it be related directly or indirectly to features of the band structure and/or crystal field levels? What type of elementary excitations and electronic transitions occur at high energies and what sort of structure do they contribute to the cross section?

The other type question concerns the order of magnitude of the scattering. Are the cross section intensities for the structure of interest large enough to be measured? This is a nontrivial matter for experiments carried out for large neutron energy transfer. In general, neutron experiments are carried out for as small a wave-vector transfer, $\vec{Q}$, as possible because of the dramatic decrease in intensity due to form factors or matrix elements decreasing with increasing $\vec{Q}$. On the other hand, $\vec{Q}$ must be large enough to satisfy the energy and momentum conservation conditions for these higher energy transfer experiments. The hope is that there will be a region of compromise where the intensities can be successfully measured.

In this paper we present some results from two different studies of the theory of inelastic neutron scattering for energy transfers of a few electron volts or less. The purpose of this work is to aid in assessing the feasibility of determining new information about electronic transitions (band structure effects) and elementary excitations at higher energy transfers. The first study is concerned with the possibility of determining band structural information about semiconductors. The second and, more ambitious, study is to investigate the scattering from ferromagnetic transition metals which results from electronic transitions as well as elementary excitations (spin waves).

The work on semiconductors represents an extension of previous calculations[1] for silicon and germanium to $\vec{Q}$'s which are more relevant to realistic experimental conditions. In this paper, we will be concerned only with calculations of the inelastic magnetic cross section for various fixed $\vec{Q}$'s as a function of neutron energy transfer. Questions related to the feasibility of such experiments have already been investigated by Allen et al.[2]

The investigation of magnetic scattering from transition metal ferromagnets also represents an extension of previous work.[3-5] This earlier work was restricted to $\vec{Q}$'s in the first Brillouin zone and primarily to energies less than 200 meV. These calculations revealed some interesting features in nickel along the [100] direction, such as the existence of an "optical" spin-wave mode and the continuation of the acoustic spin-wave branch to high energy. These predictions were subsequently confirmed by experiment[6] but the complete dispersion curves could not be completely mapped out because of the relatively high energies encountered. One of the main purposes of this work is to extend the calculations to such higher energy phenomena.

Another interesting aspect of scattering from transition metal magnets is the possibility of extracting information about the spin-polarized band structure. In principle, the cross section should contain information about the spin-splitting energy, Stoner modes, as well as non spin-flip electronic transitions. As will be shown, the calculations and analysis of this type of data are quite different from the semiconductor case because of the strong exchange and correlation effects which play a dominant role in magnetic systems.

In order to adequately assess the feasibility of obtaining information of the type referred to above one must carry out realistic calculations of the total inelastic magnetic scattering cross section for neutron energy transfers up to a few electron volts. Prior to this work, calculations of the total magnetic cross section (spin + orbit) have been limited to noninteracting systems of electrons described by the free-electron model[7-9] or by noninteracting tight-binding models of d-band transition metals.[10]

The remainder of this paper is divided into sections which relate to the theory, numerical results, and conclusions for the semiconductor and transition metal magnet problems. The general theory of neutron scattering is reviewed extensively in Ref. 11

and will not be discussed here. Instead, we will begin with expressions which are appropriate to the particular case being discussed. We will consider the semiconductor problem first.

## II. THEORY

In the one-electron approximation, the total magnetic cross section for a nonmagnetic system can be written as the sum of spin and orbital terms:

$$\frac{d^2\sigma}{d\Omega dE} = \left.\frac{d^2\sigma}{d\Omega dE}\right|_{\text{spin}} + \left.\frac{d^2\sigma}{d\Omega dE}\right|_{\text{orbital}} , \tag{1}$$

where

$$\left.\frac{d^2\sigma}{d\Omega dE}\right|_{\text{spin}} = \left(\frac{\gamma e^2}{m_e c^2}\right)^2 \frac{|\vec{k}|}{|\vec{k}_o|} \sum_{\substack{nm\\ \vec{q}}} |F^{(s)}_{nm}(\vec{q},\vec{Q})|^2 \, (f_{n\vec{q}} - f_{m\vec{q}+\vec{Q}})$$

$$\times\ \delta\left(\omega + E(n\vec{q}) - E(m\vec{q}+\vec{Q})\right) , \tag{2}$$

and

$$\left.\frac{d^2\sigma}{d\Omega dE}\right|_{\text{orbital}} = \left(\frac{\gamma e^2}{m_e c^2}\right)^2 \frac{|\vec{k}|}{|\vec{k}_o|} \frac{1}{2|\vec{Q}|^2} \sum_{\substack{nm\\ \vec{q}}} |\vec{F}^{(o)}_{nm}(\vec{q},\vec{Q})|^2 (f_{n\vec{q}} - f_{m\vec{q}+\vec{Q}})$$

$$\times\ \delta\left(\omega + E(n\vec{q}) - E(m\vec{q}+\vec{Q})\right) . \tag{3}$$

The spin and orbital form factors $F^s$ and $F^o$ respectively are defined by

$$F^{(s)}_{nm}(\vec{q},\vec{Q}) = \frac{1}{V_o} \int d^3r \; e^{i\vec{Q}\cdot\vec{r}} \, \overline{\psi}_{n\vec{q}}(\vec{r}) \psi_{m\vec{q}+\vec{Q}}(\vec{r}) , \tag{4}$$

$$\vec{F}^{(o)}_{nm}(\vec{q},\vec{Q}) = \frac{1}{V_o} \hat{Q}X \int d^3r \; e^{i\vec{Q}\cdot\vec{r}} [\overline{\psi}_{n\vec{q}}(\vec{r}) \nabla\psi_{m\vec{q}+\vec{Q}}(\vec{r}) - \psi_{m\vec{q}+\vec{Q}}(\vec{r}) \nabla\overline{\psi}_{n\vec{q}}(\vec{r})]. \tag{5}$$

In these expressions, $V_o$ is the volume of the unit cell, $E(n\vec{q})$ is the energy, $\psi_{n\vec{q}}(\vec{r})$ is the Bloch wave function, and $f_{n\vec{q}}$ is the Fermi occupation number. The notation $|\vec{F}^{(o)}|^2$ represents the scalar product $\vec{F}^{(o)}\cdot\vec{F}^{(o)}$, and the symbols n and $\vec{q}$ are band and wave-vector labels respectively. These expressions are evaluated for a semiconductor where the Fermi energy lies in the band gap. It can be seen from Eqs. (2) and (3) that the inelastic scattering results entirely from interband transitions, i.e., from the valence band to the conduction band.

The band structure used in this work was obtained using the empirical pseudopotential method of Cohen and Bergstrasser.[12] The pseudopotential method is based on the orthogonalized plane wave approach where the electronic wave function has the form

$$\psi_{n\vec{q}}(\vec{r}) = \phi_{n\vec{q}}(\vec{r}) - \sum_{\vec{G}} A_{n\vec{q}}(\vec{G}) \sum_{j} \alpha^{j}_{\vec{q}}(\vec{G}) \sum_{\ell} e^{-i(\vec{q}+\vec{G})\cdot\vec{R}_\ell}\, \theta^{j}(\vec{r}-\vec{R}_\ell) \, , \qquad (6)$$

$$\alpha^{j}_{\vec{q}}(\vec{G}) = \frac{1}{\sqrt{V_o}} \int d^3r\, \bar{\theta}^{j}(\vec{r}) e^{-i(\vec{q}+\vec{G})\cdot\vec{r}} \, , \qquad (7)$$

and $\vec{G}$ is a reciprocal lattice vector. The sum on $\ell$ is over both unit cells and sites within the unit cell, and $\theta^{j}(\vec{r}-\vec{R}_\ell)$ is the j-th core wave function associated with site $\ell$. The coefficient, $\alpha^{j}_{\vec{q}}(\vec{G})$, ensures orthogonalization of $\psi$ to the core states. The "pseudopotential" wave function, $\phi_{n\vec{q}}(\vec{r})$, is given by

$$\phi_{n\vec{q}}(\vec{r}) = \frac{1}{\sqrt{V_o}} \sum_{\vec{G}} A_{n\vec{q}}(\vec{G})\, e^{-i(\vec{q}+\vec{G})\cdot\vec{r}} \, . \qquad (8)$$

The expansion coefficient, $A_{n\vec{q}}(\vec{G})$, and the electronic energy, $E(n\vec{q})$, can be obtained by solving the Schrödinger-like equation

$$\left( - \frac{\hbar^2}{2m} \nabla^2 + V_p(\vec{r}) \right) \phi_{n\vec{q}}(\vec{r}) = E(n\vec{q})\, \phi_{n\vec{q}}(\vec{r}) \, . \qquad (9)$$

$V_p(\vec{r})$ is the pseudopotential, which can be expanded in terms of reciprocal lattice vectors

$$V_p(\vec{r}) = \sum_{\vec{G}} V_G \cos \vec{G}\cdot\vec{\tau}\, e^{-i\vec{G}\cdot\vec{r}} \, , \qquad (10)$$

where $G = |\vec{G}|$ and $\vec{\tau} = a\left(\frac{1}{8},\frac{1}{8},\frac{1}{8}\right)$. The solution of Eq. (9) can be simplified considerably by choosing a coordinate system where the two atoms in each unit cell are located at $\vec{R} \pm a\left(\frac{1}{8},\frac{1}{8},\frac{1}{8}\right)$, where $\vec{R}$ is an fcc lattice vector. This ensures that the $A_{n\vec{q}}(\vec{G})$ are real.

Numerical calculations of the cross section have shown that the contributions from the core part of the wave function in Eq. (6) are small and can, in practice, be neglected. By using this approximation, the matrix elements in Eqs. (4) and (5) become

$$F_{nm}^{(s)}(\vec{q},\vec{Q}) = \sum_{\vec{G}} A_{n\vec{q}}(\vec{G})A_{m\vec{q}+\vec{Q}}(\vec{G}) \quad , \tag{11}$$

and, with $\hat{Q} = \vec{Q}/Q$,

$$\vec{F}_{nm}^{(o)}(\vec{q},\vec{Q}) = -2i\hat{Q} \times \sum_{\vec{G}} (\vec{q}+\vec{G})\, A_{n\vec{q}}(\vec{G})A_{m\vec{q}+\vec{Q}}(\vec{G}) \quad . \tag{12}$$

## III. COMPUTATIONAL RESULTS

Calculations of the neutron scattering cross sections given in Eqs. (2) and (3) involve the determination of the band structure (electronic energies and wave functions), the spin and orbital matrix elements, and the relevant band and Brillouin zone sums. The band structure was calculated by solving Eq. (9) with a three parameter pseudopotential, where $\vec{G}$ in Eq. (10) was restricted in magnitude to 3, 8, and 11 (in units of $2\pi/a$). For completeness, the parameters are given in Table 1. Further details of the calculation are given in Ref. 1. The electronic density of states for silicon and germanium calculated using these parameters are given in Figs. 1 and 2, respectively. These results are very similar and show the valence and conduction bands separated by the band gap of about 1 eV.

Table 1. Pseudopotential parameters in Ry as defined in Eq. (10) for silicon and germanium. Only nonzero terms correspond to G = 3, 8, and 11 (in units of $2\pi/a$).

| $V_G$ | Si | Ge |
|---|---|---|
| $V_3$ | − 0.21 | − 0.23 |
| $V_8$ | 0.04 | 0.01 |
| $V_{11}$ | 0.08 | 0.06 |

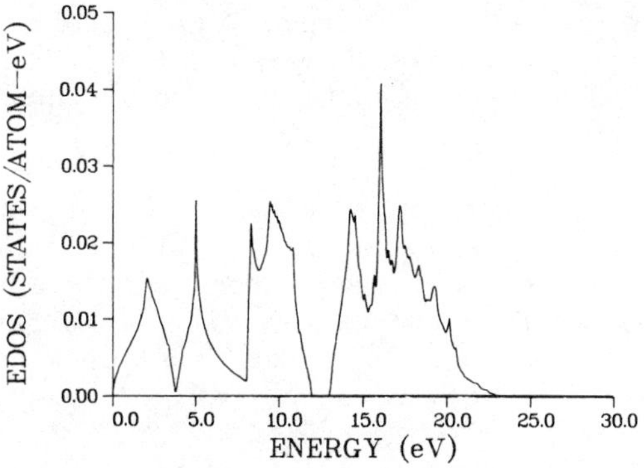


Fig. 1. Electronic density of states for silicon. The results are for four conduction and four valence bands.

The Brillouin zone sums were carried out by using symmetry arguments to reduce the full zone sum to sums over the irreducible zone and appropriate point group operations. The irreducible zone was subdivided into 1536 tetrahedra. The electronic energy, wavefunction expansion coefficients, and spin and orbital matrix elements were calculated on a mesh generated by unique corner points of these tetrahedra. The tetrahedron method[13] was then used to evaluate the Brillouin zone sums.

We have carried out an extensive series of calculations of the total cross section for $\vec{Q}$ along the [100], [111], and [110] directions. These results are based on the use of four valence and four conduction bands which should ensure reasonable accuracy

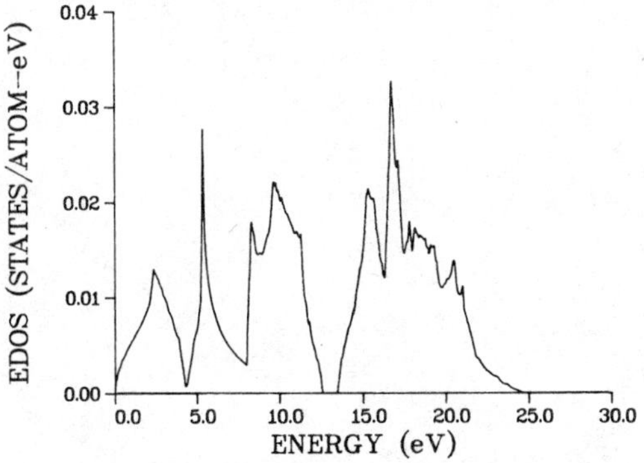


Fig. 2. Electronic density of states for germanium. The results are for four conduction and four valence bands.

for neutron energy transfers of a few eV or less. All factors in the cross section expressions have been included except for the ratio $|\vec{k}|/|\vec{k}_o|$, which depends on the particular conditions of the experiment.

As mentioned before, these calculations represent an extension of earlier work to $\vec{Q}$'s outside the first Brillouin zone. In the course of this work an error was detected in the part of the computer program which manipulated the $A_{nq}^{\rightarrow}(\vec{G})$. This resulted in some changes in the overall structure of the cross section. It turns out, however, that these changes do not affect significantly the conclusions given in Ref. 1. In addition, the intensities given in Ref. 1 were actually integrated intensities over a histogram representation 23 meV wide. These points should be kept in mind when comparing the results presented here to those given in Ref. 1.

For purposes of clarity, only a few representive examples of our results will be given. General features of these results will be pointed out when appropriate. One important result of all the numerical work is that the band and wave-vector dependence of the spin and orbital matrix elements can not be ignored. This is clearly demonstrated by the results given in Fig. 3 which represent a comparison of the joint density of states (JDOS) to the total scattering cross section for silicon with $\vec{Q} = (3.0,0.0,0.0)$. The JDOS can be obtained from Eq. (2) by replacing $F^{(s)}$ by a constant. For some $\vec{Q}$'s, the JDOS does resemble the total cross section but this is by no means a general result. The results given in Fig. 3 demonstrate another important point. Notice that there is no significant structure in the total cross section which can be

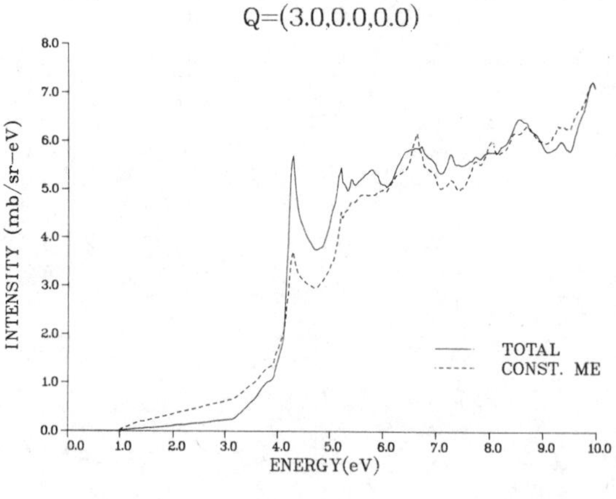


Fig. 3. Comparison of total scattering intensity and constant matrix element results (JDOS) for silicon with $\vec{Q} = (3,0,0,0)$. The dotted curve is scaled to the solid curve for comparison.

associated with the band gap (~ 1 eV). The reason for this is that the density of states for transitions across the band gap (the JDOS) is relatively small and featureless. This was found to be the case for both silicon and germanium.

Comparisons of the total scattering cross section to the orbital contribution for germanium and silicon are given by Figs. 4-7. The difference is, of course, the spin contribution. Notice that the intensity scale is in millibarns/(sr-eV). There are

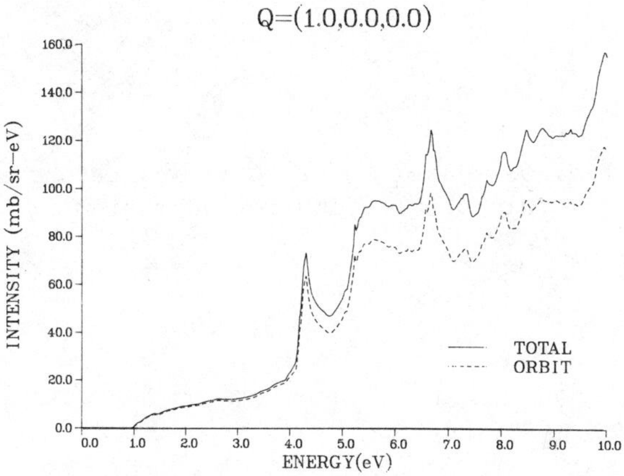


Fig. 4. Results for total scattering and orbital contribution for silicon with $\vec{Q}$ = (1.0,0,0).

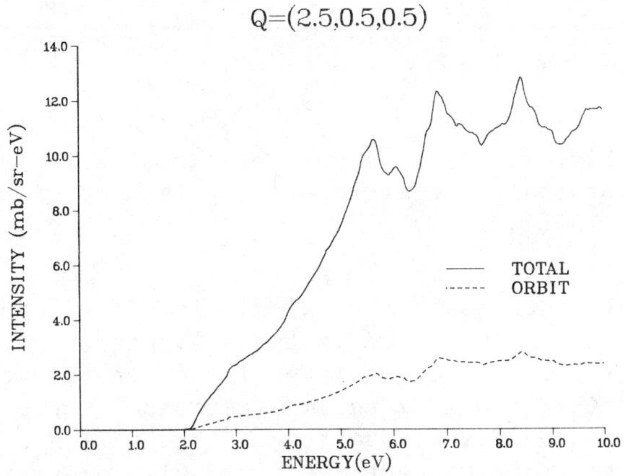


Fig. 5. Results for total scattering and orbital contribution for silicon with $\vec{Q}$ = (2.5,0.5,0.5).

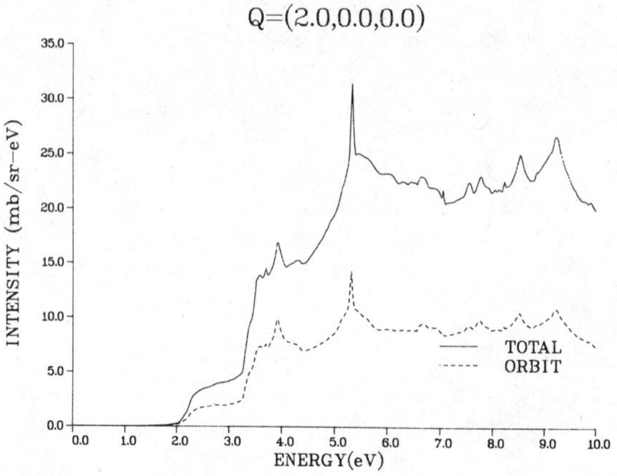


Fig. 6. Results for total scattering and orbital contribution for germanium with $\vec{Q}$ = (2.0,0,0).

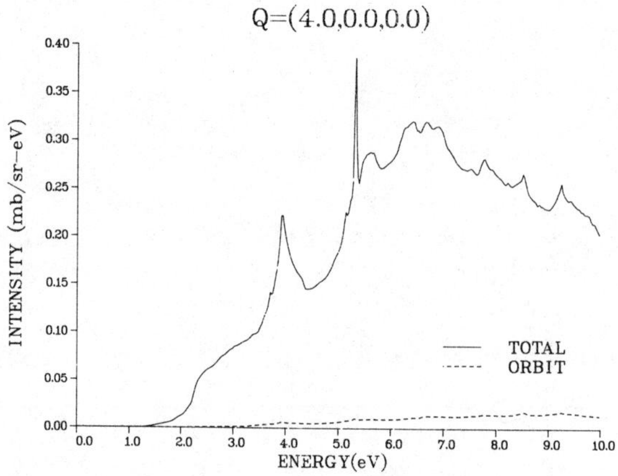


Fig. 7. Results for total scattering and orbital contribution for germanium with $\vec{Q}$ = (4.0,0,0).

several general features of the calculations for both silicon and germanium that are illustrated in these figures. For example, the orbital part of the scattering intensity dominates at small $\vec{Q}$ inside the first Brillouin zone but decreases quite dramatically with increasing $\vec{Q}$. This is due to $F^{(o)}$ and $F^{(s)}$ and is expected behavior, but the rate of decrease with $\vec{Q}$ is larger than predictions made on the basis of free electron models.[7-9]

The intensity profiles clearly have considerable energy-dependent structure which also varies with $\vec{Q}$. As noted before, there are no special features which can be related to the band gap. However, if the structure in these intensity profiles could be measured, then a "best" set of the pseudopotential parameters defined in Eq. (10) could be determined from some type of least squares fitting procedure. Another important feature of these results is that the total scattering cross section also decreases significantly with increasing $\vec{Q}$. It appears from arguments given in Ref. 2 that it is unlikely that measurements can be made for $\vec{Q}$'s inside the first Brillouin zone. This is due to the large neutron energy transfer. Therefore, the scale of intensities given in Figs. 5 and 6 is probably a reasonable upper bound for the scattering intensities to be expected in realistic experiments in the near future.

## IV. CONCLUSIONS FOR SCATTERING FROM Si AND Ge

The results presented in this paper have helped to resolve some important questions relevant to the feasibility of determining band structure information from inelastic neutron scattering experiments performed in the eV energy range. It is clear from the calculations that the band gap in silicon and germanium, and quite possibly all semiconductors, can not be determined directly from experiment. However, theoretical fits to structure obtained from measured intensity profiles could be used to determine relevant pseudopotential parameters, thereby, determining indirectly a "best" set of pseudopotential bands.

The question of whether such experiments can actually detect this structure remains unanswered. Clearly, experiments should be carried out for as small a $\vec{Q}$ as practical. The lower bound on $\vec{Q}$ is dictated by the energy and momentum conservation conditions relevant to the neutron energy transfer. The best compromise for silicon and germanium seems to be for $\vec{Q}$'s in the neighborhood of those given in Figs. 5 and 6. That is to say, the experiments must be able to detect intensities in the range of tens of millibarns/(sr-eV).

It is clear, therefore, the experiments which have been proposed for obtaining information about the band structure of semiconductors from neutron scattering experiments are going to be difficult to perform. If the challenge of measuring these relatively small cross sections can be met, then neutron scattering could become an important source of information about the band structure of semiconductors.

## V. THEORY FOR ITINERANT ELECTRON FERROMAGNETS

We turn now to the case of calculating the total inelastic magnetic scattering from itinerant electron ferromagnets. The development of the theory for this problem is much too complicated and lengthy to present here. Instead, a brief outline of the theoretical approach will be given along with the final expressions for the cross section which form the basis of the calculations.

As in the semiconductor problem, the total magnetic cross section can be written in terms of spin and orbital contributions. Unlike the semiconductor case, however, we can not use the noninteracting electron expression because of strong exchange and correlation effects which are inherent in ferromagnetic metals. These effects lead to spin-dependent electronic energy bands and wave functions and to a separation of the spin part of the scattering into spin-flip and non spin-flip parts. Because of the nature of the terms involved, spin-flip scattering is referred to as transverse scattering and non spin-flip scattering as longitudinal scattering. If spin-orbit effects are neglected, the orbital scattering does not involve a spin flip of the neutron and differs from the spin scattering in much the same way as in the semiconductor case, i.e., through differences in the form factor.

The general approach that we have used to calculate the cross section is based on an extension of previous work[3-5] which provides the details of the theory for determining the transverse part. This part contains information about spin waves which were the primary focus of that work. If we want the total scattering, the longitudinal and orbital contributions must also be included. This can be accomplished by deriving the equation of motion for the relevant Green's functions and using the same generalized RPA decoupling scheme[3-5] to make the equations tractable. An important step in solving the Green's function equations is to use the screened coulomb matrix element approximation given in Ref. 3-5. This approximation incorporates the important band and wave-vector dependence of these matrix elements and allows the Green's function equations to be converted into relatively simple matrix equations which can be solved in a straightforward manner.

The procedure results in closed form expressions for the transverse (T), longitudinal (L), and orbital (O) parts of the scattering cross section which are too lengthy and complicated to reproduce here. There are, however, a number of general features of these results which are worth noting. First we write the electronic wave functions in the KKR form

$$\psi_{n\vec{q}\sigma}(\vec{r}) = \sum_{\mu} a_{n\mu\sigma}(\vec{q})\phi_{\mu}^{\sigma}(\vec{r}) \ , \qquad (13)$$

where the $\phi_\mu$ are symmetry orbitals with symmetry index $\mu = (\ell, m)$, $a_{n\mu\sigma}(\vec{q})$ are expansion coefficients, and $\sigma$ is the spin index. Then each part of the scattering cross section can be written in terms of a complex function, $I^A$, which has the general form

$$I^A(\vec{Q},z) = \sum_{N,N'} F_N^A(\vec{Q})\, H_{N,N'}^A(\vec{q},z)\, F_{N'}^A(\vec{Q}) \;, \tag{14}$$

where A represents T, L, or O, z is a complex variable, and N is a combination of symmetry labels $(\mu, \nu)$. The wave vectors $\vec{Q}$, $\vec{q}$, and $\vec{G}$ are related by

$$\vec{Q} = \vec{q} + \vec{G} \;, \tag{15}$$

where $\vec{G}$ is a reciprocal lattice vector and $\vec{q}$ is restricted to the first Brillouin zone. The $F_N$ are form factor-like quantities which depend on $\{\phi_\mu^\sigma\}$ but are not z-dependent. The $H^A$ depend on Brillouin zone sums of a function of the electronic energy and $a_{n\mu\sigma}(\vec{q})$. The cross section is proportional to the imiginary part of $I^A$ with $z = \omega + i\varepsilon$, where $\omega$ is the neutron energy transfer and $\varepsilon$ is infinitesimally small. Notice that $H^A$ depends only on $\vec{q}$ and contains all of the $\omega$ dependence of the scattering.

The form of Eq. (14) points out one very serious problem. If we include s-, p-, and d-like symmetry terms (9 in all) in the wave function expansion, then N must run over 9x9 = 81 terms and, therefore, $H^A$ is 81x81. If we calculate the cross section on any reasonable size energy mesh (say 500 to 1000 points) from zero up to a few eV, then we clearly have a massive computer storage problem. Because of this, complete calculations of the total cross section can not be carried out at present.

It seems reasonable as a first step to neglect the s- and p-symmetry contributions to the scattering, since we expect the d-symmetry terms to dominate. This does not imply that the s- and p-symmetry terms can be neglected altogether, because they are important for the band structure calculation. This approximation reduces the calculation to a manageable size and work is currently under way to obtain numerical results for the total cross section.

As a first step we have evaluated $I^A$ in Eq. (14) by including only the terms for which the form factor terms $F_N^A$ are diagonal in their symmetry indices, i.e., only those N corresponding to $(\mu, \mu)$ combinations. These terms represent the leading contribution to the spin part of the scattering but do not adequately represent the orbital part. For this reason, we will consider only the spin part of the scattering from this point on. Methods for calculating the orbital part are currently under investigation.

With these approximations, $I^A$ assumes the following form:

$$I^A(\vec{Q},z) = |F_d^A(\vec{Q})|^2 \sum_{N,N'}' H^A_{N,N'}(\vec{q},z) \equiv |F_d^A(\vec{Q})|^2 I_o^A(\vec{q},z) \,, \qquad (16)$$

where $F_d^A$ is the corresponding d-like form factor and the sum is restricted to "diagonal" d-symmetry terms $(N = (\mu,\mu))$. It should be remembered that the expression given in Eq. (16) represents actual contributions to the scattering cross section. More accurate calculations will simply increase the number of terms in the sum on N and N' but will not alter the energy-dependent structure present in $I_o^A$.

Explicit results for $I_o^A$ are given below for the transverse and logitudinal terms.

$$I_o^T(\vec{q},z) = \sum_{\substack{\sigma \\ N,N',\overline{N}}}' [I+\Gamma^{\sigma,-\sigma}(\vec{q},z)]^{-1}_{N,N'} \, \Gamma^{\sigma,-\sigma}_{N',\overline{N}}(\vec{q},z)/U_{\overline{N}} \,, \qquad (17)$$

$$I_o^L(\vec{q},z) = \sum_{\substack{\sigma \\ N,N',\overline{N}}}' \left([I-\Gamma^{-\sigma,-\sigma}(\vec{q},z)][I-\Gamma^{\sigma,\sigma}(\vec{q},z)\Gamma^{-\sigma,-\sigma}(\vec{q},z)]^{-1}\right)_{N,N'} \Gamma^{\sigma,\sigma}_{N',\overline{N}}(\vec{q},z)/U_{\overline{N}} \,, \qquad (18)$$

where I is the unit matrix and $N = (\mu,\mu)$, $N' = (\nu,\nu)$,

$$\Gamma^{\sigma,\sigma'}_{NN'}(\vec{q},z) = \frac{U_\nu}{N_o} \sum_{\substack{nm \\ \vec{p}}}' \frac{\{a_{m\mu\sigma}(\vec{p}+\vec{q})a_{n\mu\sigma'}(\vec{p})\}\{a_{m\nu\sigma}(\vec{p}+\vec{q})a_{n\nu\sigma'}(\vec{p})\}}{z-E(m\vec{p}+\vec{q}\sigma)+E(n\vec{p}\sigma')} \times \{f_{n\vec{p}\sigma'} - f_{m\vec{p}+\vec{q}\sigma}\} \,, \qquad (19)$$

$$\{a_1a_2\} = a_1a_2 + a_2a_1 \,, \qquad (20)$$

and $U_{N'} = U_{(\nu,\nu)} = U_\nu$. The $U_\nu$ are constants which are part of the screened coulomb matrix element approximation discussed above. There are only two independent $U_\nu$, one for $t_{2g}$ and the other for $e_g$ symmetry.

The scattering expressions in Eqs. (16), (17), and (18) are more complicated than the ones used for the semiconductor case. The exchange and correlation effects included in the calculation for itinerant ferromagnets have introduced enhancement factors into the scattering cross section. These are the matrices which multiply the $\Gamma^{\sigma,\sigma'}_{N',\bar{N}}/U_{\bar{N}}$ term in Eqs. (17) and (18). If these enhancement factors are replaced by the unit matrix then it can be shown that the scattering cross sections reduce to the noninteracting result examined in Section II. The enhancement factors are very important since they significantly affect the behavior of the scattering cross section and incorporate spin waves into the theory. If the band and wave-vector dependence of the screened coulomb matrix elements used in the derivation of Eqs. (17) and (18) are neglected, we obtain the familiar enhancement factors of Izuyama, Kim, and Kubo.[14] As will be shown in the next section, this is not a good numerical approximation.

## IV. NUMERICAL RESULTS FOR NICKEL - SPIN ONLY

The numerical evaluations of the cross section expressions for the transition metal ferromagnets nickel and iron is currently in progress. Thus far, we have obtained preliminary results only for nickel for scattering along the [100] direction. This part of the problem was done first because it appeared to be the most interesting from the point of view of scattering at energies above 100 meV. These results, together with a brief review of previous work, will be given in this section.

The starting point of the cross section evaluation is the band structure calculation. The method of calculation is outlined in Refs. 3-5. In general terms, the calculation uses as input a "realistic" set of paramagnetic bands and the two parameters $U_\nu$ (for $e_g$ and $t_{2g}$ symmetries) mentioned in the previous section. A self-consistent calculation is then carried out to generate the ferromagnetic band structure. The two parameters are then fixed to give the measured moment and symmetry character of the moment (from form factor data). Results of such a calculation are shown in Fig. 8. There are a number of interesting features of this band structure which are worthy of note. The energy difference between opposite spin bands at the same wave vector is called the spin-splitting energy and it is clearly wave-vector dependent. The primary reason for this is that the spin-splitting energy for pure $e_g$ and $t_{2g}$ symmetry states differ significantly (e.g., the rigidly spin-split d-band states along Γ to X). Because of the relatively small $e_g$ splitting we find only one d-like band above the Fermi energy at the X point. The spin-splitting energy of the d-bands nearest the Fermi energy at the L point was found to be ~ 300 meV. These features are in excellent agreement with experiment and are

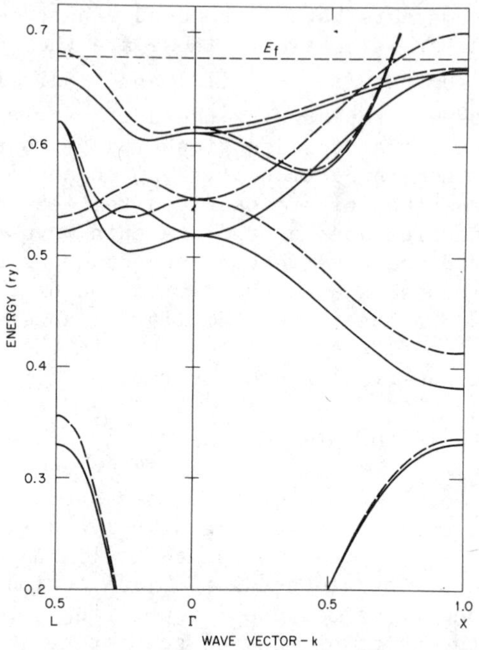


Fig. 8. Energy bands for ferromagnetic nickel.

significantly different from those obtained from most other published band structures for nickel.

The inelastic scattering cross section calculated from this band structure was also found to have some unusual features. Previous calculations[3,5] were restricted to energies less than 200 meV, which is well below the range of interest for electronic transitions. A summary of the spin-wave spectra obtained from the position of peaks in the transverse cross section is given in Fig. 9. The spin-wave spectra obtained from the band structure is shown in Fig. 8 and is the curve labeled CD1. Notice that the acoustic branch appears to cross and interact with an "optical" branch. The calculated intensity in the part of the lower branch which can be associated with this optical mode was found to be relatively small and this part of the curve ($|\vec{q}|>0.5$) has not been observed experimentally. However, the rest of the spectra, including the behavior in the region where the two modes interact and the continuation of the acoustic branch to high energies, was subsequently verified by experiment.[6] Notice also the excellent agreement between theory and experiment below 100 meV. The curve labeled CD2 represents only the lower part of the spin-wave spectra obtained from a slightly different, but equally "reasonable," set of input paramagnetic bands.

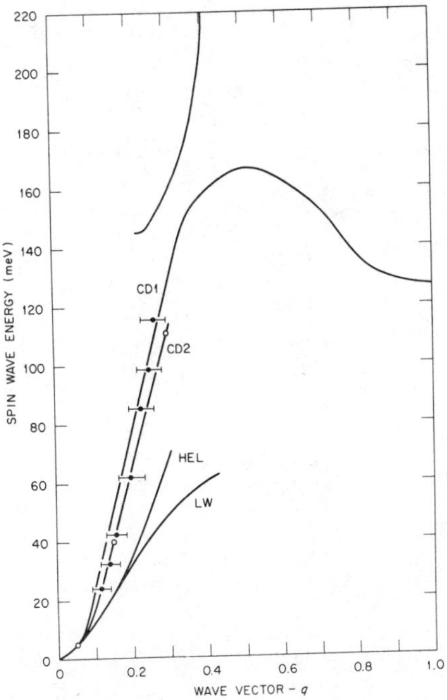


Fig. 9. Spin-wave dispersion curves for ferromagnetic nickel. Experimental data are represented by the bars and are from Ref. 15.

The curve labeled HEL was generated by a set of ferromagnetic bands which are known to be incorrect. Finally, the curve labeled LW was obtained using the same set of bands used to generate the CD1 spectra but neglecting the band and wave-vector dependence of the screened coulomb matrix element in the band structure and cross section expressions. This result demonstrates the need for including these effects in the calculation.

In the remainder of this section we will present some numerical results for the spin part of the zero temperature scattering for fixed $\vec{Q}$ along [100] and neutron transfer energies ranging from zero to approximately 3 eV. These results do not include the form factor which, at the level of approximation of Eq. (16), is simply a scale factor. This work represents an extension of that presented in Refs. 3-5 to include the longitudinal scattering and a much higher range of neutron energy transfer.

Results for several different $\vec{Q}$'s are given in Figs. 10-11. Each figure gives the longitudinal, transverse, and total spin part of the scattering which has been scaled to unity for each $\vec{Q}$ in order

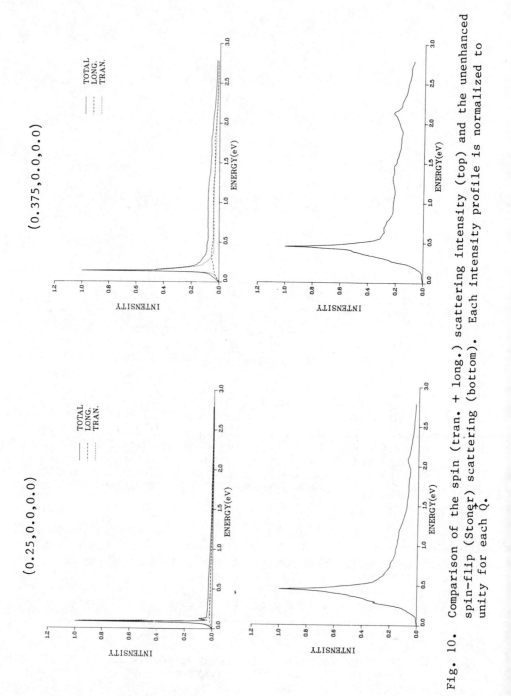

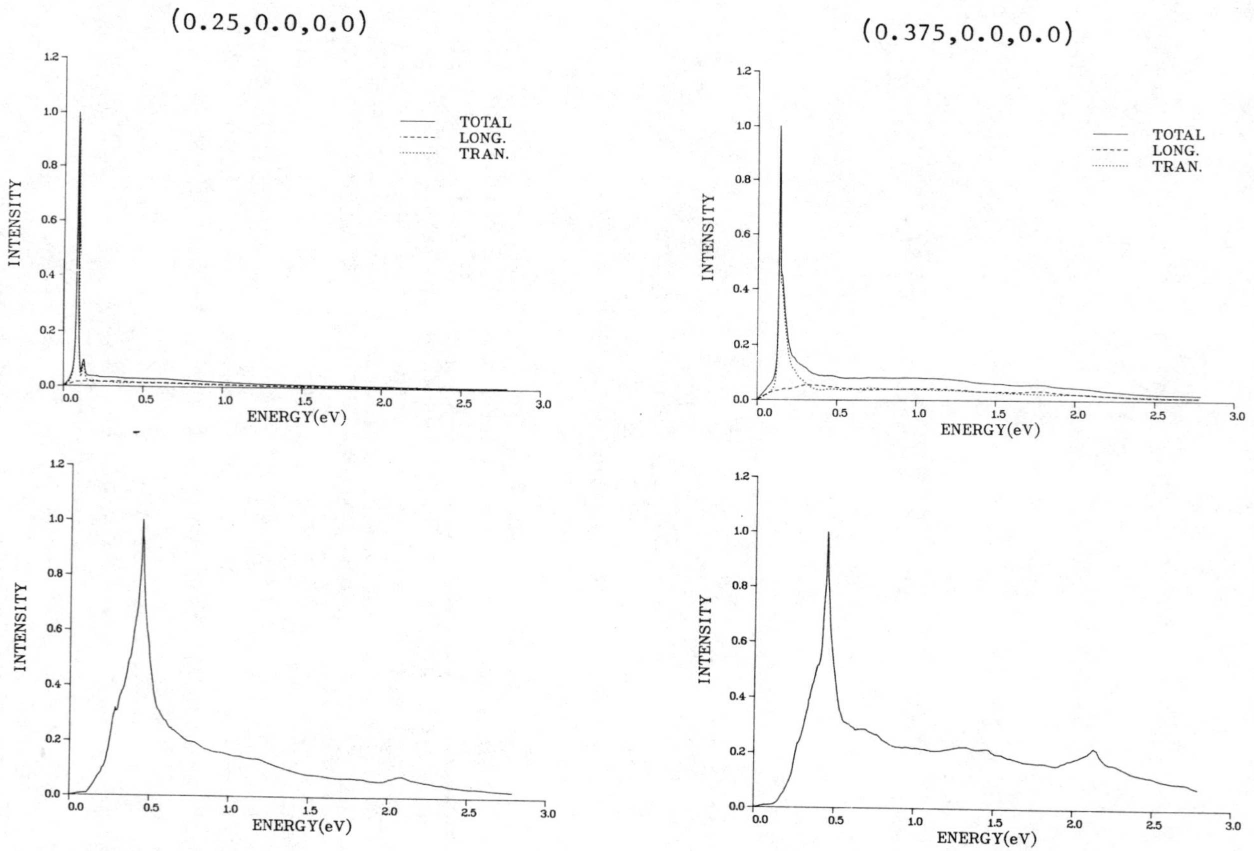


Fig. 10. Comparison of the spin (tran. + long.) scattering intensity (top) and the unenhanced spin-flip (Stoner) scattering (bottom). Each intensity profile is normalized to unity for each $\vec{Q}$.

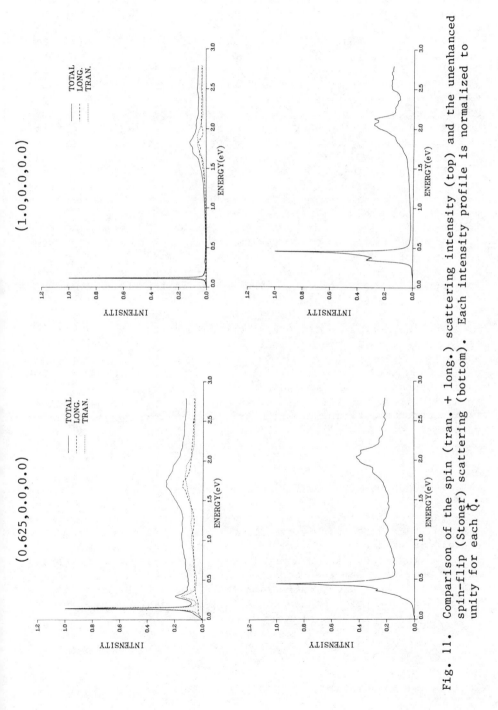

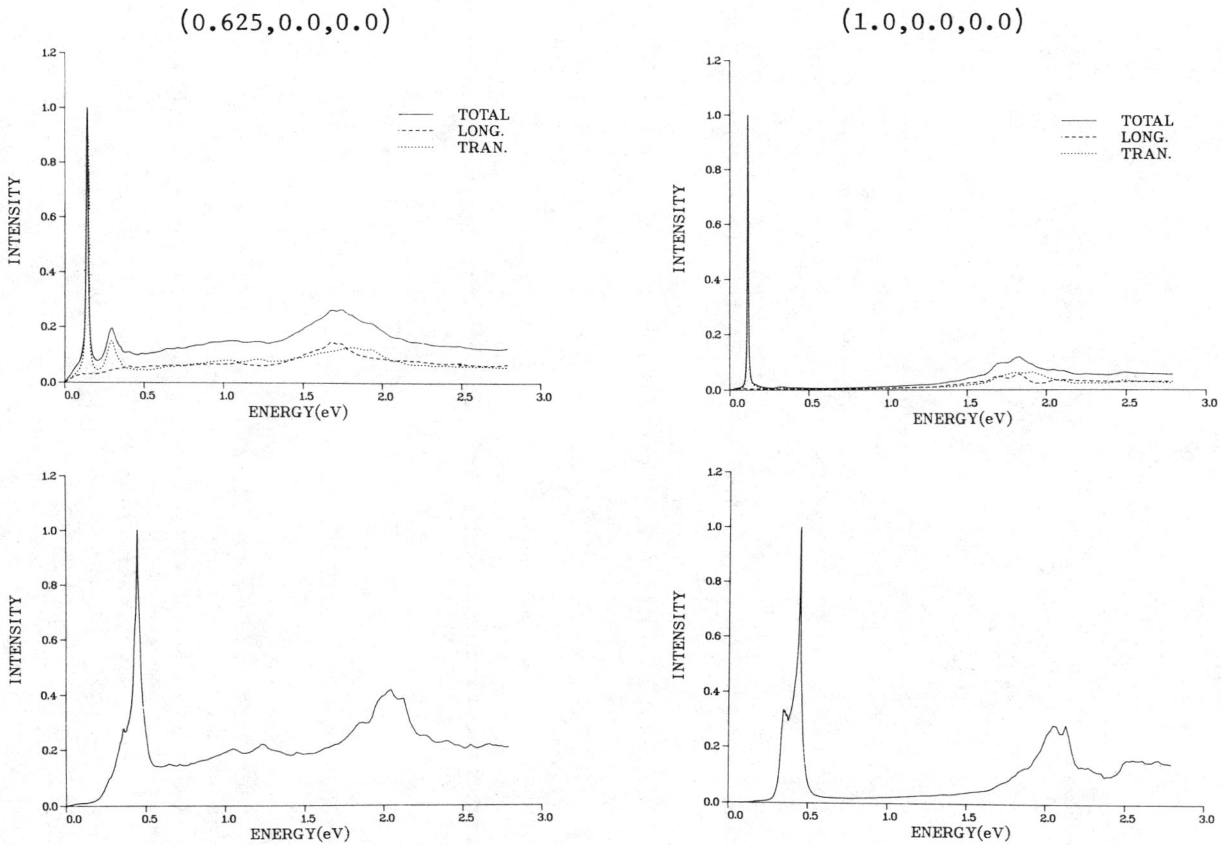


Fig. 11. Comparison of the spin (tran. + long.) scattering intensity (top) and the unenhanced spin-flip (Stoner) scattering (bottom). Each intensity profile is normalized to unity for each $\vec{Q}$.

to ease comparison. In reality the absolute intensities fall off considerably as $|\vec{Q}|$ increases. As an aid to interpreting the structure in the spin part of the scattering, we have calculated the transverse cross section with the enhancement factor replaced by the unit matrix. This yields the transverse part of the noninteracting cross section which results from spin-flip electronic transitions. Excitations of this type are called Stoner excitations. Some of these results are also shown in Figs. 10-11. Comparison of the total spin with the Stoner scattering results given in these figures for the same $\vec{Q}$ show that the structure at the lower energies comes from the enhancement factors (spin waves) while the structure at the higher energies (and high $\vec{Q}$) is generated by both spin-flip (Stoner) and non spin-flip single particle transitions.

There are clearly two spin-wave branches which persist out to the zone boundary ($\vec{Q} = (1,0,0)$ in units of $2\pi/a$). A plot of the position of the spin-wave peaks as a function of $\vec{Q}$ for energies below 200 meV yield the dispersion curves given in Fig. 9 (the curve labeled CD1). The intensity of the lower branch decreases significantly in the region where the two branches "cross" and remains relatively small out to the zone boundary. The intensity of the upper branch decreases and the width of the peak increases as $\vec{Q}$ approaches the zone boundary. The energy of this upper branch exceeds 300 meV at the zone boundary.

The large peak in the unenhanced part of the spin-flip scattering (Figs. 10-11) occurs in the range of spin-splitting energies shown in Fig. 8. Notice that this peak is completely suppressed by the enhancement factor for all $|\vec{q}|$ out to the zone boundary. This is unfortunate because it indicates that information about the spin-splitting energy for nickel can not be determined directly from neutron scattering experiments. On the other hand, the structure seen in the 1.5 to 2.0 eV energy range at the higher $\vec{Q}$'s could provide some useful band structure information.

## VII. CONCLUSIONS FOR SCATTERING FROM NICKEL

Based on the rather limited results given in the previous section it appears that important information about ferromagnetic nickel can be determined from neutron scattering experiments performed using a spallation source. Experiments at higher neutron energy could map out the complete spin-wave spectra and provide information about electronic transitions above 1 eV. Unfortunately, it appears that the enhancement factors suppress any structure in the scattering cross section below 1 eV which is generated by electronic transitions. In particular, there appears little hope that information about the spin-splitting energy for nickel could be determined directly from such experiments.

There is still a great deal of work to be done. We intend to carry out extensive calculations of the total cross section for nickel and iron in order to determine the complete spin-wave spectra along the major symmetry directions. Detailed analysis of the results of these calculations should provide a complete picture of the type of information which can be obtained about the band structure. There is some hope that the spin-splitting energy can be measured in iron since it is about a factor of five larger than in nickel while the scale of spin-wave energies appears to be roughly the same. Accurate calculations of the cross section will also help to assess the feasibility of detecting the structure which has been predicted.

ACKNOWLEDGEMENT

We are grateful to S. W. Lovesey of the Rutherford Appleton Laboratory for his continued support of the research.

REFERENCES

1. J. F. Cooke and J. A. Blackman, Calculation of neutron cross sections for interband transitions in semiconductors, Phys. Rev. B 26:4410 (1982).

2. D. R. Allen, E. W. J. Mitchell, and R. N. Sinclair, A resonance detector spectrometer for neutron inelastic scattering in the eV region, J. Phys. E 13:639 (1980).

3. J. F. Cooke and H. L. Davis, k-dependent exchange and the dynamic susceptibility of ferromagnetic nickel, p. 1218 in: "Proceedings of the Eighteenth Conference on Magnetism and Magnetic Materials," C. D. Graham, Jr. and J. J. Rhyne, eds., AIP Conf. Proc. No. 10, AIP, New York (1975).

4. J. F. Cooke, Neutron scattering from itinerant-electron ferromagmets, Phys. Rev. B 7:1108 (1973).

5. J. F. Cooke, J. W. Lynn, and H. L. Davis, Calculations of the dynamic susceptibility of nickel and iron, Phys. Rev. B 21:4118 (1980).

6. H. A. Mook and D. Tocchetti, Neutron-scattering measurements of the generalized susceptibility, Phys. Rev. Lett. 43:2029 (1979).

7. R. J. Elliott, Theory of neutron scattering by conduction electrons in a metal and on the collective-electron model of a ferromagnet, Proc. R. Soc. London Ser. A 235:289 (1956).

8. S. Doniach, Theory of inelastic neutron scattering in nearly ferromagnetic metals, Proc. Phys. Soc. London 91:86 (1967).

9. J. E. Hebborn and N. H. March, Orbital and spin magnetism and dielectric response of electrons in metals, Adv. Phys. 19:175 (1970).

10. S. W. Lovesey and C. G. Windsor, Neutron orbital cross section for a tight-binding model of paramagnetic nickel, Phys. Rev. B 4:3048 (1971).

11. W. Marshall and S. W. Lovesey, "Theory of Thermal Neutron Scattering," Oxford University Press, New York (1971).

12. M. L. Cohen and T. K. Bergstresser, Band structures and pseudopotential form factors for fourteen semiconductors of the diamond and zinc-blende structures, Phys. Rev. 141:789 (1966).

13. G. Lehmann and M. Taut, On the numerical calculation of the density of states and related properties, Phys. Status Solidi 54:469 (1972).

14. T. Izuyama, D. Kim, and R. Kubo, Band theoretical interpretation of neutron diffraction phenomena in ferromagnetic metals, J. Phys. Soc. Jpn. 18:1025 (1963).

15. H. A. Mook, J. W. Lynn, and R. M. Nicklow, Temperature dependence of the magnetic excitations in nickel, Phys. Rev. Lett. 30:556 (1973).

# STUDY OF LOCALISED ELECTRONIC EXCITATIONS BY NEUTRON SCATTERING

R.J. Elliott

Department of Theoretical Physics
1 Keble Road
Oxford OX1 3NP, U.K.

## 1. INTRODUCTION

Neutron scattering allows the spectroscopic study of the energy levels associated with both nuclear and electronic motion. With the advent of neutron sources with higher incident energy the upper limit of the range covered will be extended from say 0.1 eV to perhaps 1 eV. It is therefore appropriate to consider what types of experiment will be profitable in this area. The brief for these lectures is to consider localised electronic excitations.

Experiments to date have been largely concerned with crystal field levels in rare earth ions. The technique is particularly useful in metals where other spectroscopic methods cannot be used. A wide range of such levels, for example in some transition metal and actinide compounds, including spin-orbit splittings, will soon be accessible. It is also possible that low lying electronic levels of non-magnetic impurities in crystals and of atoms and molecules in gases might be studied.

The first part of the lectures will be concerned with the theory of such transitions between electronic energy levels, assuming fixed nuclei. Such transitions can be observed in principle at all $\underline{Q}$, but the form factors cause the cross sections to fall off rapidly at large Q. In addition to strictly localised levels a brief discussion of possible transitions to exciton states will be given. These are co-operative excitations with well defined momentum $\underline{q}$ which must be conserved in the transition, but are not being considered elsewhere in this meeting, although co-operative magnetic excitations (spin waves) will be discussed by Lindgard, while electronic band states will be considered by Cooke.

The coupling between the electronic and nuclear motion can give rise to important effects. Because of the direct coupling, a neutron scattering event can give a direct impulse to the nuclei which causes it to move relative to the electrons. Such a virtual process can result in transitions between electronic states. This process has been considered theoretically for atoms, molecules, impurities in crystals and excitons, and these results will be reviewed.

## 2. Scattering cross sections

The differential cross section for inelastic scattering may be written, for energy transfer $\omega$ and momentum transfer Q (Marshall and Lovesey 1971, called ML hereafter)

$$\frac{d^2\sigma}{d\Omega dE} = \frac{k}{k_o} \sum_f \Big| <0\Big| \sum_\ell (b + b'(\underline{\sigma}.\underline{I}_\ell))\exp(i\underline{Q}.\underline{R}_\ell)$$

$$+ \sum_i \beta(\underline{\sigma}.\underline{M}_i)\exp(i\underline{Q}.\underline{r}_i)\Big|f>\Big|^2 \delta(E_0 + E_1 - \omega) \qquad (1)$$

where we have assumed that initially the crystal is in the ground state $|0>$ and is excited into final state $|f>$. The first term in the matrix element is the coherent and incoherent interaction with a nucleus with spin $\underline{I}_\ell$ at $\underline{R}_\ell$ and the second is the magnetic interaction with an electron at $\underline{r}_i$; b, b' and $\beta = (e^2\gamma/m_e c^2)$ are the appropriate scattering lengths. The operator $M_i$ takes the form

$$\hat{\underline{M}}_i = \underline{Q} \times (s_i \times \hat{\underline{Q}}) - (i/\hbar Q)(\hat{\underline{Q}} \times \underline{p}_i) = \hat{\underline{Q}} \times (\underline{\mu}_i \times \hat{\underline{Q}}) \qquad (2)$$

where $\hat{\underline{Q}} = \underline{Q}/Q$ and

$$\underline{\mu}_i = \underline{s}_i - (i/\hbar Q)(\hat{\underline{Q}} \times \underline{p}_i). \qquad (3)$$

$\underline{\sigma}$ represents the neutron spin and for unpolarised neutrons the electronic part of (1) depends on

$$\sum_{\alpha\beta}(\delta_{\alpha\beta} - \hat{Q}_\alpha \hat{Q}_\beta)\beta <0|\sum_i \mu_{i\alpha}\exp(i\underline{Q}.\underline{r}_i)|f><f|\sum_i \mu_{i\beta}\exp(-i\underline{Q}.\underline{r}_i)|0>. \qquad (4)$$

## 3. Atomic transitions

The matrix elements of $\underline{\mu}\exp(i\underline{Q}.\underline{r})$ between atomic states have recently been considered in detail by Lovesey (1978). For a many-electron atom these reduce, since the operator is a sum over the electron operators, to an appropriate sum over one electron matrix elements

$$<\ell, m, m_s|\exp(i\underline{Q}.\underline{r})\underline{\mu}|\ell', m', m'_s> \qquad (5)$$

where the electron changes from angular momentum state $|\ell,m,m_s>$ to $|\ell',m',m_s>$ in the transition. For the purpose of subsequent discussion we will choose the direction of m quantisation along $\hat{\underline{Q}}$. Expanding in terms of spherical harmonics of the unit direction vector $\hat{\underline{r}}_i$

$$\exp(i\underline{Q}.\underline{r}_i) = 4\pi \sum_\lambda j_\lambda\ (Qr_i)\ Y^o_\lambda(\hat{\underline{r}}_i) \tag{6}$$

where $j_\lambda(x)$ is a spherical Bessel function. The only components of $\underline{\mu}_i$ which enter into (4) are those perpendicular to the direction $\hat{\underline{Q}}$. Hence in these axes there is a selection rule in (7) which requires

$$m-m' = 0;\quad m_s-m'_s = \pm 1 \tag{7a}$$

for the spin part of $\mu$ and

$$m-m' = \pm 1 \quad m_s-m'_s = 0 \tag{7b}$$

for the orbital part.

The actual values of the matrix elements are very complicated and have been written in terms of Clebsch-Gordon coefficients in ML. Since the convenient axis of quantisation for the atomic states will not in general coincide with $\hat{\underline{Q}}$, there will be a complicated dependence on the direction of scattering relative to the crystal axes.

In addition to the angular factors the matrix element includes the radial integral over the atomic wave function f in the form

$$\int f_\ell(r)\ j_\lambda(Qr) f_{\ell'}(r)\ r^2 dr \tag{8a}$$

for the spin part and

$$\frac{1}{Q} \int f_\ell(r) j_\lambda(Qr)[\frac{\partial}{\partial r} f_{\ell'}(r)] r^2 dr \tag{8b}$$

for the orbital part, which give a strong dependence on the magnitude of Q.

As $Q \rightarrow 0$, the orbital part of $\underline{\mu}$ which varies like $\underline{p}_i/Q$ can become large but only has matrix elements between atomic states of opposite parity. The next term in the exponential expansion is shown by ML to give, after some manipulation,

$$e^{i\underline{Q}.\underline{r}}\underline{\mu} \rightarrow \tfrac{1}{2}(\underline{\ell} + 2\underline{s}) \tag{9}$$

the operator of the electronic magnetic moment. In the presence of a magnetic field, Elliott and Kleppmann (1975) show that $\underline{\ell}$ should be replaced by

$$\underline{\ell} \Rightarrow \underline{\ell} + (e/2\hbar c)\underline{r} \times (\underline{H}\times\underline{r}) \qquad (10)$$

in the symmetric gauge $\underline{A} = \frac{1}{2}(\underline{H}\times\underline{r})$. Thus at low $\underline{Q}$ and therefore low energy transfer $\omega$, atomic transitions between states of the same parity will obey magnetic dipole selection rules. At intermediate values $Qa \sim 1$ where a is the atomic radius all the harmonics $\lambda$ in (6) play a significant role and the selection rules are much less restrictive.

At large Qa the form factors fall off rapidly - for hydrogenic wave functions like $(Qa)^{-(4+\ell+\ell')}$. For the simplest case of a hydrogenic s→p transition with a p-state orbital in the direction of a unit vector $\underline{n}$, with wave function $(\underline{n}.\underline{r})\phi_1(r)$, Elliott and Shukla (1979) show that the matrix element

$$\langle 0,0|e^{i\underline{Q}.\underline{r}}\,\mu|1,n\rangle \sim \frac{8\pi}{Q^5}\,[(\hat{\underline{Q}}.\underline{n})\;4\underline{s}\;\frac{\partial}{\partial r}\,[f_o(r)\phi_1(r)]_{r=0} -$$

$$- (\hat{\underline{Q}}\times\underline{n})\;[\phi_1(r)\,\frac{\partial}{\partial r}\,f_o(r)]_{r=0}\}. \qquad (11)$$

A somewhat more complicated atomic transition, $6s^2 \rightarrow 6s6p$ in KCl:Tl is considered by Lovesey (1978). He finds that the transition $^1S_0 \rightarrow {}^1P_1$ involves the orbital part of $\mu$, and has intensity $\frac{1}{3}|F^o|^2$ while the transitions to $^3P_1$ and $^3P_2$ involve the spin part and have magnitude $\frac{3}{2}|F^s|^2$. The transition $^1S_0 \rightarrow {}^3P_0$ is forbidden. Here

$$F^o(Q) = 2m_e(E_p-E_s)\;F_1^o(Q) + F_2^o(Q) \qquad (12a)$$

where $F_1^o(Q) = \int r^3 dr f_s(r) f_p(r)[j_o(Qr) + j_2(Qr)]/Q$

and $F_2^o(Q) = \int r dr f_s(r) f_p(r)[2j_o(Qr)-j_2(Qr)]/Q$

while $F^s(Q) = \int r^2 dr f_s(r) f_p(r) j_1(Qr)$. (12b)

The form factors involved in d→d and f→f transitions have been studied by Trammell (1953), Balcar (1970) and in detail in ML. The orbital and spin components are slightly different from each other. Those contributions involving the smaller $\lambda$ are larger at low Q where they vary as $Q^\lambda$ and all fall off rapidly at large Q.

These form factor properties mean that with present and forseeable techniques, atomic transitions will only have observable cross sections for $Qa \lesssim 1$. However for transitions of energy $\omega \sim 1$ eV say, geometrical conditions greatly restrict the available range of Q (Loewenhaupt 1984). For incident energy of 2 eV, $Q_{min} \sim 9 \times 10^8$ cm$^{-1}$ and this value does not fall to $1 \times 10^8$ cm$^{-1}$

until $E^o \sim 100$ eV. Under these conditions the cross section is only appreciable at very small angles $\sim 1^o$ and very good angular and energy resolution will be required.

## 4. Exciton states

Since these are not considered elsewhere we shall briefly consider the collective atomic like excitations called Frenkel excitons. Exciton states can be conveniently classified into two extreme types, though many are intermediate between them. In the Frenkel type the excitation is effectively confined to a single atom or molecule, and passes from one site to another across the crystal. Thus if $\Psi_\ell$ represents the crystal with atom $\underline{R}_\ell$ excited, the exciton state with wavevector q takes the form

$$N^{-1/2} \sum_\ell \exp(i\underline{q}.\underline{R}_\ell)\Psi_\ell . \tag{13}$$

In the Wannier exciton the excitation volume covers many atoms and the entity is best regarded as a bound electron-hole pair. The electron is in a conduction band state with wavevector $\underline{k}_e$ and the hole represents an electron missing from the valence band state $-\underline{k}_n$ so that the exciton state may be represented by the form

$$\sum_{k_e,k_n} A(\underline{k}_e,\underline{k}_n)\,|c,\underline{k}_c;\ v,\ \underline{k}_n> \tag{14}$$

where $\underline{k}_e+\underline{k}_n=\underline{q}$. For a weakly bound exciton with large spatial extent $A(\underline{k}_e,\underline{k}_n)$ will be large only over a narrow range of $\underline{k}_e$ and $\underline{k}_n$. The relative electron-hole motion can then be approximately represented by the wavefunction

$$\Phi(\underline{r}_e,\underline{r}_n) = \sum_{k_e k_n} A(\underline{k}_e,\underline{k}_n)\exp(i\underline{k}_e.\underline{r}_e)\exp(i\underline{k}_n.\underline{r}_n) = \phi(\underline{q},\underline{r}_e-\underline{r}_n) \times \exp[i\tfrac{1}{2}\underline{q}.(\underline{r}_e+\underline{r}_n)] \tag{15}$$

which is modulated by the periodic parts u of the Bloch functions.

The cross section for the creation of an exciton with state (13) from the ground state by scattering a neutron through $\underline{Q}$ is given by the square of the matrix element of $\Sigma\underline{\mu}_i \exp(i\underline{Q}.\underline{r}_i)$ between the atomic ground state and $\Psi_i$. As in (5) these are related to electronic matrix elements provided the momentum conservation is satisfied ie $\underline{Q}=\underline{q}+\underline{\tau}$. The cross section for a typical process is proportional to $\beta^2$(~0.1 barns) together with the form factor. For scattering within the first zone ($\tau=0$) this is comparable with the cross section for spin-wave scattering in a magnetic material. Detailed calculations for some atomic transitions are discussed above. For large Q the form factor falls off at at least as rapidly as $(aQ)^{-8}$.

For $Q = 3 \times 10^8 cm^{-1}$ and $a = 1 \times 10^{-8} cm$ this gives a reduction to $10^{-4}$ barns. As before it therefore seems essential to work at high incident energy and low $\underline{Q}$. Even in this regime however there may be difficulties due to the broad linewidth since it is known from optical experiments that these excitons interact strongly with phonons.

Many Frenkel excitons correspond to atomic transitions $p^6 \to p^5 s$ which can be written in terms of holes as $s^2 \to sp$, like the atomic transition considered earlier and discussed for small and intermediate Q by Lovesey (1978). This is a good approximation in rare gases and a moderate one in alkali halides. Using the approximate form (11) for large Q we see that spin scattering can only take to longitudinal excitons, while orbital scattering occurs to transverse excitons. Because of the singlet nature of the ground state the latter will be confined to singlet excitons. These selection rules will be broken down by spin-orbit coupling. White (private communication) has considered the experimental aspects of this problem and has suggested that it might be possible to detect the excitons produced in neutron scattering by fluorescence. Unfortunately the direct optical decay together with phonon generation of an exciton at $\underline{q}$ is likely to be much slower than the exciton-phonon scattering which transfers the exciton to its lowest $\underline{q} = 0$ state. Coincidence counting fluorescent photons may, however, give useful information.

The cross section for Wannier excitons is described in terms of band states which will be considered in other lectures.

## 5. Electronic excitation via the nuclear cross section

The possibility of observing electronic excitations by neutron scattering from the nuclear terms in (1) via the non-adiabatic nucleus-electron coupling was first considered by Kashcheev and Krivoglaz (1962) for impurity centres in crystals. These considerations were extended to excitons by Krivoglaz and Rashba (1980) and by Elliott and Shukla (1979). Lovesey et al. (1982) and Ruijgrok et al. (1983) have more recently considered the case of such scattering in gases of atoms and molecules, and we begin with these simpler cases.

The atomic wave function can be written

$$\psi(\underline{R}_\ell - \underline{r}_i)\, e^{i\underline{K}.\underline{R}} \tag{16}$$

where $\underline{r}_i - \underline{R}_\ell = \underline{r}$ is the distance between the electron and the nucleus and $\underline{R}$ is the centre of mass

$$\underline{R} = (m\underline{r}_i + M\underline{R}_\ell)/(m+M) \tag{17}$$

The matrix element for scattering, form (1) involves $e^{i\underline{Q}.\underline{R}}$ and using $\underline{R}_\ell = \underline{R} + m\underline{r}/(m+M)$, the matrix element is approximately

$$b\int\psi_o^*(r)\ \exp\ [im(\underline{Q}.\underline{r})/M]\psi_i(r)d\underline{r}\ \ \delta(\underline{K}_0-\underline{K}_1-\ \underline{Q}) \tag{18}$$

The smallness of (m/M) makes the dipole-approximation valid and the cross section for single electron atoms is then

$$b^2(m/M)^2|\langle\ell,m|\underline{Q}.\underline{r}|\ell'm'\rangle|^2\ \ \delta(\underline{K}_0-\underline{K}_1-\underline{Q})\ \ \delta(E_0-E_1\ -\omega) \tag{19}$$

where the energy conservation term involves both the centre of mass and the electronic motion. The cross section is small because of the term in the mass ratio. However in this process the form factor is negligible and it would be advantageous to work at large Q.

Another simple case which is instructive, is to consider a shell model impurity centre. In this model, used extensively in the theory of lattice vibrations, the atom is replaced by a central core which moves with the nucleus, and a light electron shell coupled together by a simple harmonic force. Thus the relative electron-nucleon motion is here approximated by a harmonic oscillator. We further assume that the effects of the crystal can be replaced by a harmonic coupling of the atom to a fixed centre. Writing U as the core displacement and u as the shell displacement, the equations of motion can be written generally in the form

$$\begin{aligned} M\omega^2U &= AU + Bu \\ m\omega^2u &= Cu + BU \end{aligned} \tag{20}$$

The coupled motion has a low frequency 'phonon' mode at $\omega_1$ say and a high frequency 'electronic' mode at $\omega_2$. The matrix element for the nuclear scattering again involves $e^{i\underline{Q}.\underline{U}}$ which must be divided between the two normal modes. To lowest order in m/M

$$M^{\frac{1}{2}}U \sim u_1 - \frac{B}{C}\left(\frac{m}{M}\right)^{\frac{1}{2}} u_2 \tag{21}$$

The matrix element for transitions with one electronic excitation and n phonon excitations is then

$$b\langle 0|\exp(i\underline{Q}.\underline{u}_1/M^{\frac{1}{2}})|n\rangle\langle 0|i\underline{Q}.\underline{u}_2|1\rangle(Bm^{\frac{1}{2}}/CM) \tag{22}$$

The second term can be expanded because of the small (m/M) factors. The matrix element of $u_2$ in this model is $(\hbar/\omega_2)^{\frac{1}{2}}$ and $(\hbar/m\omega_2)^{\frac{1}{2}} \sim$ a an electronic radius$^2$. The result is similar to that in (19) apart from the (B/C factor. The first term is the usual vibrational matrix element. At large Q the terms contain the Debye Waller factor:

$$\exp[-\tfrac{1}{2}Q^2\langle U^2\rangle]\quad \text{where } U^2 \sim \hbar/M\omega_1 \tag{23}$$

which acts like a form factor, though the cut-off occurs at much larger Q than in the atomic case.

A more complete treatment of the problem involves a better description of the electronic wave function and a treatment of the electron-lattice coupling arising from corrections to the adiabatic approximation. Krivoglaz and Rashba (1980) define an electron phonon interaction in the form

$$\mathcal{H}_{ep} = V_{\ell\ell'}(c^+ + c)(\gamma^+_{\ell\ell'} + \gamma_{\ell\ell'}) \tag{24}$$

where $c^+$, c create and destroy phonons, while $\gamma^+$, $\gamma$ do the same for the electronic transition $\ell \to \ell'$. This term is then used in first order perturbation theory in processes where the neutron-nuclear scattering creates or destroys a virtual phonon and $\mathcal{H}_{ep}$ transform this into an electron excitation. The effective matrix element for the process is

$$\frac{<0,\ell|e^{i\underline{Q}\cdot\underline{U}}|1,\ell><1,\ell|V_{\ell\ell'}c^+\gamma_{\ell\ell'}|0,\ell'>}{E_{\ell\ell'} - \hbar\omega_1} - \frac{<0,\ell|V_{\ell\ell'}c\gamma_{\ell\ell'}|1,\ell'><1,\ell'|e^{i\underline{Q}\cdot\underline{U}}|0,\ell'>}{E_{\ell\ell'} + \hbar\omega_1} \tag{25}$$

which gives

$$b\ Q\left(\frac{\hbar}{M\omega_1}\right)^{\frac{1}{2}} \exp\left[-Q^2<U>^2\right] \frac{V\hbar\omega_1}{E_{\ell\ell'}^2} \tag{26}$$

If we make the connection $B\left(\frac{\hbar}{M\omega_1}\right)^{\frac{1}{2}} \left(\frac{\hbar}{m\omega_2}\right)^{\frac{1}{2}} = V$ and $C = m\omega_2^2$, $E_{\ell\ell'} = \hbar\omega_2$, this agrees with the estimate (22).

The electron phonon coupling term (24) can be derived from the adiabatic approximation correction as shown by Lovesey et al. (1982). Using the Born-Oppenheimer wave functions for the electronic and vibrational parts.

$$\psi_\ell(r,R)\ \chi_\ell(R). \tag{27}$$

The corrections are in the form

$$\frac{1}{M} \nabla\chi_\ell(R) \sum_{\ell'} \psi_{\ell'}(r,R)\ (G_{\ell\ell'}/E_{\ell\ell'}) \tag{28}$$

where

$$G_{\ell\ell'} = \int dr\ \psi_{\ell'}(r,R)\nabla_R\ \psi_\ell\ (r,R).$$

This detailed analysis of the $H_2^+$ ion gives results in qualitative agreement with the above estimates. It also points out that electronic excitations in molecules and atomic centres would satisfy the Frank-Condon approximation. The short time of the neutron-nuclear interaction means that the transition takes place while the heavy nuclei are fixed, ie vertically on the configuration co-ordinate diagram. Thus the broadening of the lines would be comparable with those seen in optical transitions in similar systems. By analogy phonon and rotational side bands will occur on the electronic transitions. Systems which have not been investigated in any detail to date are those showing the Jahn-Teller effect. Here the strong vibronic interaction and the small electronic excitation energies suggest that it would be a favourable case for mixed excitations to be investigated via the nuclear coupling.

## 6. Conclusions

The cross section for electronic excitations in atomic centres at higher energy are comparable to those already observed at low energy for spin waves and crystal field excitations, provided they can be observed where Qa is of order unity. For larger Q however the electronic form factor leads to much smaller values. The cross section for excitations via the nuclear scattering is basically smaller by a factor $(m/M)^2 \sim 10^{-6}$. However there is no form factor in this case apart from the Debye-Waller factor and hence it is relatively larger at higher Q. The most favourable cases are probably those involving H because of its large cross section, or Jahn-Teller systems because of the favourable electron-phonon coupling.

## References

Balcar, E., Lovesey, S.W. and Wedgewood, F.A., 1970, J.Phys.C3, 1292.
Elliott, R.J., and Kleppmann, W.G., 1975, J. Phys.C8, 2737.
Elliott, R.J., and Shukla, P., 1979, J. Phys. C12, 5463.
Kashcheev, V.N., and Krivoglaz, M.G., 1962, Sov. Phys. Solid State 3, 2301.
Krivoglaz, M.A., and Rashba, E.I. 1980, Sov. Phys. Solid State 21, 1705.
Loewenhaupt, M., 1984, 'Proc. Los Alamos Meeting on High Energy Excitations in Condensed Matter' (to be published).
Lovesey, S.W., 1978, J. Phys. C11, 3971.
Lovesey, S.W., Bowman, C.D., and Johnson, R.G., 1982, Z.Phys.B47, 137.
Marshall, W., and Lovesey, S.W., 1971, Theory of Thermal Neutron Scattering (O.U.P.).
Ruijgrok, Th.W., Nijboer, B.R.A., and Hoare, M.R., 1983, Physica 120A, 537.
Trammell, G.T., 1953, Phys. Rev. 92, 1387.

# DYNAMIC PROPERTIES OF QUANTUM SOLIDS AND FLUIDS

Henry R. Glyde

Physics Department
University of Delaware
Newark, Delaware 19716

ABSTRACT

The study of excitations in solidified $^3$He and $^4$He by inelastic neutron scattering provides the experimental tool to test our microscopic understanding of quantum solids. A review of this study, with emphasis on its relation to our current microscopic picture, is presented. A similar review of elementary excitations in liquid $^3$He, since they were first observed by neutrons in 1974, is also presented. Some recent new results on excitations, in the substantially larger field, of liquid $^4$He are highlighted again with an attempt to relate these to a microscopic picture of Bose liquids.

## 1. INTRODUCTION

Helium was first liquified in 1908 by Kamerlingh Onnes and first solidified in 1926 by Keesom. The much rarer isotope $^3$He was first liquified in 1947 by Sydoriak, Hammel and Grilly and first solidified in 1951 by Osborne, Abraham and Weinstock.

The aim in these lectures is to highlight some interesting features of excitations in these remarkable liquids and solids emphasing new work since previous excellent and exhaustive reviews; by Woods and Cowley[1], Stirling,[2] Cowley,[3] Zawadowski[4] and Price[5] for liquid He and by Domb and Dugdale[6], Guyer[7], Koehler[8], Horner[9], Glyde[10] and by Varma and Werthamer[11] for solid He. These reviews provide a complete historical perspective. General books on condensed helium are by Keller[12] and Wilks[13] and edited by Bennemann and Ketterson.[14]

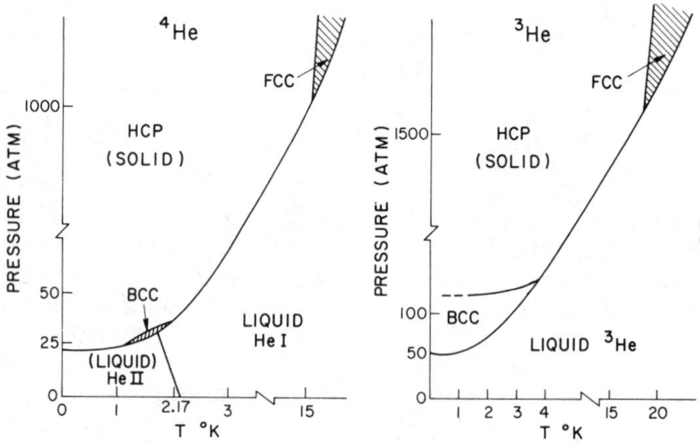


Fig. 1. Phase diagram of helium.

In Fig. 1 we show a phase diagram of helium at low temperatures. Both $^3$He and $^4$He remain liquid down to T $\approx$ 0 K under their own saturated vapor pressure (svp). This is because the interatomic potential is weak and the mass is small leading to a highly kinetic, weakly bound liquid. The interatomic potential may be approximately represented by the Lennard-Jones expression

$$v(r) = 4\varepsilon\{(\sigma/r)^{12} - (\sigma/r)^{6}\}$$

with parameters $\varepsilon = 10.22$ K and $\sigma = 2.556$ Å. The most accurate representation of $v(r)$ at present is the HFDHE2 potential of Aziz et al.[15] Both liquid $^4He$ and liquid $^3He$ become superfluid at low T; at T = 2.17 K at svp and at T = 1 mK at p $\approx$ 30 atm., respectively. Under applied pressure both $^3He$ and $^4He$ crystalize into bcc, hcp and fcc structures depending upon the temperature and pressure. The bcc phase is generally regarded as the most open structure favoring large amplitude vibration at low pressure. However, a return to the bcc phase in $^4He$ at very high pressures $\sim$ 150 kbar has recently been observed.[16] In the low density bcc phase the nuclear spins (½) order[17] into an antiferromagnet below 1 mK.

The excitations in both superfluid (He II) and normal (He I) $^4He$ and in all structures of solid $^4He$ have been studied by inelastic neutron scattering. Because of the high absorption cross section of the $^3He$ nucleas for neutrons (interacting in the singlet state) neutron studies of $^3He$ are very difficult. Studies of liquid $^3He$ began only in 1974 with the pioneering measurements of Scherm et al[18] and only preliminary elastic scattering studies of bcc $^3He$ have been reported.[19]

For an unpolarized neutron beam and $^3He$ sample, and with effects of absorption removed, the combined dynamic form factor

$$S(Q,\omega) = S_c(Q,\omega) + (\sigma_i/\sigma_c) S_I(Q,\omega) \tag{1}$$

is observed.[18,20] Here $\sigma_i$ and $\sigma_c$ are the incoherent and coherent[21] scattering cross sections and $\sigma_i/\sigma_c \approx 0.20$ (Ref. 22). The

$$S_c(Q,\omega) = \frac{1}{2\pi} \int dt\, e^{i\omega t} \frac{1}{N} \langle\rho(Q,t)\rho(-Q,0)\rangle \tag{2}$$

is the usual coherent dynamic form factor proportional to the $Q^{th}$ Fourier component of density waves in the sample $\rho(Q) = \sum_\ell e^{-iQ\cdot r(\ell)}$. In liquid $^4He$, where $\sigma_i = 0$, only coherent scattering takes place and $S(Q,\omega) = S_c(Q,\omega)$. The

$$S_I(Q,\omega) = \frac{1}{2\pi} \int dt\, e^{i\omega t} \frac{1}{NI(I+1)} \langle\vec{I}(\vec{Q},t)\cdot\vec{I}(-\vec{Q},0)\rangle \tag{3}$$

is the spin dependent form factor proportional to the $Q^{th}$ Fourier component of the spin density waves, $I(Q,t) = \sum_\ell I(\ell,t)e^{-iQ\cdot r(\ell,t)}$. In $^3He$ $S_I(Q,\omega)$ is important since in the range $0.5\ \text{Å} \leq Q \leq 1.0\ \text{Å}$ $S_I(Q)$ is approximately twice $S_c(Q)$. Rather than evaluate $S(Q,\omega)$ itself it is more convenient to evaluate time-ordered expectation values such as

$$\chi_c(Q,t) = -\, i(1/N)\langle T\rho(Q,t)\rho(-Q,0)\rangle \tag{4}$$

which is related to $S_c(Q,\omega)$ by $S_c(Q,\omega) = - 1/\pi\ [n(\omega)+1]\ \mathrm{Im}\ \chi(Q,\omega)$.

As noted above we highlight some recent results with an emphasis on featuring the similarities in the excitations of liquid and solid helium and the relation to microscopic pictures. We begin with solid helium because it is more straightforward and better understood on a microscopic basis.

## 2. SOLID HELIUM

To introduce solid helium we begin with bcc $^4$He which is most like liquid $^4$He. There the Debye-Waller temperature is[23] $\theta_{DW}$ = 22.5 K. Assuming an effective harmonic model this suggests a zero point energy, $E_{zp} \approx (9/8)\theta_D \approx 25$ K and a mean square amplitude of vibration

$$\langle u\rangle^2 = \frac{9}{4}\ \frac{1}{Mk\theta_{DW}} = 109\ \frac{1}{M\theta_{DW}}\ \mathring{A}^2 \approx 1\ \mathring{A}^2. \tag{5}$$

Thus $E_{zp}$ is comparable to the potential energy so that the atoms are highly kinetic and not well localized at lattice points. Also $E_{zp}$ is much greater than the thermal energy ($T \lesssim 2$ K) so that quantum effects are most important and T = 0 K is a good approximation. The volume of the small sliver of bcc $^4$He shown in Fig. 1 is V = 21 cm$^3$/mole giving an interparticle spacing R = 3.56 Å ($a_o$ = 4.120 Å). Thus the Lindemann ratio is $\delta = (\langle u^2\rangle/R^2)^{1/2} \approx 0.3$. Thus we expect a weakly bound, highly anharmonic solid composed of light atoms executing large amplitude vibrations.

To describe solids S(Q,ω) is usually expanded[24,25] in terms of scattering from single phonons, pairs of phonons and so on. This is an expansion of S(Q,ω) in powers of $\langle[Q\cdot u]^2\rangle$ and from (5) $\langle[Q\cdot u]^2\rangle \sim 1$ for $Q \sim 1$ Å$^{-1}$ in bcc $^4$He. Hence we expect single phonon excitation to dominate S(Q,ω) only at very low $Q \lesssim 1$ Å$^{-1}$. At intermediate Q, 1 Å$^{-1} \leq Q \leq 3$ Å$^{-1}$, both single and many phonons are created in the scattering and there is substantial interference between the single and multiphonon creation. For $Q \geq 3$ Å$^{-1}$ so many phonons are created that the scattering is like that from single atoms in a weakly interacting gas. This range of excitation creation, from single phonons at low Q through multiphonon (and interference) to single particle like excitations for $Q \gtrsim 3$ Å$^{-1}$, is common to both liquid[1] and bcc $^4$He.

In bcc helium the $\langle u^2\rangle$ is so large and the vibrations so anharmonic that a harmonic theory predicts imaginary frequencies. The dynamics must be described by a self consistent phonon (SCP) theory.[7-11] in which the phonon frequencies and the vibrational amplitudes are evaluated iteratively until consistent. Otherwise the SCP theory, from the point of view of $S(Q,\omega)$, is much like conventional anharmonic lattice dynamics.[24,25] In bcc $^3$He there is also rapid quantum mechanical exchange of atoms, at a rate $\lesssim 2\times10^7$ Hz. However, since the exchange frequency is much less than the vibrational frequencies $\sim 10^{12}$ Hz (and the exchange energy $J\sim10^{-3}$ K is much less than the ground state energy) exchange and statistics can be largely neglected in treating the dynamics. Thus solid $^3$He and $^4$He can be treated using the same SCP theory ignoring the difference in quantum statistics. As pressure is applied and R is reduced the $\langle u^2\rangle$ is reduced and the solid becomes less anharmonic. In the fcc phase at V = 9.03 cm$^3$/mole ($p \approx 4.7$ kbar), the smallest volume studied to date by neutrons, the SCP theory is still needed but an explicit treatment of short range correlations[26] in the atomic motion is not required.

## 2.1. Phonon Expansion of $S(Q,\omega)$

To express $S(Q,\omega)$ in terms of scattering from phonons we substitute $\rho(Q)$ in (4) so that $(r = R+u)$

$$\chi(Q,t) = -i \sum_{\ell} e^{-iQ\cdot R(\ell 0)} \langle T e^{-iQ\cdot u(\ell t)} e^{iQ\cdot u(00)} \rangle . \tag{6}$$

Taking advantage of the time ordering operator we expand the exponentials in 6). There are two key points. Firstly, the Debye-Waller factor $d(Q) = \langle e^{iQ\cdot u}\rangle \approx e^{-2W(Q)}$ is extracted so that each term contains $d^2(Q)$. Secondly, following Ambegoakar, Conway and Baym (ACB),[27] all the terms containing the single phonon green function $D_{\alpha\beta}(\ell 0,t) = -i\langle T u_\alpha(\ell t) u_\beta(00)\rangle$ are collected into a single term. Expressing D in phonon coordinates $(D(q\lambda,\omega))$ leads to

$$S_c(Q,\omega) = S_o(Q) + S_p(Q,\omega) + S_m(Q,\omega). \tag{7}$$

Here $S_o(Q)$ is the elastic scattering term. The $S_p(Q,\omega)$ contains all the terms having $D(q\lambda,\omega)$ as a factor

$$S_p(Q,\omega) = -\frac{1}{\pi}[n(\omega)+1]\,\mathrm{Im}[R(Q,q\lambda,\omega)D(q\lambda,\omega)R(Q,q\lambda,\omega)]. \tag{8}$$

It represents the scattering which excites a single phonon plus the scattering which contains a single phonon as an intermediate state. That is, the incoming neutron could excite two phonons which coalesce via three phonon anharmonic processes into a single phonon during the scattering process. This process is referred to as a one-two phonon anharmonic interference process and is included in $S_p$ in addition to the simple one phonon scattering term $S_1(Q,\omega)$. The R is a general structure factor

$$R = F(Q,q\lambda) - E(Q,q\lambda,\omega).$$

which includes the simple one phonon structure factor $F(Q,q\lambda) = d(Q)(2M\omega_{q\lambda})^{-\frac{1}{2}}[Q\cdot e_{q\lambda}]$ and the term $E(Q,q\lambda,\omega)$ which couples the one phonon scattering to multiphonon excitation via anharmonic terms. If F only is retained, (11) reduces to (no interference)

$$S_1(Q,\omega) = \frac{1}{2\pi}[n(\omega)+1)[F(Q,q\lambda)]^2 A(q\lambda,\omega)\Delta(Q-q-\tau) \quad (9)$$

where $A(q\lambda,\omega) = -2\text{Im } D(q\lambda,\omega)$ is the one phonon response function,

$$A(q\lambda,\omega) = \frac{8\omega_{q\lambda}^2\Gamma(q\lambda,\omega)}{[-\omega^2+\omega_{q\lambda}^2+2\omega_{q\lambda}\Delta(q\lambda,\omega)]^2+[2\omega_{q\lambda}\Gamma(q\lambda,\omega)]^2}\,. \quad (10)$$

Here $\Delta$ and $\Gamma$ are the phonon frequency shift and inverse lifetime due to three phonon processes, i.e.

$$\Gamma = -\frac{\pi}{2}\sum_{1,2}|V(q\lambda,1,2)|^2\{(n_1+n_2+1)\delta(\omega-(\omega_1+\omega_2))$$

$$+ (n_2-n_1)[\delta(\omega+\omega_2-\omega_1) - \delta(\omega-\omega_2+\omega_1)]\}. \quad (11)$$

where $V(q\lambda,1,2)$ is the cubic anharmonic coefficient and $1 = q_1\lambda_1$. The $S_m(Q,\omega)$ is the remaining multiphonon scattering which contains tw phonon or higher processes.

Anharmonic interference effects take place in any moderately anharmonic solid[29-32] and were first observed by Cowley et al[28] in alkali halides. To date interference between one and two phonon processes only has been evaluated explicitly in $S_p(Q,\omega)$. In this case $E(Q,q\lambda,\omega)$ is also proportional to $V(q\lambda,-1,-2)$ and if $E(Q,q\lambda,\omega)$ is not greatly frequency dependent we may write $S_p$ in the form

$$S_p(Q,\omega) = S_1(Q,\omega) + \alpha(Q)S_1(Q,\omega) + \beta(Q)[\frac{\omega^2-\omega_{q\lambda}^2}{\omega_{q\lambda}}]S_1(Q,\omega). \quad (12)$$

The interference contains two terms; one proportional to $S_1(Q,\omega)$ (on energy shell term) which will increase or decrease the apparent magnitude of $S_1$ depending upon the sign of $\alpha(Q)$ and one (off energy shell-term) which will distort the shape of $S_1(Q,\omega)$. If $S_1(Q,\omega)$ is sharply peaked at $\omega_{q\lambda}$, the second term can affect the shape of $S_p(Q,\omega)$ only little.

We now sketch neutron scattering studies of solid helium using the above as a basis.

## 2.2. Low Q, Single Phonon Excitation

In Fig. 2 is shown the one phonon frequencies observed in bcc $^4$He by Osgood et al[33] compared with the dispersion curves calculated by Glyde and Khanna.[34] The observed "one phonon" peak positions for longitudinal phonons along [q,0,0] at high q are distorted upward by interference effects and certainly lie above the true one phonon frequencies.[23] The calculations are microscopic using only the interatomic potential and lattice constant as input. The dashed line is the self consistent harmonic (SCH) frequencies and the solid line is the SCH frequencies with the cubic term added as a perturbation (SCH+C). A T-matrix method is used to describe the short range correlations in the atomic motion induced by the hard core of the potential. The agreement between the SCH+C and observed frequencies is reasonable and typical of SCP calculations in highly anharmonic crystals.

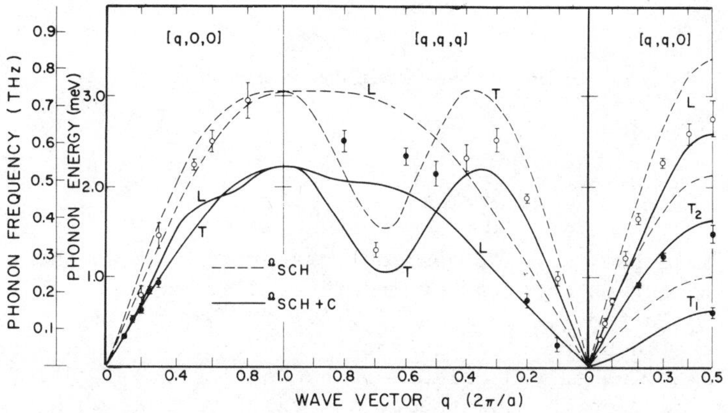


Fig. 2. Phonon dispersion in bcc $^4$He, $a_o$ = 4.111 Å. Points are observed values by Osgood et al (Ref. 33). From Glyde and Khanna, Ref. 34.

In Fig. 2 are the full width at half maximum (FWHM), W = 2Γ, of the phonon groups used to identify the phonon frequencies shown in Fig. 3. There we see that for a typical longitudinal phonon (i.e. having wave vector Q ∿ 0.7 Å$^{-1}$ in the middle of the Brillouin zone) ω ∿ 1 meV ∿ 0.25 THz ∿ 12 K. The Γ is therefore large and ∿ 1/3 of the phonon energy. The calculations considered three phonon processes only via the cubic anharmonic term (11). The calculated $\Gamma$ is dominated entirely by the temperature independent term of (11) ($\delta(\omega-\omega_1-\omega_2)$) and we do not expect $\Gamma$ to be temperature dependent in bcc $^4$He. The calculated $\Gamma$ is clearly substantially less than the observed value which may be due to using harmonic like, infinite lifetime phonons in (11).

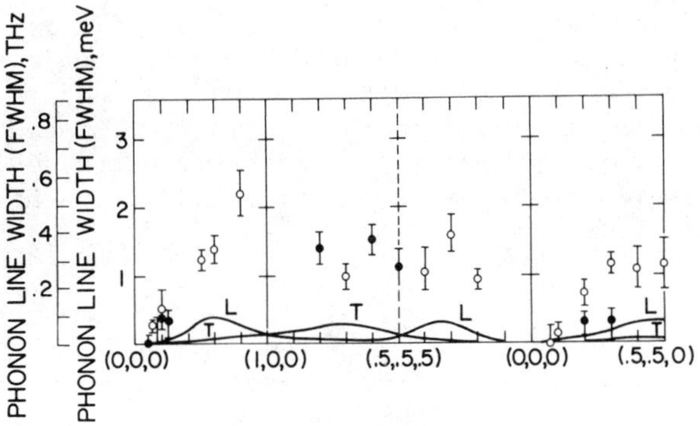


Fig. 3. Phonon line widths in bcc $^4$He as Fig. 2.

To span the range of neutron scattering data we show the dispersion curves observed by Eckert et al[35] in fcc $^4$He at V = 9.03 cm$^3$/mole and T = 38 K in Fig. 4. This is the most compressed helium studied by neutrons to date. The solid lines are the SCH+C dispersion curves calculated by Collins and Glyde.[36] The agreement between the SCP theory and experiment is much better because fcc $^4$He at 9.03 cm$^3$/mole is much less anharmonic and an explicit account of short range correlations does not seem to be required. In fcc $^4$He at 11.07 cm$^3$/mole[37] including a specific treatment of short range correlations does improve agreement between theory and experiment somewhat. At 9.03 cm$^3$/mole a typical phonon half width is $\Gamma \sim$ 1/10th of the phonon energy and the SCP calculations lie much closer to the observed values. Between 22 K and 38 K at V = 9.03 cm$^3$/mole $\Gamma$ increases by at most 15%. This confirms that the $\Gamma$ is dominated by the temperature independent three phonon process as noted above. Clearly the temperature dependence of the width ($\Gamma(T)$) of the excitations in solid helium and liquid $^4$He are quite different, since $\Gamma(T)$ of phonons in liquid $^4$He varies over two orders of magnitude between 1 K and 2.17 K.

## 2.3. Intermediate Q, One-Multiphonon Interference

To display the nature of interference contributions to $S_p(Q,\omega)$

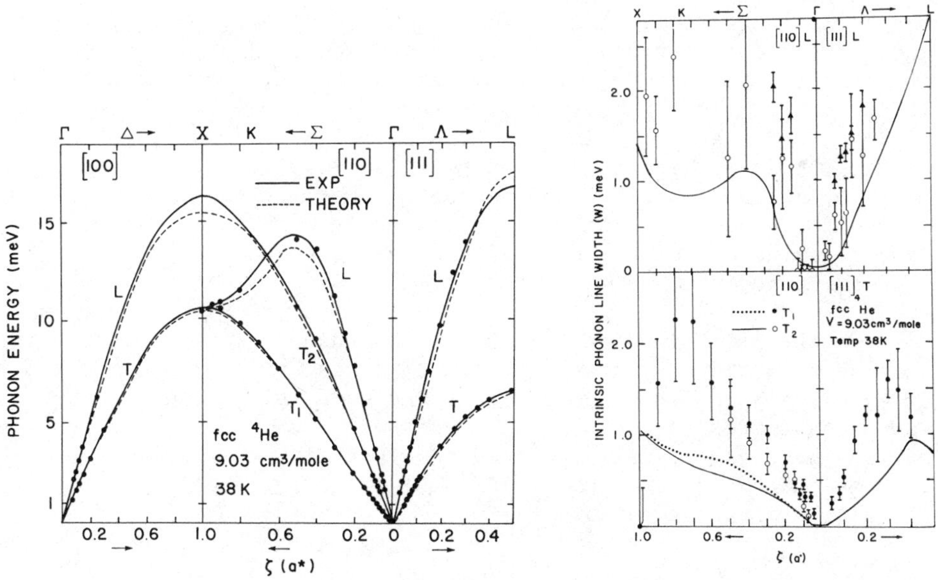


Fig. 4. Phonon dispersion and lifetimes in fcc $^4$He. Points are observed values of Eckert at al (Ref. 35). From Collins and Glyde, Ref. 36.

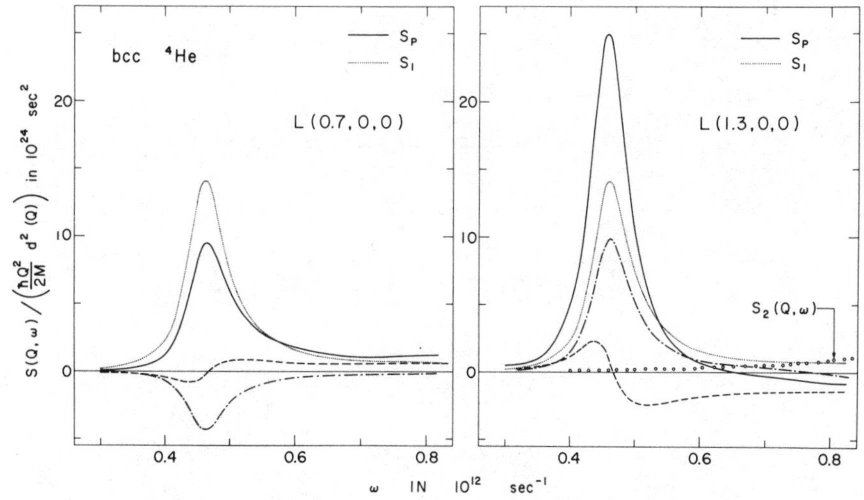


Fig. 5. Interference contributions; ····· $S_1(Q,\omega)$, —— - —— - —— on energy shell term, ----- off energy shell term, —— $S_p(Q,\omega)$. From Glyde, Ref. 31.

we show a calculation of $S_p(Q,\omega)/F(Q,q\lambda)$ for a longitudinal phonon at $Q = (2\pi/a)(0.7,0,0)$ and $Q = (2\pi/a)(1.3,0,0)$ in bcc $^4$He in Fig. 5.

The simple one phonon contribution, $S_1(Q,\omega)/F(Q,q\lambda)$, is the same for these two Q values which are equidistant from $Q = (2\pi/a)(100)$. The $\alpha(Q)S_1$ and $\beta(Q)S_1$ in (12) are proportional to $|Q|^3V(q_\lambda,-1,-2)$ and vanish at recirprocal lattice points and at the midpoint between two lattice points. $V(q,-1,-2)$ is an odd function of q and oscillates in sign about these symmetry required zeros. In the example in Fig. 5 $\alpha(Q)$ and $\beta(Q)$ and have opposite signs at (0.7,0,0) and (1.3,0,0). The on energy shell term $\alpha(Q)S_1(Q,\omega)$ clearly has the most impact on $S_p(Q,\omega)$ in the one phonon peak region. If we focussed on the one phonon peak region only of $S_p(Q,\omega)$ we would find it quite different in magnitude than expected for simple one onon scattering.

On the left side of Fig. 6 we show phonon groups observed in bcc $^4$He. In the one phonon approximation the pair of phonon groups in each graph should be essentially the same, since the Q's are equidistant from the Brillouin zone edge. The interference contributions have dramatically changed the intensity of the pairs. This is due chiefly to the $\alpha(Q)S_1(Q,\omega)$ on energy shell term. If the phonon group is narrow, the off energy shell term is not able

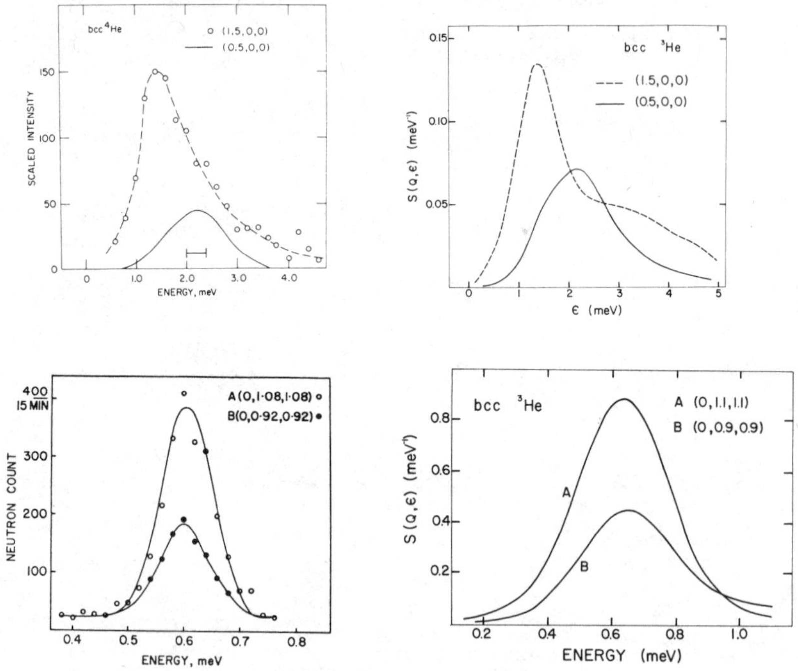


Fig. 6. Phonon groups observed in bcc $^4$He by Osgood et al (Ref. 33) and by Minkiewicz et al (Ref. 23). Calculations are for bcc $^3$He. From Glyde and Hernadi, Ref. 37a.

to displace the peak position or distort the peak shape. When the phonon width Γ is large, as is the case in the L (0.5,0,0) phonon the interference can significantly distort the peak shape as well as the peak height. On the right side of Fig. 6 are calculations of $S_p(Q,\omega)$ in bcc $^3He$.

ACB[27] show that both $S_1$ and $S_p$ satisfy the sum rule

$$\int_{-\infty}^{+\infty} d\omega\ \omega\ S_{p,1}(Q,\omega) = d^2(Q)\ \frac{[Q\cdot\varepsilon]^2}{2M}\ . \tag{13}$$

The interference terms do not contribute to the ACB sum rule. The sum rule, however, is difficult to implement in practice because $S_p(Q,\omega)$ has a long tail reaching to high frequency. The tail portion makes an important contribution to the ACB sum rule and in practice is impossible to separate from the multiphonon background. If we evaluate (13) by integrating $S_p(Q,\omega)$ over the one phonon peak (OPP) region only i.e. evaluate

$$M_p^{+}(Q) = \int_{OPP} d\omega\ \omega\ S_p(Q,\omega) \tag{14}$$

we will not capture the whole sum rule and the interference effects will show up as an oscillation of $M_p^{+}(Q)$ about the expected sum rule value (13). Similarly if we evaluate the one phonon contribution to the static structure factor from the observed one phonon group in the same way,

$$S_p^{+}(Q) = \int_{OPP} d\omega\ S_p(Q,\omega) \tag{15}$$

we expect $S_p^{+}(Q)$ will oscillate about the anticipated one phonon value $[Q^2/2M\omega_{q\lambda}]d^2(Q)$ due to interference. Fig. 7 shows the values of $S_p^{+}(Q)/[Q^2/2M\omega_{q\lambda}]$ in fcc $^4He$. Interference contributions manifest themselves as an apparent oscillation in the one phonon intensity, as first clarified in Helium by Horner.[29]

## 2.4. High Q, Single Particle Excitations

For $Q > 3$ Å$^{-1}$, Kitchens et al[33] find the scattering intensity from low density solid He is broad and similar to that expected from atoms in a weakly interacting gas. That is, the peak is approximately Gaussian in shape with a peak position just below the free atom recoil energy, $\omega_R = Q^2/2M$. Indeed at $Q \approx 4.5$ Å$^{-1}$, Kitchens et al find $S(Q,\omega)$ is approximately the same in bcc $^4He$, in hcp $^4$He at the same volume and in liquid $^4$He at svp.[38] At Q $\approx$ 4.5 Å$^{-1}$ the energy transfer at the peak of $S(Q,\omega)$ is $\sim$ 10 meV $\sim$ 100 K. This is $\sim$ 10 times the well depth of the interatomic potential. Thus we would expect the scattering to be predominantly from a single atom and that the recoiling atom would be insensitive to details of crystal structure. These ideas are discussed in more detail in section 3.5.

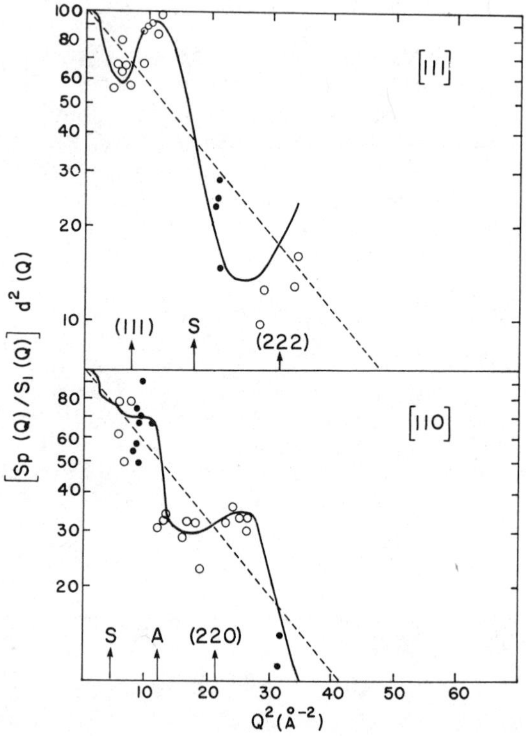


Fig. 7. $S_p^+(Q)/(Q^2/2M\omega_{q\lambda})$ in (15) as obtained by integrating over the observed one phonon peak region (points) (Eckert et al, Ref. 35) and as calculated (Ref. 36) including interference contributions (solid line). The dashed line is the result without interference. From Collins and Glyde, Ref. 36.

In Fig. 8 is shown the peak and half width positions of $S(Q,\omega)$ calculated by Horner[39] for bcc $^4$He. There we see that for Q>3 Å$^{-1}$ the calculated peak position of $S(Q,\omega)$ lies somewhat below but close to $\omega_R$. Shown as dots are the peak positions of the scattering intensity observed by Kitchens et al.[33] The scattering at

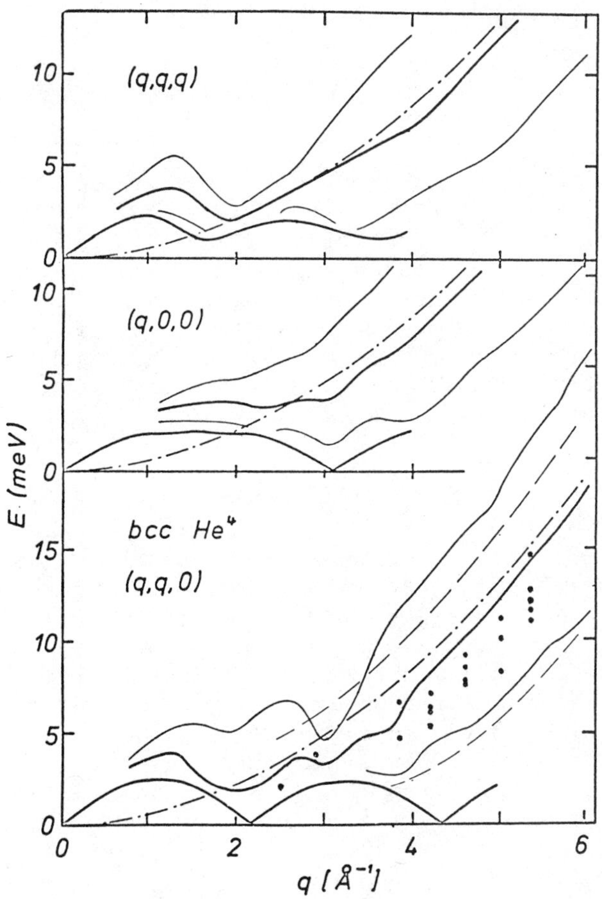


Fig. 8. Energy versus Q of scattering from bcc $^4He$ calculated by Horner (Ref. 44). Dark solid line shows peak of $S(Q,\omega)$ and one phonon dispersion curve. Lighter lines denote one half height positions. Dots are observed peak positions of Kitchens et al (Ref. 33). From Horner, Ref. 39.

$Q > 3$ Å$^{-1}$ is clearly largely independent of the symmetry direction considered. It is interesting that along the [q,q,q] direction, the predicted one phonon excitation curve and total scattering intensity is almost identical to that observed by Cowley and Woods[38] in liquid $^4He$ (see Fig. 11).

Most recently, Hilleke et al[40] have measured neutron scattering from hcp $^4He$ at high momentum transfer, 12 Å$^{-1} \leq Q \leq 22$ Å$^{-1}$, using the pulsed IPNS1 neutron source at Argonne National Laboratory. The aim in these studies, as in liquid $^4He$, is to measure the momentum distribution, n(p). The momentum distribution may be obtained from the observed $S(Q,\omega)$ within the impulse approximation (IA) (see section 3.5)

$$S_{IA}(Q,\omega) = \int d^3p\; n(p)\; \delta(\omega - \frac{Q^2}{2M} - \frac{Q \cdot p}{M}). \qquad (16)$$

If we assume a Gaussian distribution $n(p) = (2\pi\langle p_\alpha^2\rangle)^{-\frac{1}{2}} e^{-p^2/2\langle p_\alpha^2\rangle}$ then $S_{IA}(Q,\omega) = (2\pi\sigma)^{-\frac{1}{2}} e^{-(\omega-\omega_R)^2/2\sigma}$ where $\sigma = \langle p_\alpha^2\rangle Q^2/M^2$. From $\sigma$ we may obtain the KE = $3\langle p_\alpha^2\rangle/2M = (3/2)\sigma M/Q^2$. This represents an important new tool for studying quantum crystals. For example, the calculated KE lies 15% below those obtained from the observed $\sigma$. While there remains issues such that the energy distribution of the incident neutron beam and the validity of the impulse approximation to be fully clarified, these studies promise to be an exciting new area of research.

## 2.5. Solid $^3$He

Measurements of elastic scattering from bcc $^3$He are currently in progress at the CENG, Grenoble[19] and jointly at Argonne National Laboratory and Studsvik.[19] The aim is to determine the magnetic structure in the spin ordered phases of bcc $^3$He. Below 1 mK and in zero magnetic field bcc $^3$He orders into an antiferromagnetic phase. The structure is not definitively established but that it is believed to be the up-up-down-down structure in which two planes of atoms have spin up followed by two of spin down and so on. In a modest magnetic field the uudd structure transfers into a spin-flop ferromagnetic phase.[17] The elastic scattering measurements will provide important information on the nature of exchange in solid helium which cannot be represented by a Heisenberg Hamiltonian and which involves three and four atom exchange processes.

Glyde and Hernadi[37a] have evaluated $S(Q,\omega)$ for paramagnetic bcc $^3$He. The coherent part (2) is like that of bcc $^4$He discussed above. Since the exchange constant is small ($J \sim 1$ mK) compared to the usual energy transfer $\omega$, the spins in bcc $^3$He can be regarded as free and independent. In this case $S_I(Q,\omega)$ reduces to the usual incoherent $S_i(Q,\omega)$. Only if very low energy transfers or extremely high resolution ($\sim 1$ mK $\sim 0.1$ μeV) were used could the dynamics of the spins be observed. A still difficult, but somewhat easier, quasielastic measurement to determine exchange constants in bcc $^3$He has been proposed.[41]

Isotropic mixtures of solid $^3$He and $^4$He is an interesting open area for exploration.

# 3. LIQUID $^4$He

Neutron scattering from liquid $^4$He has a long and rich[1-5] history. Perhaps most interesting, new experiments continue to reveal new physics in this complex and facinating liquid. It is fair to say that much is not yet understood about the basic nature of the excitations and their interactions in liquid $^4$He. There is not universal agreement on the interpretation of measurements.

We attempt here to relate new results to a microscopic picture of liquid $^4$He. To set the stage we begin with a brief sketch of the microscopic picture referring along the way to some well established experimental results.

### 3.1. Background

Landau proposed that single excitations in He II follow a continuous dispersion curve, $\omega(Q)$, having an acoustic 'phonon' form at $Q \to 0$ ($\omega(Q) = c\,Q$) and a minimum at higher Q of form, $\omega(Q) = \Delta + (Q-Q_R)^2/2M^*$. The excitations in the minimum region he called 'rotons'. This dispersion curve is shown in Fig. 9. A fit to $\omega(Q)$ yields $\Delta \approx 8.7$ K at $Q_R \approx 1.9$ Å$^{-1}$ with $M^* = 0.15$ M.

Feynman[53] provided the first microscopic derivation of the phonon-roton dispersion curve. He argued that the excited state containing a single excitation above the ground state $\phi$ should be of the form $\psi = \sum_\ell f(r_\ell)\phi$. This involves, in principle, all atoms in the fluid. He derived a variational equation for $\psi$ and found a minimum energy for the excitation of

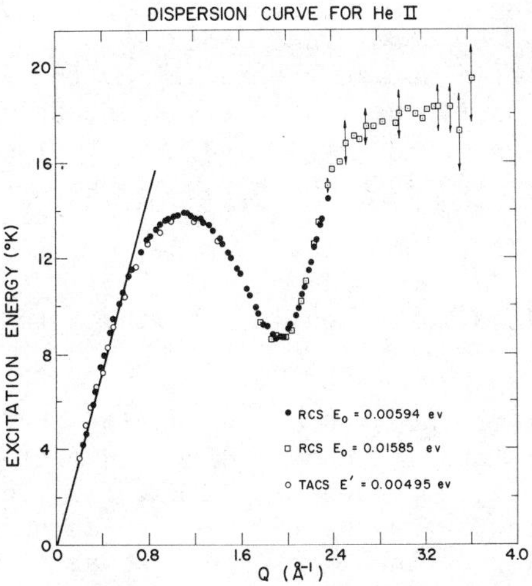


Fi.g 9. Elementary excitation dispersion curve in He II at 1.1 K and svp. From Cowley and Woods, Ref. 38.

$$\omega(Q) = \frac{Q^2}{2M} S^{-1}(Q) \tag{17}$$

if $f(r) = e^{+iQ\cdot r}$. Since $\sum_{\ell} e^{+iQ\cdot r_\ell} = \rho^+(Q)$, the excited state is $\psi = \rho^+(Q)\phi$ and the elementary excitation is a density wave of wave vector Q. This density excitation Feynman called a 'phonon'. Feynman showed, in analogy with phonons, that (at T = 0 K)

$$\lim_{Q\to 0} S(Q) = \frac{Q}{2Mc} \tag{18}$$

so that $\omega(Q) = cQ$ at $Q \to 0$. Since S(Q) has a maximum at $Q \sim 2$ Å$^{-1}$, $\omega(Q)$ has a minimum in the 'roton' region as proposed by Landau. The $\omega(Q)$ in (17) is an upper bound to the true excitation energy. Feynman and Cohen[43] proposed an improved wave function $\psi$ which localized the excitation to involve only a few atoms. This reduced $\omega(Q)$ into substantially better agreement with experiment.[1]

Cohen and Feynman[44] proposed that the phonon-roton excitations could be observed by inelastic neutron scattering the that the scattering could create many as well as a single excitation. Miller et al[45] articulated this by proposing $S(Q,\omega)$ be written as

$$S(Q,\omega) = Z(Q)\delta(\omega-\omega(Q)) + S_{II}(Q,\omega) \tag{19}$$

where Z(Q) is the weight of the single phonon scattering (with response approximated by a delta function) and $S_{II}(Q,\omega)$ is the multiphonon creation contribution. This is the form used to analyze $S(Q,\omega)$ today. The bottom half of Fig. 10a shows Z(Q) at 1.1 K as determined by Cowley and Woods.[38] Fig. 10b shows the fraction, $H_I$, of the f-sum rule taken up by the one phonon component in (19). The top half of Fig. 10 is explained below. In Fig. 11 is shown the energy versus wave vector dependence of the scattering at 1.1 K for intermediate Q values. From Fig. 10 and Fig. 11 we see that for $Q \gtrsim 2.5$ Å$^{-1}$ that the scattering is almost entirely in the multiphonon component $S_{II}(Q,\omega)$ which peaks at an energy just below the free atom recoil energy. The scattering shown in Fig. 11 is almost identical to that predicted for bcc $^4$He by Horner[39] (see Fig. 8).

Feynman's theory focussed on evaluating the excitation wave function $\psi = \rho^+(Q)\phi$ and its energy directly (without explicit reference to a possible condensate in He II). The correlated basis function (CBF) method[46] may be regarded as an outgrowth of this approach. Beginning in the late 1950's a parallel theory was developed[47-53] based on direct evaluation of response functions using many-body techniques. We sketch the response function

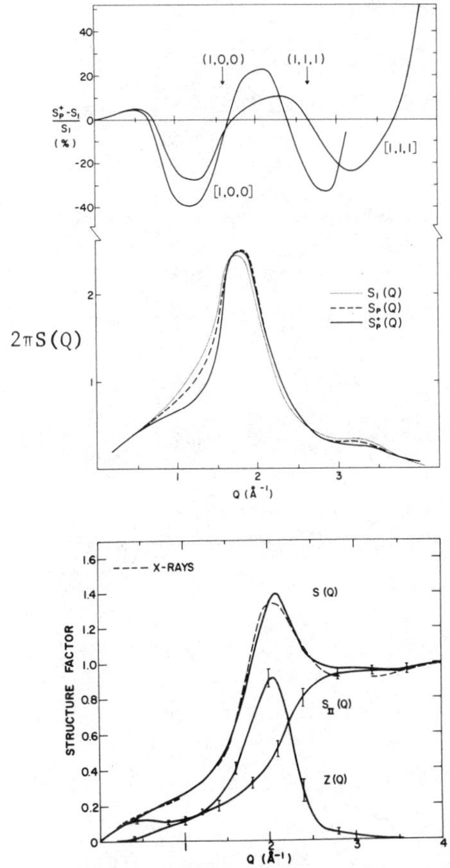


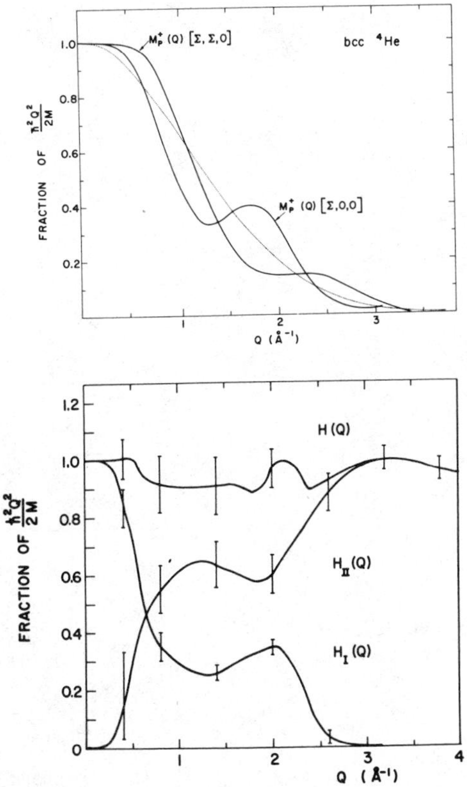


Fig. 10a. Bottom half: observed one phonon ($Z(Q)$) and multiphonon ($S_{II}(Q)$) contributions to $S(Q)$, see Eq. (19). From Ref. 38. Upper half, calculated one phonon contribution to $S_p{}^+(Q)$ to $S(Q)$ along the [1,0,0] direction in bcc $^4$He. From Ref. 31.

Fig. 10b. Bottom half: fraction of f-sum rule taken up by observed one phonon peak region ($H_I(Q)$) and multiphonon regions ($H_{II}(Q)$) in He II. From Ref. 38. Upper half: Calculation one phonon fraction of f-sum rule from (14) in bcc $^4$He. From Ref. 31.

approach since it contains explicit reference to the condensate and therefore to possible changes in $S(Q,\omega)$ at $T_\lambda$ which have been proposed by recent measurements.[54-56] Also the formulation suggests a possible coupling and interference between one and multiphonon excitations in liquid $^4$He which we discuss in section 3.3.

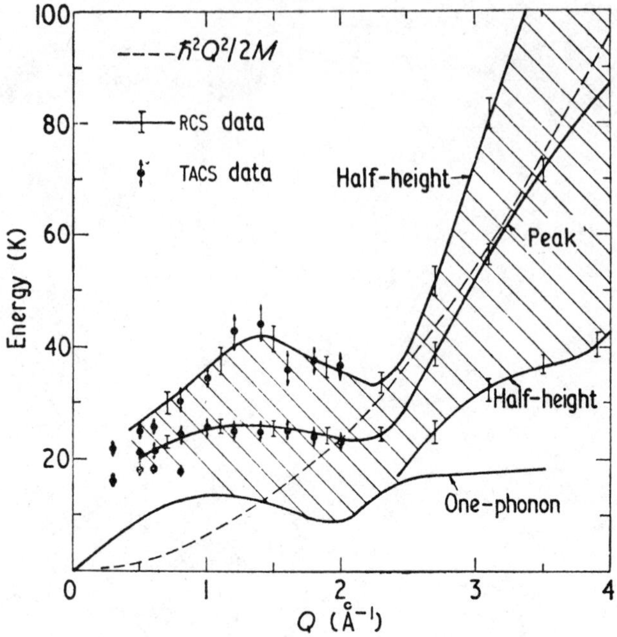


Fig. 11. Energy versus Q of scattering at intermediate Q in He II at 1.1 K and p ≈ 0. From Cowley and Woods, Ref. 43.

In this technique the dynamic susceptibility (4) is formally divided into a singular and regular part,[49]

$$\chi(Q,\omega) = \chi_S(Q,\omega) + \chi_R(Q,\omega). \tag{20}$$

The key in this separation is that $\chi_S$ is made up of all terms containing the single particle Green function $G_{\alpha\beta}$ as a factor, i.e.

$$\chi_S = \Lambda_\alpha G_{\alpha\beta}\Lambda_\beta \tag{21}$$

where $\Lambda_\alpha(Q,\omega) = n_o^{\frac{1}{2}}[1 + \bar{P}(Q,\omega)]$ is the 'structure factor' multiplying $G_{\alpha\beta}$. Here $n_o$ is the condensate fraction and $\bar{P}(Q,\omega)$ is a factor containing two Green functions and a vertex coupling these two to $G_{\alpha\beta}$ in (21). Thus

$$S(Q,\omega) = -\frac{1}{\pi}[n(\omega)+1]\ \mathrm{Im}\{\Lambda_\alpha(Q,\omega)G_{\alpha\beta}(Q,\omega)\Lambda_\beta(Q,\omega)+\chi_R(Q,\omega)\}. \tag{22}$$

Note the formal similarity of the first term of (22) to $S_P(Q,\omega)$ in (11). Here the single particle Green function $G_{\alpha\beta}$ plays the role of D, $n_o$ of $F(Q,q\lambda)$ and $\bar{P}(Q,\omega)$ of the frequency dependent $E(Q,q\lambda,\omega)$.

The forms (21) and (22) suggest[49] that any poles in $G_{\alpha\beta}$ will appear as poles in $\chi$ and be observed in $S(Q,\omega)$, provided $n_o$ is finite. This is quite clear in the Bogoliubov approximation (dilute Bose gas) where $n_o$ can be large and $\Lambda_\alpha \approx n_o^{\frac{1}{2}}$. This is important since in the presence of the condensate the observed peaks in $S(Q,\omega)$ can be identified with the poles in $G_{\alpha\beta}$. The peaks of $S(Q,\omega)$ therefore identify the quasiparticle energies. The quasiparticles are the elementary excitations which determine the thermodynamic properties. Since the $\Lambda(Q,\omega)$ are also functions of $\omega$ the spectral shape of $\chi(Q,\omega)$ and $G_{\alpha\beta}(Q,\omega)$ need not be the same but if the poles are sharp enough they should occur at the same energy in $\chi$ and G.

We now indicate that for finite $n_o$, $\chi$ and G have two identical poles. To do this we introduce the irreducible part of $\chi$, $\tilde{\chi}$, which is the sum of all contributions (diagrams) to $\chi$ which cannot be broken into two parts by breaking a single interaction line. With this definition of $\tilde{\chi}$,

$$\chi = \tilde{\chi}/[1 - v(Q)\tilde{\chi}] \equiv \tilde{\chi}/\varepsilon \tag{23}$$

where $v(Q)$ is the Fourier transform of $v(r)$ and $\varepsilon = [1 - v(Q)\tilde{\chi}]$. Similarly we may write $\chi_R = \tilde{\chi}_R/\varepsilon_R$ and $\Lambda = \tilde{\Lambda}/\varepsilon_R$ where $\varepsilon_R = [1-v(Q)\tilde{\chi}_R]$ is a "regular part" dielectric function and $\tilde{\Lambda}$ is assumed to have no singularities. $G_{\alpha\beta}$ is formally written in terms of a numerator ($\tilde{N}$) and denominator (D); $G_{\alpha\beta} = \tilde{N}/D$ and $\tilde{G}_{\alpha\beta} = \tilde{N}/\tilde{D}$. It is not obvious here but Ma and Woo,[51] Kondor and Szépfalusy[51] show that

$$G_{\alpha\beta} = \tilde{N}/D = \varepsilon_R\tilde{N}/[\varepsilon_R\tilde{D} - \tilde{\Lambda}v(Q)\tilde{\Lambda}\tilde{N}]. \tag{24}$$

Using (23) and (21) $\chi$ may be rearranged as[50-52]

$$\chi = \tilde{\chi}\tilde{D}/[\varepsilon_R\tilde{D} - \tilde{\Lambda}v(Q)\tilde{\Lambda}\tilde{N}]. \qquad (T < T_\lambda) \tag{25}$$

Comparing (24) and (25) shows G and $\chi$ will have identical poles; at the zeros of their common denominator. This identity of poles appears as a result of the 'coupling' term $\tilde{\Lambda}v(Q)\tilde{\Lambda}\tilde{N}$ and which is proportional to $n_o$ through $\tilde{\Lambda}$.

To explore the physical nature of these poles let us go above $T_\lambda$ where $n_o = 0$. In this case $\chi_S = 0$ and

$$\chi = \tilde{\chi}_R/\varepsilon_R \qquad (T > T_\lambda).$$

The $\chi$ then has a pole when $\varepsilon_R(Q,\omega_1(Q)) = 0$. This pole coresponds to the excitation in the density-density response function and is a zero sound-like excitation. The energy is denoted[50,52] $\omega_1(Q)$. Similarly where $n_o = 0$, $G_{\alpha\beta} \to \tilde{G}_{\alpha\beta} = \tilde{N}/\tilde{D}$ $(T > T_\lambda)$ which has a pole when $\tilde{D}(Q,\omega_2(Q)) = 0$. The poles in G correspond to the quasiparticle excitations[50,52] and their energies, $\omega_2(Q)$, identify the quasiparticle energies. Clearly, above $T_\lambda$ the quasiparticle and zero sound excitations are quite separate and specifically the quasiparticle poles would not be observed in a neutron scattering study of $\chi$. Below $T_\lambda$, where $n_o$ is finite, though small, the two excitation energies $\omega_1(Q)$ and $\omega_2(Q)$ are shifted in position and both poles appear in $\chi$ and $G_{\alpha\beta}$. The new pole positions, denoted $\tilde{\omega}_1(Q)$ and $\tilde{\omega}_2(Q)$ are given by zeros in the denominators of (24) and (25). Setting the common denominator equal to zero leads to an equation[52,53] like that for the two frequencies of two coupled oscillators, with coupling proportional to $n_o$.

The predictions above may be summarized as follows. In $S(Q,\omega)$ only the poles of $\chi$ are observed. Above $T_\lambda$ there is no connection between $\chi$ and $G_{\alpha\beta}$ so that $S(Q,\omega)$ does not provide any information on the quasiparticle energies. Below $T_\lambda$ the poles of $\chi$ and $G_{\alpha\beta}$ become identical so that a measurement of the peaks or poles in $\chi$ can be used to identify quasiparticle energies. The spectral distribution of $\chi$ and $G_{\alpha\beta}$ are not necessarily the same and, if the peak is broad (as it is at higher Q), the frequency dependence of $\bar{P}(Q,\omega)$ could shift the peak position in $\chi$ away from the pole in $G_{\alpha\beta}$. Below $T_\lambda$, $\chi$ and $G_{\alpha\beta}$ each have two identical poles, at $\tilde{\omega}_1(Q)$ and $\tilde{\omega}_2(Q)$. The $\tilde{\omega}_1(Q)$ pole corresponds to the zero sound mode with energy shifted from its value $\omega_1(Q)$ above $T_\lambda$ due to the "coupling term" $\tilde{\Lambda}v(Q)\tilde{\Lambda}\tilde{N}$ in (25). Since $n_o$ is small the shift should be small. The $\tilde{\omega}_2(Q)$ corresponds to the pole of $G_{\alpha\beta}$ that above $T_\lambda$ was independent of $\chi$. There is now evidence[57] that this $\tilde{\omega}_2(Q) \approx Q^2/2M$ "pole" can be seen in $S(Q,\omega)$ at low Q in He II with weight which increases as T is reduced.

Before passing to new experimental results we must refer to the new polarization potential theory developed by Aldrich, Pethick and Pines[58] since reviews of liquid $^4$He. This theory has been recently summarized by Pines[59] with full reference to the literature. They use the general expression (23) to evaluate the total $\chi$ in the form

$$\chi = \chi_{SC}/[1 - (f_q^s + f_q^v(\omega^2/q^2))\chi_{SC}]. \tag{26}$$

The $\chi_{SC}$ (short for $\chi$ screened) is expressed as

$$\chi_{SC}(q,\omega) = \alpha_q \frac{N_q^2/M_q^*}{\omega^2 - \varepsilon_q^2} + \chi_M(q,\omega) \tag{27}$$

which is the sum of a single excitation component of weight $\alpha_q$ and a multiexcitation component $\chi_M(q,\omega)$. The single excitations have effective mass $M^*_q$ related to $f_q^v$ and $\chi_M(q,\omega)$ is approximated by its static limit $\chi_M(q,\omega) = -NA_q$. The $f^s{}_q$, $f^v{}_q$, $\alpha_q$ and $A_q$ are q dependent parameters determined by fits to S(q), neutron data, the sound velocity as a function of pressure and by sum rules. Particularly, $f^s_q$ is determined as the Fourier transform of a model potential $f_s(r)$ in r space. The $f_s(r)$ has a soft core at small r to simulate short range correlations between atoms in the fluid induced by the hard core of v(r) and a long range part as expected for two isolated He atoms. This theory has been most successful in correlating and predicting neutron scattering intensities, single excitation dispersion curves and their variation with pressure.[59]

## 3.2. A New Interpretation

Woods and Svensson[54] and Svensson et al[55] have proposed an exciting new interpretation of the total $S(Q,\omega)$. They propose that the observed $S(Q,\omega)$ can be fitted at all temperatures, both above and below $T_\lambda$, by the form

$$S(Q,\omega) = \frac{\rho_s(T)}{\rho} S_s(Q,\omega) + \frac{\rho_n(T)}{\rho} S_n(Q,\omega) \ . \tag{28}$$

Here $\rho_s(T)$ ($\rho_n(T)$) is the superfluid (normal) density which entirely accounts for the temperature dependence of $S(Q,\omega)$ ($\rho = \rho_s + \rho_n$). $S_s(Q,\omega)$ consists of a one-phonon peak plus a broad multiphonon background characteristic of the superfluid. The $S_n(Q,\omega)$ consists of a single broad 'background peak' characteristic of the normal fluid. The $S_n(Q,\omega)$ is shown as the dashed line in Fig. 12 at T = 2.15K for Q = 1.13 Å$^{-1}$ which is in the 'maxon' region of the phonon-roton dispersion curve. The $(\rho_s(T)/\rho)S_s(Q,\omega)$ is given by the difference between the total observed intensity at any T in Fig. 12 and the dashed line. This interpretation suggests that the very sharp one-phonon peak in $S_s(Q,\omega)$ disappears entirely above $T_\lambda$ leaving only a much broader distribution above $T_\lambda$. It also imples that $S_s(Q,\omega)$ is a special characteristic of a superfluid. Further support for (28) comes from accurate measurements[60,61] of S(Q) and g(r) which Svensson et al[56] find can be fitted by analogous relations for S(Q) and g(r).

Can support for (28) be found in theory? Griffin and Talbot[62] have developed relationships for $S(Q,\omega)$ similar to (28). However, they[63] now prefer the general formulation presented above. Eqs. (21) and (22) do suggest there will be a component of $S(Q,\omega)$ proportional to $n_o$. However, $n_o$ must be related to $\rho_s$ and since $n_o$ is small we would not expect a superfluid contribution to dominate $S(Q,\omega)$ even at low T. At T $\sim$ 1 K, $\rho_s(T) \approx \rho$ so that $S(Q,\omega) \approx S_s(Q,\omega)$ in (28). We would expect the zero-sound excitation energy $\tilde{\omega}_1(Q)$, below $T_\lambda$, to differ from its value $\omega_1(Q)$ above $T_\lambda$. But again since $n_o$ is always small the difference between

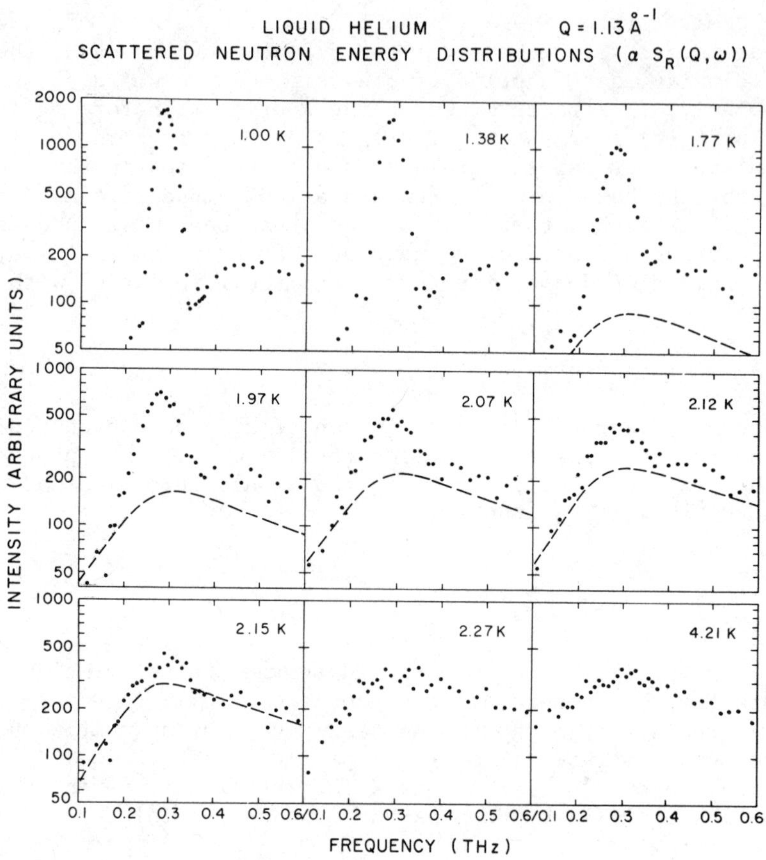


Fig. 12. $S(Q,\omega)$ observed in liquid $^4He$ at $Q = 1.13$ Å$^{-1}$. From Woods and Svensson, Ref. 54.

$\tilde{\omega}_1(Q)$ and $\omega_1(Q)$ should be small. Thus while theory does suggest $S(Q,\omega)$ will be different above and below $T_\lambda$, it does not suggest a general form (28).

It does seem from (22) that if the excitation energy corresponding to $G_{\alpha\beta}$ is desired, only the sharp component appearing below $T_\lambda$ should be used. Svensson[64] notes that roton energies obtained using only the sharp component $S_s(Q,\omega)$ rather than the whole $S(Q,\omega)$ agrees better with quasiparticle energies inferred from thermodynamic measurements. Clearly, (28) raises a new and interesting view which, while perhaps too simple, has foundation in experiment. Equally clearly any theory which attempts to explain it must incorporate $n_o$ explicitly in some way. Deriving a relation of the general form (28) presents an exciting challange for theorists.

## 3.3. One-Multiphonon Interference?

Following Cowley and Woods,[38] Werthamer[30] and substantial discussion by Svensson et al[65], we draw attention to the possibility of interference between one phonon and multiphonon scattering in liquid $^4$He. This is based on an analogy with the one and many phonon interference via anharmonic effects identified in solid $^4$He (section 2.3).

If single phonon scattering in liquid $^4$He were like that from a harmonic solid, the one phonon structure factor (Z(Q)) would have the form (T = 0 K)

$$S_1(Q) = \left( \frac{Q^2}{2M\omega(Q)} \right) e^{-\alpha Q^2} .$$

Here $\alpha Q^2 = 2W(Q)$ is the Debye-Waller factor. At $Q \to 0$ ($\omega \to cQ$) this has the same limit, $S_1(Q) = Z(Q) = (Q/2Mc)$ as (18) in liquid $^4$He. In liquid $^4$He $\omega(Q)$ is approximately linear up to $Q \approx 0.7$ Å$^{-1}$ (see Fig. 9). The observed Z(Q) shown in Fig. 10a, however, lies nearly a factor of two below (Q/2Mc) at $Q \simeq 0.7$ Å$^{-1}$. The observed Z(Q) is obtained by integrating the observed intensity over the one phonon peak region only. This corresponds to evaluating $S_p^+(Q)$ in (15). In the top of Fig. 11 we show $S_p^+(Q)$ evaluated in bcc $^4$He. Clearly $S_p^+(Q)$ and Z(Q) have a similar shape. This suggests a similar interference effect may be operating in liquid $^4$He which takes intensity out of the one phonon peak region of $S(Q,\omega)$ and transfers it to higher $\omega$, which is regarded as the multiphonon region. This transfer is possible only if $S_1(Q,\omega)$ has a high frequency tail and Bartley and Wong[50] suggest that it does in liquid $^4$He.

Similarly, the fraction, $H_I(Q)$, of the f-sum rule $(Q^2/2M)$ taken up by scattering in the one phonon peak region of $S(Q,\omega)$ in He II is shown in Fig. 10b. $H_I(Q)$ corresponds to evaluating $M_P^+$ in (14) and dividing it by $(Q^2/2M)$ to present it as a fraction of the f-sum rule. In the upper half of Fig. 10b is shown calculations of $M_P^+$ along two symmetry directions in bcc $^4$He. For a harmonic solid $M_P^+$ is given by $d^2(Q)$ shown as the dotted line in Fig. 10b. Clearly $M_P^+(Q)$ along the [1,0,0] direction and $H_I(Q)$ show a similar dependence on Q with a substantial dip at $Q \approx 1.3$ Å$^{-1}$ and a hump at $Q \approx 2$ Å$^{-1}$. Again this suggests similar interference effects may be transferring intensity from the apparent one phonon region into the apparent multiphonon region in liquid $^4$He.

In the presence of interference it would be more representative to replace (19) by

$$S_1(Q,\omega) = Z_1(Q)\, A(Q,\omega) + Z_{INT}(Q,\omega)\, A(Q,\omega) + S_M(Q,\omega). \qquad (29)$$

Here $Z_1(Q)$ would represent the 'pure' one phonon contribution and $Z_{INT}(Q,\omega)$ would be responsible for the deviations from $Z_1(Q)$, due to interference. Indeed Wong and Gould[50] have obtained $S(Q,\omega)$ in the form (29) following the theory sketched in section 3.1. They find $Z_{INT}(Q,\omega)$ is proportional to $Q^3$ as is the case for one-two phonon interference in the solids. Again, we note the formal similarity of the singular part of $S(Q,\omega)$ for liquid $^4He$ in (22) and of $S_P(Q,\omega)$ for solid helium in (8). The analogy between the liquid and solid cannot be complete, however, since three phonon processes which provide the dominant interference contribution in solids are largely excluded in liquid $^4He$.

In the 'maxon' region ($Q \approx 1.1$ Å$^{-1}$) Svensson et al.[65] find the apparent interference effect increases with pressure. That is, the $Z(Q)$ shown in Fig. 10a, which is already low at $Q \simeq 1$ Å$^{-1}$, decreases still further at $p \approx 25$ atm and the intensity in the multiphonon region increases. This "further" redistribution of intensity from the one to multiphonon region as pressure is applied has also been explained in terms of hybrydization between two-roton bound states and the single phonon-roton excitations by Zawadowski et al.[66]

Articulation of possible interference contributions and their pressure and temperature dependence presents an interesting problem in liquid $^4He$.

### 3.4. Phonon and Roton Lifetimes

Recently, Mezei[67] and Mezei and Stirling[68] have made accurate measurements of roton and phonon lifetimes, $\tau$, in liquid $^4He$ using a combination of Neutron Spin Echo and Triple Axis Spectrometer techniques. Their results are shown in Fig. 13 which, combined with earlier roton line widths determined by Greytak and Yan[69] and by Tarvin and Passell[70], display several interesting features.

Firstly, the half width at half maximum (HWHM) $\Gamma = \tau^{-1}$ of both phonons and rotons in liquid $^3He$ at $T \approx 1$ K is $\Gamma \approx 0.02$ K. This is roughly two orders of magnitude smaller than the $\Gamma$ observed for a typical longitudinal phonon in solid $^4He$ or for the zero sound mode in liquid $^3He$. Secondly, the $\Gamma(T)$ in liquid $^4He$ is strongly temperature dependent increasing by over two orders between 1 K and $T_\lambda = 2.17$ K. In contrast the phonon line widths in solid $^4He$ and the zero sound mode width in liquid $^3He$ are approximately independent of temperature. Clearly, very different phonon decay processes are taking place in superfluid $^4He$. Is this unique feature related to the condensate? The answer is probably not, at least not directly.

In a solid, phonons decay chiefly via three phonon processes, given by (11). In solid $^4He$ the temperature independent process in

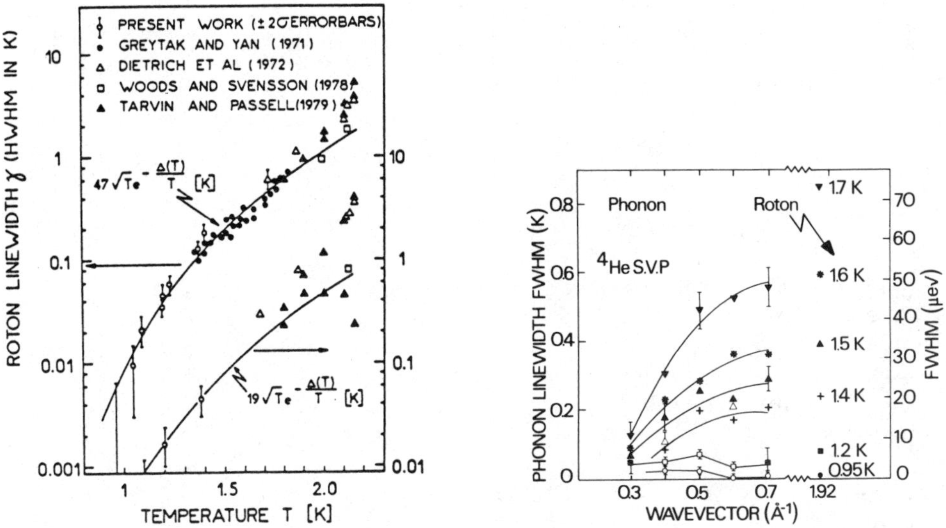


Fig. 13. Phonon and roton linewidths in He II. From Mezei and Stirling (Ref. 68) and Mezei (Ref. 67), respectively. Right figure on roton widths (and shifts) with T includes data from Refs. 69,70. The lines are calculated values of Γ and Δ.

which the excited phonon spontaneously decays into two other phonons, dominantes; the first term in (11). At $T \sim 1$-2 K, the Bose factors in (11) are negligible. Three phonon decay processes can take place in a solid because there are many phonon branches and the solid is anisotropic. By contrast in liquid $^4$He there is only a single, isotropic dispersion curve. Unless there is upward dispersion, three phonon processes cannot take place. Maris[71] has clearly set out the conditions required for three phonon processes. He shows that for $Q \lesssim 0.5$ Å$^{-1}$, where there is indeed upward dispersion, three phonon processes are allowed but are cut off for higher Q. This seems to account for the drop in Γ at $Q \gtrsim 0.5$ Å$^{-1}$ at T = 0.95 K in Fig. 13 and is the first direct evidence of three phonon processes in liquid $^4$He. Even at low Q the phase space available for three phonon processes is small. In liquid $^3$He the zero sound mode can decay to pairs of particle-hole excitations which seems large enough[72] to account for the large and temperature independent width in liquid $^3$He.

The dominant decay processes in liquid $^4$He are therefore of higher order and complicated[1]; either four phonon, five phonon or three phonon processes taking account of the finite width of the phonons. All these processes require the existence of thermal phonons or rotons. The number of thermal rotons is $N_R(T) \propto \sqrt{T}\, e^{-\Delta/T}$ and this temperature dependence agrees with that of $\Gamma(T)$ well.[67,69] Bedell et al[73] have recently evaluated roton-roton scattering (a four "phonon" process) in detail using a T-matrix to represent the scattering amplitude. Again the temperature dependence of $\Gamma(T)$ is given largely by $N_R(T)$. Their work contains a full discussion of this and other interesting features of He II.

### 3.5. The Condensate Fraction

In 1938 London proposed that a substantial fraction, $n_o$, of the atoms in superfluid $^4$He were condensed in the zero momentum state, in analogy with Bose Einstein Condensation in a dilute Bose gas. In the last five years neutron scattering measurements have confirmed unambiguously this controversial proposal. Writing the momentum distribution as

$$n(p) = n_o \delta(p) + (1-n_o) n^*(p), \tag{30}$$

where $n^*(p)$ is the momentum distribution of the excited atoms in $p \neq 0$ states, the observed condensate fraction $n_o(T)$ can be expressed as

$$n_o(T) = n_o(0)\, [1 - (\frac{T}{T_\lambda})^\alpha]. \tag{31}$$

The observed values of $n_o(T)$ are shown in Fig. 14 and give[74] $n_o(0) = 13.3 \pm 1.2$ % and $\alpha = 4.7 \pm 1.2$. Superfluidity and $n_o$ are related!

There are basically four methods for determining[74] $n_o(T)$. Hohenberg and Platzman[75] proposed measurements of $S(Q,\omega)$ at large Q to determine $n_o(T)$. At high momentum transfer they argued, firstly, that a neutron would scatter predominantly from a single nucleus (atom) in the fluid (single particle scattering). In this case the $\ell \neq 0$ terms in (6) can be neglected and $S_c(Q,\omega)$ reduces to the incoherent $S_i(Q,\omega)$. Secondly, the energy transfer is large, much larger than the interatomic potential. In this case the scattering would be from "nearly free particles". If the potential is neglected, $S_i(Q,\omega)$ reduces to $S_{IA}(Q,\omega)$ given by (16). Equivalently, when the energy transfer is large the scattering time is short and the single atom receives an impulse from the incoming neutron. In a short time approximation $S_i(Q,\omega)$ reduces to $S_{IA}(Q,\omega)$.

While these arguments are asymptotically correct[76] at $Q \to \infty$, at finite Q there has been complications; (1) from interference between scattering from more than one atom ($\ell \neq 0$ terms in $S_c(Q,\omega)$) and (2) from interactions between the scattered atom and the remainder of the system[77] (final state interactions). The

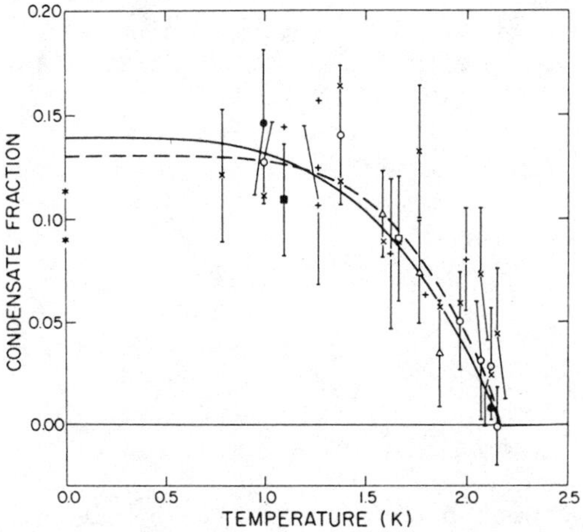


Fig. 14. The observed condensate fraction in He II at svp from Ref. 80 (solid circles and square and solid curve), Ref. 82 (open circles), Ref. 82 (open triangles), Ref. 84 (open square) and Ref. 88 (solid triangle). The *'s are calculations: upper (Ref. 87 and 113), lower (Ref. 86). From Svensson, Ref. 80.

'interference' ($\ell \neq 0$ terms) cause oscillations[77] in both the peak position and the width of $S(Q,\omega)$ as a function of Q of period $\sim 2$ Å$^{-1}$. The final state interactions distorts the shape of $S_{IA}(Q,\omega)$ and displaces its peak to below the recoil energy $\omega_R = Q^2/2M$. The problem in the past has been to correct for these distortions and a detailed account of how this is now done has recently been presented by Sears.[78]

Using (16) Woods and Sears[79] and Sears et al[80] have obtained reliable values of $n_o(T)$ based on measurements of $S(Q,\omega)$ at several large Q values and removing the effects of interference and final state interactions. This work yields $n_o(0) = 12.8 \pm 1.8\%$.

Hyland, Rowland and Cummings[81] (HRC) proposed a second method based on measuring the pair correlation function g(r) as a function of T and the relation

$$[g(r,T)-1] = (1-n_o(T))^2[g^*(r,T^*)-1]. \tag{32}$$

Here $g^*(r,T^*)$ is g(r) due to the excited atoms and it is assumed that $g^*(r,T^*)$ can be obtained from the observed g(r) at a temperature just above $T_\lambda$, i.e. $g^*(r,T) = g(r,T_\lambda)$. It is also assumed in (32) that $g^*$ is not substantially temperature dependent nor influenced by an increasing condensate fraction as T decreases below $T_\lambda$. Although the foundations of (32) have been critisized, observed values[82-84] of $g(r,T)$ appear to obey the HRC relation with an $n_o(T)$ independent of r and give an $n_o(0) = 13.9 \pm 2.7$.

Sears[78] has proposed a method based on determining the kinetic energy K(T) and the relation

$$n_o(T) = 1 - \frac{K(T)}{K(T_\lambda)} . \tag{33}$$

Two determinations of K(T) in (33) lead to $n_o(0) = 13.5 \pm 2.6$ and $12.2 \pm 4.5$, respectively. Finally, Campbell[85] has proposed a method based on measured values of surface tension which gives $n_o(0) = 13\%$.

All these methods provide mutually consistent $n_o(T)$ values as shown in Fig. 14. Calculated values of $n_o(0)$ by Kalos et al[86] (9%) and by Whitlock et al.[87] (11.3%) agree well with the observed values. These results all confirm that superfluid $^4$He has a condensate fraction, although one much smaller than for a weakly interacting Bose gas ($n_o \approx 100\%$). Recent measurements by Mook[88] suggest that $n_o$ decreases with applied pressure, as is expected since interactions increase in strength with pressure.[87,89]

## 4. LIQUID $^3$He

### 4.1. Experimental Results

The study of excitations in liquid $^3$He by inelastic neutron scattering opened in 1974 with the pioneering experiments of Scherm et al.[18,90] and those of Sköld, Pelizzari and collaborators[22,91] in 1976. These experiments are made difficult by the large absorption cross-section of $^3$He for neutrons, $\sigma_a = 5327$b compared to a scattering cross-sections of $\sigma_c = 4.1$ b and $\sigma_i \approx 1.2$ b. All results to date are from these two groups and are for unpolarized neutron beams scattering from unpolarized $^3$He samples.

The $S(Q,\omega)$ (1) observed by Sköld and Pelizzari[22] is shown in Fig. 15, for T = 40 mK and T = 1.2 K. Beginning with T = 40 mK and low $Q \approx 0.4$ Å$^{-1}$, we see the observed $S(Q,\omega)$ has two peaks. The peak at low energy is identified as the 'paramagmon' resonance in the spin density fluctuations, $S_I(Q,\omega)$ of (4). The existence of a peak in $S_I(Q,\omega)$ at low $\omega$ suggests that the spin-antisymmetric interaction entering $S_I(Q,\omega)$ is negative and that the liquid is nearly ferromagnetic. The peak at higher energy is identified as the zero sound excitation predicted in 1966 by Pines[92] in the density fluctuations, $S_c(Q,\omega)$ of (2). It's energy is identified as E in Fig. 15. A peak at high energy suggests the spin-symmetric

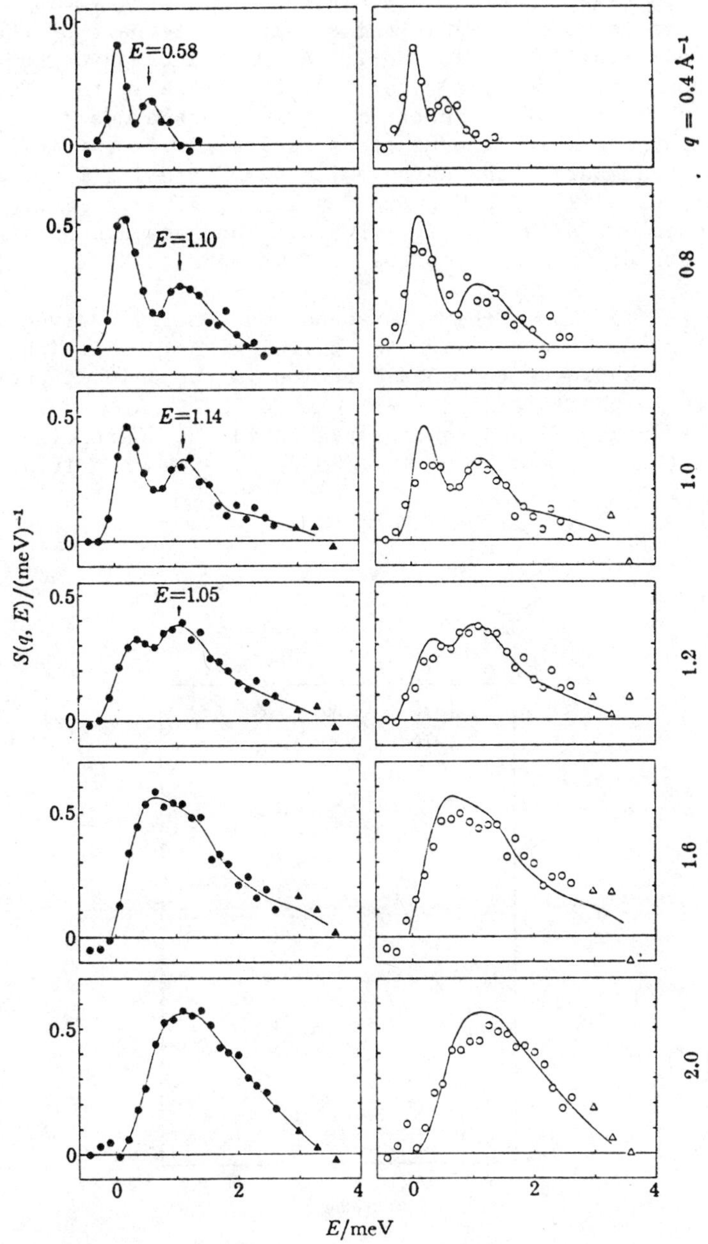


Fig. 15. S(Q,ω) observed at T = 40 mK (left) and at 1.2 K (right). The solid lines are fitted by eye to the T = 40 mK values. From Sköld and Pelizzari, Ref. 22.

interaction entering $S_c(Q,\omega)$ is positive. This zero (zero atomic collisions) sound mode (ZSM) is the direct counterpart of the $\omega_1$ mode observed[93] in $S_c(Q,\omega)$ in $^4He$ I. As Q is increased the two peaks broaden until they disappear. At $Q \geq 1.6$ Å$^{-1}$, the $S(Q,\omega)$ in Fig. 15 is characterist of scattering from a gas a weakly interacting Fermions in which the neutrons excite single particle-hole excitations. Comparing the data at T = 40 mK and 1.2 K we see that the ZSM remains well defined up to T = 1.2 K. The Fermi temperature of liquid $^3He$ at svp is $T_F \simeq 1.5$ K. The paramagmon peak appears to broaden somewhat between 40 mK and 1.2 K.

Shown in Fig. 16 is the zero sound mode energy derived from $S(Q,\omega)$ in Fig. 15 compared with the calculations of Aldrich and Pines.[94] Also shown as the shaded region is the allowed energy band for single particle-hole (P-H) excitations. At $Q \approx 1.3$ Å$^{-1}$ the ZSM enters the P-H band and decays rapidly to particle-hole excitations (Landau damping). The ZSM terminates when it hits the P-H band.

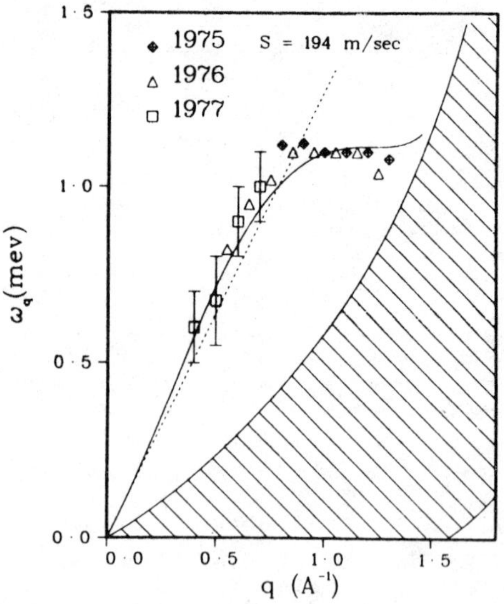


Fig. 16. The zero sound mode energy as calculated by Aldrich and Pines compared with observed values: (1975) Sköld et al (Ref. 91), (1976, 1977) Sköld and Pelizzari (Ref. 22). From Aldrich and Pines, Ref. 94.

In Fig. 17 is shown the zero sound mode peak observed by Hilton et al[90] at three Q values at T = 20 mK. The widths (FWHM) of the zero sound mode derived from the meaurements are shown in Fig. 18. Since at the low Q values shown in Fig. 18 the ZSM cannot decay to single P-H excitations, decay by other energically allowed processes must be taking place. The open squares show the ZSM widths calculated by Glyde and Khanna[72] for decay to pairs of particle-hole excitations.

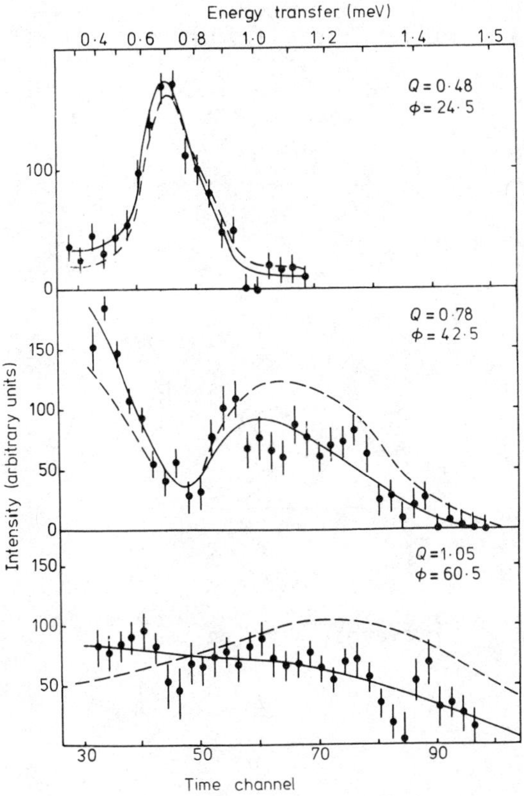


Fig. 17. Scattered intensity from liquid $^3$He observed at T = 20 mK. Full curves are guides to the eye and the dashed curves are these guides converted to $S(Q,\omega)$. From Hilton et al (1980) Ref. 90.

Figs. 15-18 briefly summarize existing data which are now mutually consistent. The zero sound mode has an energy $\sim$ 30% lower than that in $^4$He and exists up to Q $\approx$ 1.3 Å$^{-1}$ only beyond which it is heavily Landau damped. The short length of the ZSM is certainly due to Fermi statistics. At low T ($T \lesssim 1$ K) the ZSM width in $^3$He is at least two orders of magnitude broader than the corresponding phonon-roton excitation in He II. At low T, decay mechanisms clearly exist in $^3$He that do not exist in He II. At high T (near $T_\lambda$ in $^4$He) the widths of 'zero sound' in $^3$He and $^4$He become comparable. At high $Q \gtrsim 2$ Å$^{-1}$ there are no collective excitations in $^3$He or $^4$He and the observed $S(Q,\omega)$ is very similar, as is shown in Fig. 19. In Fig. 19 the $S(Q,\omega)$ observed in $^4$He at T = 4.2 K has been adjusted to T = 0 K by adjusting the Bose factor $[n(\omega) + 1]$ in (22). This affects $S(Q,\omega)$ at low $\omega$ only. With this general background we go on to interpret $S(Q,\omega)$ in $^3$He.

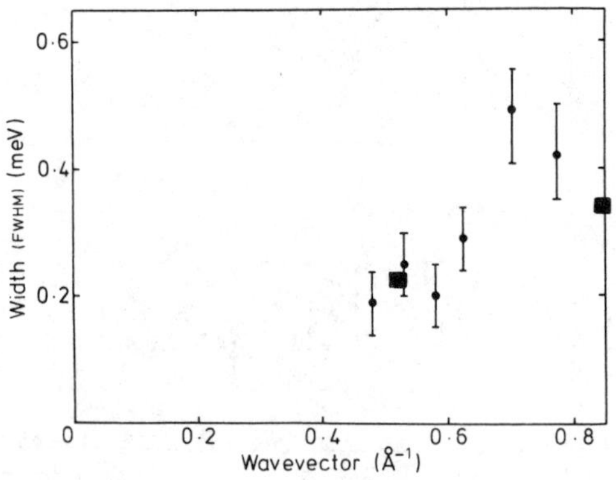


Fig. 18. Widths of the zero sound mode (observed Ī) and as calculated (□) by Glyde and Khanna (Ref. 72). From Hilton et al (1980), Ref. 90.

## 4.2. Interpretation

### (a) Free Fermions

Van Hove[96] in his classic papers apparently first deduced $S_o(Q,\omega)$ for a gas of non-interacting Fermions. This simple case provides much insight on $S(Q,\omega)$ in liquid $^3$He. To follow standard developments, we normalize $\chi(Q,t)$ in (4) with the volume V rather than N ($V_o = V/N$).

The ground state of a Fermi gas has all the lowest momentum states occupied, two particles per state, up to the Fermi momentum $k_F$. The density is uniform. In second quantization, the density wave creation operator is $\rho^+(Q) = \sum_k a^+_{k+Q} a_k$. A density wave is

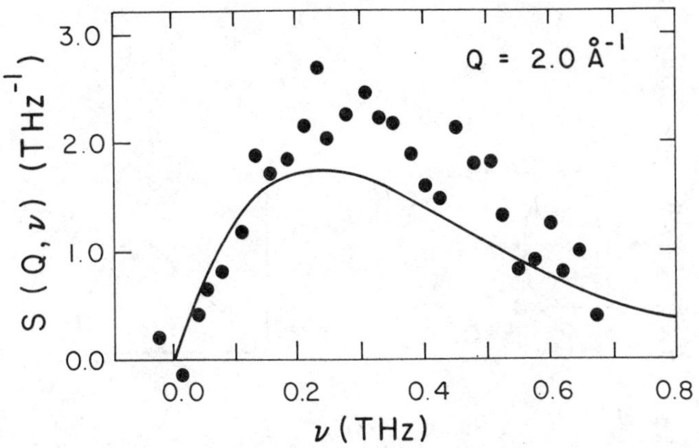


Fig. 19. $S(Q,\nu)$ observed in liquid $^3$He (points) by Sköld et al at T = 15 mK and in liquid $^4$He at T = 4.2 K adjusted to T = 0 K. From Woods, Ref. 95.

therefore induced in the gas by annihilating a particle from within the Ferm sea ($a_k$, $k < k_F$), leaving a hole, and creating a particle above the Fermi sea ($a^+_{k+Q}$, $|\vec{k}+\vec{Q}| > k_F$). A density-wave of wave vector $\vec{Q}$ is a superposition of particle ($\vec{k}+\vec{Q}$) - hole (k) excitations all separated by wave vector $\vec{Q}$. Two examples of particle-hole (P-H) excitations (A and B) are depicted in Fig. 20. The energy transferred, $\omega = \varepsilon(\vec{k}+\vec{Q}) - \varepsilon(k)$, by a neutron in creating the P-H pair (A and B) is also depicted in Fig. 20. The width of the allowed energy band of P-H excitations reflects the diameter of the Fermi sea.

When $\rho^+(Q) = \rho(-Q)$ and $\rho(Q)$ are substituted into $\chi$ a straightforward evaluation leads to the Lindhard function

$$\chi^\circ(Q,\omega) = \frac{1}{V}\sum_{k\sigma}\frac{n(k) - n(k+Q)}{\omega-[\varepsilon(k+Q)-\varepsilon(k)]+i\eta} \tag{34}$$

where n(k) is the Fermi function. The Lindhard function may be evaluated analytically both at T = 0 K and at finite T. The resulting $S_o(Q,\omega)$ is[97]

$$S_o(Q,\omega) = -\frac{V_o}{\pi}[n(\omega)+1]\left(\frac{dn}{d\varepsilon}\right)\left(\frac{\pi z}{4x}\right)\ln\left[\frac{1+e^{-a_1}}{1+e^{-a_2}}\right]$$

where $a_1=[(y/x-x^2)/4-\mu/\mathring{\varepsilon}_F]/z$, $(dn/d\varepsilon)=3/(2\mathring{\varepsilon}_F\mathring{V}_o)$ is the density of states per unit volume at the Fermi energy $\mathring{\varepsilon}_F$, $a_2 = [(y/x+x^2)/4-\mu/\mathring{\varepsilon}_F]/z$ and

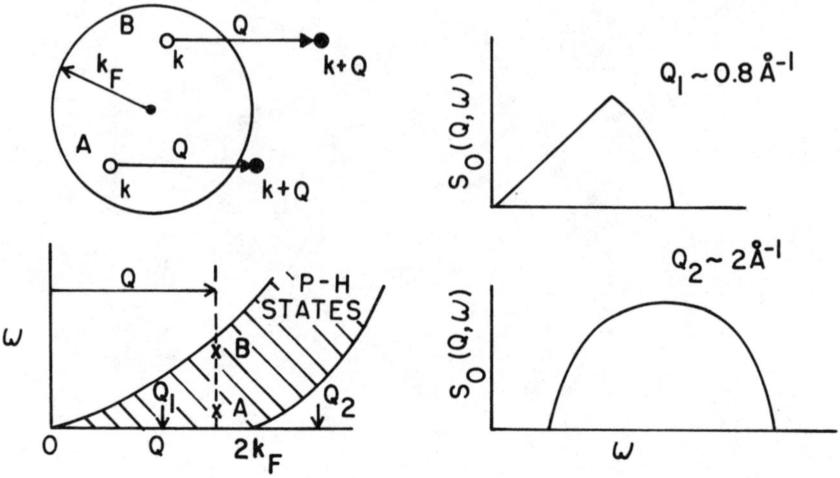


Fig. 20. Fermi sea, particle-hole excitations and $S_o(Q,\omega)$ for a Fermi gas.

$y = \omega/\varepsilon_F^\circ$ and $z = T/\varepsilon_F^\circ$. $S_o(Q,\omega)$ is sketched for two representative values of Q in Fig. 20 for $^3$He at svp where $k_F \approx 0.78$ Å$^{-1}$ and $\varepsilon_F^\circ \approx 5.0$ K. In the absence of interaction both $S_c$ and $S_I$ reduce to $S_o$.

(b) Interactions

In the presence of interactions, the full $\chi(Q,\omega)$ may be rigorously expressed in the form[98,99]

$$\chi_{C,I}(Q,\omega) = \chi^\circ(Q,\omega)/[1 - V_{s,a}(Q,\omega)\chi^\circ(Q,\omega)]. \qquad (35)$$

Here $\chi^\circ$ is a convenient zero order value for a system of independent Fermions such as (34). The $V(Q,\omega)$ represents all effects of interaction between the P-H pairs. Eq. (35) is, for example, a rearrangement of Dyson's equation for the P-H propagator, $\chi^{-1} = \chi^{\circ -1} - V(Q,\omega)$, and $V(Q,\omega)$ is the P-H self energy.[98] The $V_s = \frac{1}{2}(V^{\uparrow\uparrow}+V^{\uparrow\downarrow})$ and $V_a = \frac{1}{2}(V^{\uparrow\uparrow}-V^{\uparrow\downarrow})$ are the spin-symmetric and spin-antisymmetric P-H interactions, respectively. The full $\chi$ may also be decomposed in the form (23). If we replace $V(Q,\omega)$ by $v(Q)$ in (35) we have the simple random phase approximation (RPA).

Almost all calculations of $S(Q,\omega)$ use the form (35). At low Q and low $\omega$, (35) reduces[100] to the collisionless Landau-Boltzmann equation for the response of the density (or spin density) to a perturbation. In this limit, and if we keep only spherically symmetric components of $V$, the $V$ reduce to the Landau parameters $V_s=(dn/d\epsilon)^{-1}F_o^s$ and $V_a=(dn/d\epsilon)^{-1}F_o^a$. $F_o^s$ is found empirically[100] to be $F_o^s\sim10.7$. This strong positive interaction in (35) pushes $\chi_C(Q,\omega)$ up to higher frequencies leading to the ZSM. $F_o^a$ is found to be $F_o^a\sim-0.67$ and this negative interaction in (35) pushes $\chi_I(Q,\omega)$ down into a resonance at low $\omega$ within the P-H band. This broadly explains the origin of the two peaks in the observed $S(Q,\omega)$ in Fig. 16 as $Q \to 0$.

In general the P-H (or particle-particle) interaction is very complicated. It has three chief components. There is a strong repulsion at close approach due to the hard core of the potential (plus a weak attraction at long range). There is the Fermi statistical repulsion. Thirdly there are the interactions between pairs induced via the density and spin density excitations. The third component appears to dominate at low Q and $\omega$ while the hard core dominates at high Q and $\omega$.

## 4.3. Theory

Akhiezer et al[101] apparently first calculated $S(Q,\omega)$ for interacting $^3$He. They derived the form (1) and evaluated $S_c(Q,\omega)$ at low Q assuming it contained a zero sound excitation. They approximated $S_I(Q,\omega)$ by $S_o(Q,\omega)$. An early calculation of $S_c(Q,\omega)$ by Lovesey[102] at high Q used continued fraction methods and found good agreement with the high Q results of Scherm et al.[18]

Following the prediction by Pines[92] that zero sound should exist at $Q \sim k_F$, Aldrich, Pethick and Pines[94] developed a polarization potential theory for liquid $^3$He along similar lines discussed in section 3.1 for $^4$He. They express $\chi$ in the general form (35),

$$\chi^{C,I}(q,\omega) = \chi_{SC}^{C,I}/[1-(f_q^{s,a} + g_q^{s,a}(\omega^2/q^2))\chi_{SC}^{C,I}] \qquad (36)$$

where to follow their notation we write q = Q. The $\chi_{SC}^{C,I} = \alpha_q^{C,I}\chi^{\circ*}(Q,\omega)+\chi_M^{C,I}$ where $\chi^{\circ*}$ is the Lindhard function (35) for free quasiparticles having an effective mass $M^*(q)$ and $\chi_M(q)$ is a static multiparticle-hole component. For the coherent part, $\alpha_q^C$ and the form of $\chi_M(q)$ were determined by fitting to the high q neutron scattering data of Scherm et al[18] while $\alpha_q^I = 1$ and $\chi_M^I = 0$. The $f_q^{s,a}$ are Fourier transforms of an effective quasi-particle interaction. This model potential has a spin dependent soft core

$$f^{\sigma\sigma'}(r) = a^{\sigma\sigma'}[1 - (r/r_c^{\sigma\sigma'})^8] \qquad\qquad r < r^{\sigma\sigma'}$$

and the remainder of f(r) $(r > r^{\sigma\sigma'})$ is much like v(r). The

$f_q^{s,a} = \frac{1}{2}(f^{\uparrow\uparrow} \pm f^{\uparrow\downarrow})$ and the parameters $a^{\uparrow\uparrow}$, $a^{\uparrow\downarrow}$, $r_c^{\uparrow\uparrow}$ and $r_c^{\uparrow\downarrow}$ where chosen so that the $q \to 0$ limits reproduce the Landau parameters. The $Ng_q^s$ was chosen to give the correct effective mass at $q = 0$ and $Ng_q^s$ was assumed to fall off reasonably as q increased. The $M^*$ in $\chi^{o*}(q,\omega)$ is $M_q^* = M_o(1 + Ng_q^s)$ and $Ng_q^a$ was set to zero.

While this theory contains many q dependent adjustable parameters it is able to provide a complete description of $S(Q,\omega)$ up to $Q \lesssim 1.6$ Å$^{-1}$. The polarization potential theory has also been used to evaluate scattering amplitudes, transport properties and other properties[59] in liquid $^3$He. Some problems in consistently establishing the parameters in the polarization potential theory have recently been discussed by Béal-Monod and Valls.[103]

Glyde and Khanna[72] have developed a simple phenomenological theory based on (35) and Landau theory. In the $Q \to 0$ limit (35) reduces to the Landau-Boltzmann equation for $\chi$ provided $V(Q,\omega)$ is related the parameterized interaction (F) used by Landau as[104],

$$\lim_{Q\to 0} V_{s,a}(Q,\omega) = \left(\frac{dn}{d\varepsilon}\right)^{-1}\left[F_o^{s,a} + \frac{F_1^{s,a}}{1+F_1^{s,a}/3}\left(\frac{\omega}{Qv_F}\right)^2\right] \qquad (37)$$

where $v_F = k_F/M^*$. In making this equality only the $\ell = 0$ and $\ell = 1$ angular momentum components of $V(Q,\omega)$ have been retained (as is usual in Landau theory). This limit also sets the effective mass in $\chi^{o*}$ as $M^* = M_o(1 + F_1{}^s/3)$. To allow some contribution to $S_c(Q,\omega)$ from multiparticle-hole excitations at larger Q, without violating the f-sum rule, $V_s(Q,\omega)$ was multiplied by a function $\beta(Q)$. The $\beta(Q)$ contained one adjustable parameter and $\beta(Q) \approx 1$ for $Q \lesssim 1$ Å$^{-1}$. Since $S_I(Q,\omega)$ does not satisfy an f-sum rule $V_a$ in (37) was used at all Q.

As shown in Fig. 21 this simple picture agrees reasonably well with experiment up to $Q \approx 1$ Å$^{-1}$. The calculated ZSM peak in Fig. 21 also includes a decay term to pairs of P-H excitations with the P-H interaction again approximated by (37). This model predicts a substantial pressure dependence of the ZSM energy (Fig. 21) which follows from the strong pressure dependence of $F_o{}^s$.

With their measurements, Stirling et al[90] presented an RPA calculation of $S(Q,\omega)$. This had the interesting feature of a first principles calculation of the energy spectrum $\varepsilon(k)$ appearing in $\chi_o$.

Takeno and Yoshida[99] derived (35) rigorously finding ($\alpha$ = C or I)

$$V_\alpha(Q,\omega) = V_\alpha(Q,0) + [V_\alpha(Q,\infty) - V_\alpha(Q,0)]\omega M_{\alpha 2}(Q,\omega).$$

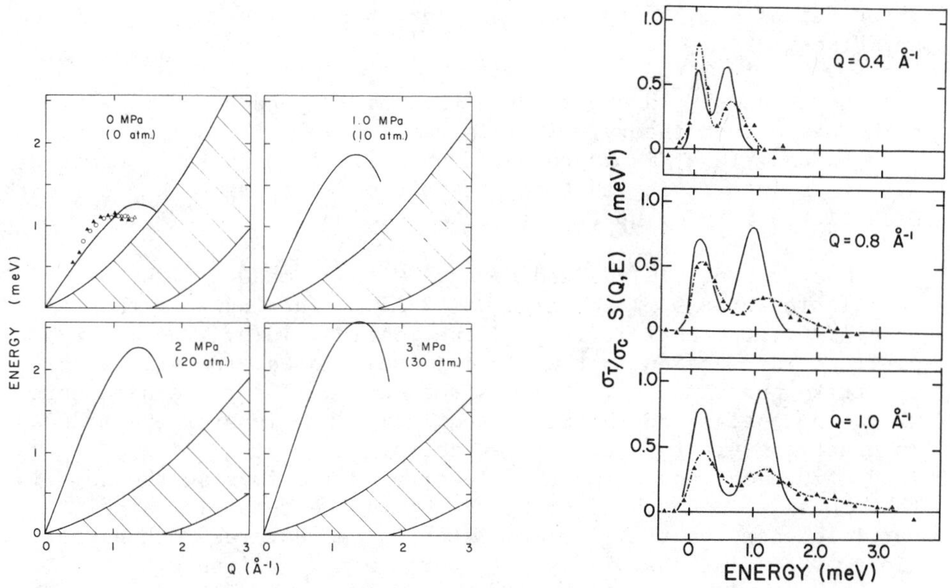


Fig. 21. Right: $S(Q,\omega)$ in liquid $^3He$ at svp. Left: Zero sound mode versus pressure. Data are from Sköld and Pelizzari, Ref. 22. From Glyde and Khanna, Ref. 72.

Here $M_{\alpha 2}(Q,\omega)$ is the interacting part of the second order memory function and $V(Q,0)$ and $V(Q,\infty)$ are static limits of $V(Q,\omega)$. They firstly expand $V(Q,\omega)$ in powers of $\omega$ keeping terms up to $\omega^2$ only to get $V_\alpha(Q,\omega)=V_\alpha(Q,0)+b_\alpha\omega^2$. This shows that the interactions of the form (36) and (37) correspond to low frequency as well as low Q approximations. Using the compressibility and f-sum rules for both $S_C(Q,\omega)$ and $S_I(Q,\omega)$ they obtain the interaction (37) with $F_1^s$ replacing $F_1^a$ in $V_I(Q,\omega)$. This is not correct, however, since $S_I$ does not satisfy an f-sum rule. Next they keep only the static parts $V_C(Q,0)$ and $V_I(Q,0)$ and determine these by fitting to the observed $S_C(Q)$ and $S_I(Q)$. Finally, they determine $V_\alpha(Q,\infty)$ from third moment sum rules and represent the imaginary part of $M_{\alpha 2}(Q,\omega)$ by a Gaussian in $\omega$. The last, dynamic interaction provides damping of zero sound and a broadened $S(Q,\omega)$ in reasonable agreement with experiment.

Valls et al[105] have also derived a microscopic expression for $\chi_\alpha(Q,\omega)$ using the memory function formulation. They analyze this in detail and show that it can be reduced to the form (36) provided the part of the interaction corresponding to collisions between particles is ignored (i.e., a zero collision approximation). By fitting to $S_C(Q)$ and $S_I(Q)$ they obtain polarization potential interactions similar to those of Aldrich and Pines[94] and of Bedell and Pines.[106]

Béal-Monod[107] has evaluated the spin-dependent part $\chi_I(Q,\omega)$ using paramagnon theory. In this case $\chi^\circ$ in (35) is (34) with $M^* = M_O$ and $V(Q,\omega)$ is represented by a single adjustable constant I. It is interesting that this simple model appears to provide the best fit to $S_I(Q,\omega)$ up to $Q \gtrsim 1.2$ Å$^{-1}$ so far!

Most recently Glyde and Hernadi[108] have made a first principles calculation of $S(Q,\omega)$ for high $Q \gtrsim 2$ Å$^{-1}$. The input is the Aziz et al[15] pair potential v(r) and the liquid volume. The inter-particle interaction in the fluid is calculated from v(r) within a T-matrix approximation.[109] This accounts well for the hard core of the potential and the Fermi statistics but ignores the induced interactions via the collective excitations. At $Q \gtrsim 1.5$ Å$^{-1}$ liquid $^3$He does not support collective excitations so the hope is that these will not be important. Using the T-matrix, single particle energies $\varepsilon(k)$ are evaluated in the Hartree-Fock approximation. The relation (35) is used for $\chi$ with $\chi_O$ given by (34) representing independent Fermions have Hartree-Fock energies $\varepsilon(k)$. The interaction $V_{s,a}(Q,\omega)$ is represented the spin-symmetric and spin-antisymmetric T-matrices.

The resulting $S(Q,\omega)$ predicts the peak position well but the intensity in the peak region is too low with too much intensity at higher $\omega$. If the imaginary part of the T-matrix is ignored, the resulting $S(Q,\omega)$ agrees very well with experiment. This suggests there is as yet too much damping in the theory but that the real part of the T-matrix has the correct sign and magnitude at high Q. Since $S(Q,\omega)$ in $^3$He and $^4$He are similar at $Q \gtrsim 2$ Å$^{-1}$ this approach may also be applicable to He I.

## 4.4. Future Prospects

There remain many interesting avenues to explore in $^3$He. It would be interesting to measure the pressure dependence of the scattering to see whether the ZSM energy increases with pressure or not. It would be interesting to extend measurements to higher Q values, particular to measure the momentum distribution n(p). Can the step in the momentum distribution of the particles (as distinct from the quasiparticles) at $p_F$ be observed in $S(Q,\omega)$? It would be interesting to measure $S(Q,\omega)$ more precisely at low Q and low $\omega$ to see whether a 'paramagmon' model really fits better than a Fermi liquid model. Higher temperature measurements would also be interesting to see at what T the paramagnon resonance and zero sound mode disappears. These lectures have also ignored the whole area of mixtures[5] entirely.

While it is now possible to describe $S(Q,\omega)$ accurately using soundly based theory,[94] its implementation requires many parameters fitted to experiment. A challange is to develop microscopic theories having few or no free parameters. At low Q this will be difficult. Also a consistent theory at finite temperature is needed. For example, a remaining puzzle is the temperature dependence of the effective mass. Specific heat measurements suggest $M^*$ decreases strongly with T, from $M^* \gtrsim 3$ at T = 0 K to $M^* \lesssim 1$ at T = 150 mK. However, if $M^* \lesssim 1$ the zero sound mode would lie in the P-H band and be damped. From Fig. 1, the ZSM clearly exists up to T = 1.2 K and this apparent discrepancy needs resolution.

Finally, perhaps further in the future scattering from polarized $^3He$ using polarized neutron beams may be possible. In this case the neutron and $^3He$ would interact in the spin triplet state only and the large absorption would be removed.

ACKNOWLEDGEMENTS

It is a pleasure to acknowledge valuable discussions with K. Bedell, R. A. Cowley, L. Passell, R. Scherm, K. Sköld, W. G. Stirling and most recently with A. Griffin and E. C. Svensson.

REFERENCES

1. A. D. B. Woods, and R. A. Cowley, Rept. Prog. Phys. 36, 1135, (1973).
2. W. G. Stirling, J. Phys. (Paris), 39, C6-1334, (1978).
3. R. A. Cowley, in: Conf. on "Neutron Scattering", vol. II, ed., R. M. Moon, Nat. Tech. Inf. Service, Springfield, Virginia (1976); in "Quantum Fluids" ed., J. Ruvalds and T. Regge, North Holland, Amsterdam (1978).
4. A. Zawadowski, ibid, (1978).
5. D. L. Price, in: Ref. 14, Vol. II (1978).
6. C. Domb and J. S. Dugdale, in: "Progress in Low Temperature Physics",Vol. 2, ed., J. Gorter, North Holland, Amsterdam, (1957).
7. R. A. Guyer, in: "Solid State Physics", ed., F. Seitz and D. Turnbull, Academic, N.Y., 23, 413 (1969).
8. T. R. Koehler, in: "Dynamical Properties of Solids", Vol. II, ed., G. K. Horton and A. A. Maradudin, North Holland, Amsterdam, (1975).
9. H. Horner, ibid, Vol. 1, (1974).
10. H. R. Glyde, in: "Rare Gas Solids", Vol. I, ed., M. L. Klein and J. A. Venables, Academic Press, New York, (1976).
11. C. M. Varma and N. R. Werthamer in Ref. 14, Vol. I, (1976).

12. W. E. Keller, "Helium-3 and Helium-4", Plenum, New York, (1969).
13. J. Wilks, "The Properties of Liquid and Solid Helium", Oxford University Press, London, (1967).
14. "The Physics of Liquid and Solid Helium", ed. K. H. Bennemann and J. B. Ketterson, Wiley, New York, Vol. I, (1976), Vol. II (1978).
15. R. A. Aziz, V. P. S. Nain, J. Carley, W. L. Taylor, and G. T. McConville, J. Chem. Phys., 70, 4330, (1979).
16. P. Loubeyre, J. M. Besson, J. P. Pinceau and J. P. Hansen, Phys. Rev. Lett. 49, 1172, (1982); D. Levesque, J. J. Weis, and M. L. Klein, Phys. Rev. Lett. 51, 670, (1983).
17. J. H. Hertherington, M. Roger and J. M. Delrieu, Rev. Mod. Phys., 55, 1, (1983).
18. R. Scherm, W. G. Stirling, A. D. B. Woods, R. A. Cowley, and G. J. Coombs, J. Phys. C. 1, L341, (1974).
19. K. Sköld (private communication), J. Floquet (private communication).
20. V. F. Sears, J. Phys. C. 9, 409, (1976).
21. T. A. Kitchens, T. Oversluizen, L. Passell, and R. I. Schermer, Phys. Rev. Lett. 32, 791 (1974).
22. K. Sköld and C. Pelizzari, Phil. Trans. Roy. Soc. London, Ser. B, 290, 305, (1980).
23. V. J. Minkiewicz, T. A. Kitchens, G. Shirane, and E. B. Osgood, Phys. Rev. A 8, 1513 (1973).
24. W. Marshall and S. W. Lovesey, "Theory of Thermal Neutron Scattering", Oxford University Press, London, (1971).
25. R. A. Cowley, Adv. Phys. 12, 421, (1963); Rep. Prog. Phys. 31, 123 (1968).
26. L. H. Nosanow, Phys. Rev. 146, 120 (1966).
27. V. Ambegaokar, J. Conway, and G. Baym, in "Lattice Dynamics", ed., R. F. Wallis, Pergamon, New York, 261, (1965).
28. R. A. Cowley, E. C. Svensson, and W. J. L. Buyers, Phys. Rev. Lett. 23, 325 (1969); R. A. Cowley and W. J. L. Buyers, J. Phys. C. 2, 2262, (1969).
29. H. Horner, Phys. Rev. Lett. 29, 556 (1972).
30. N. R. Werthamer, Phys. Rev. Lett. 28, 1102 (1972).
31. H. R. Glyde, Can. J. Phys. 52, 2281 (1974).
32. J. Meyer, G. Dolling, R. Scherm, and H. R. Glyde, J. Phys. F 6, 943 (1976).
33. E. B. Osgood, V. J. Minkiewicz, T. A. Kitchens and G. Shirane, Phys. Rev. A 5, 1537 (1972); T. A. Kitchens, G. Shirane, V. J. Minkiewicz and E. B. Osgood, Phys. Rev. Lett. 29, 552 (1972).
34. H. R. Glyde and F. C. Khanna, Can. J. Phys. 50, 1152 (1972).
35. J. Eckert, W. Thomlinson, and G. Shirane, Phys. Rev. B 18, 1120 (1978).
36. W. M. Collins and H. R. Glyde, Phys. Rev. B 18, 1132 (1978).
37. C. Stassis, G. Kline, W. A. Kamitakahara, and S. K. Sinha, Phys. Rev. B 17, 1130 (1978).

37a. H. R. Glyde and S. I. Hernadi, Phys. Rev. B 25, 4787 (1982).
38. R. A. Cowley and A. D. B. Woods, Can. J. Phys. 49, 177 (1971).
39. H. Horner in: "Proceedings of the Thirteenth International Conference on Low Temperature Physics", ed. by K. D. Timmerhaus, W. J. O'Sullivan, and E. F. Hammel, Plenum, New York, (1974).
40. R. O. Hilleke, P. Chaddaha, R. O. Simmons, D. L. Price and S. K. Sinha, Phys. Rev. Lett. (in press).
41. M. Roger and H. R. Glyde, Phys. Lett. A 89, 252 (1982).
42. R. P. Feynman, Phys. Rev. 94, 164 (1954).
43. R. P. Feynman, and M. Cohen, Phys. Rev. 102, 1189, (1956).
44. M. Cohen and R. P. Feynman, Phys. Rev. 107, 13 (1957).
45. A. Miller, D. Pines and P. Nozières, Phys. Rev. 127, 1452 (1962).
46. C. W. Woo, in Ref. 14, Vol. I (1976) and references there.
47. S. T. Beliaev, Sov. Phys. - JETP 7, 289 (1958).
48. N. Hugenholtz and D. Pines, Phys. Rev. 116, 489 (1959).
49. J. Gavoret and P. Nozières, Ann. Phys. NY 28, 349 (1964).
50. V. K. Wong and H. Gould, Ann. Phys. NY 83, 252 (1974); D. L. Bartley and V. K. Wong, Phys. Rev. B 12, 3775 (1975).
51. P. Szépfalusy and I. Kondor, Ann. Phys. NY 82, 1 (1974); S. K. Ma and C. W. Woo, Phys. Rev. 159, 165 (1967).
52. T. H. Cheung and A. Griffin, Can. J. Phys. 48, 2135 (1970); A. Griffin and T. H. Cheung, Phys. Rev. A 7, 2086 (1973).
53. E. Talbot and A. Griffin, Ann. Phys. 151, 71 (1983).
54. A. D. B. Woods and E. C. Svensson, Phys. Rev. Lett. 41, 974 (1978).
55. E. C. Svensson, R. Scherm, and A. D. B. Woods, J. de Phys. 39, C6-211 (1978).
56. E. C. Svensson, V. F. Sears and A. Griffin, Phys. Rev. B 23, 4493 (1981).
57. N. M. Blagoveschchenskii, E. B. Dokukin, Zh. A. Kozlov and V. A. Parfenov, JETP Lett. 31, 4 (1980).
58. C. H. Aldrich, III and D. Pines, J. Low Temp. Phys., 25, 677 (1976); C. H. Aldrich, III, C. J. Pethick and D. Pines, J. Low Temp. Phys. 25, 691 (1976).
59. D. Pines in: International School of "Enrico Fermi" - Varenna, Italy (1983).
60. E. C. Svensson, V. F. Sears, A. D. B. Woods, and P. Martel, Phys. Rev. B 21, 3638 (1980).
61. E. C. Svensson and A. F. Murray, Physica 108B, 1317 (1981).
62. A. Griffin, Phys. Rev. B 19, 5946 (1979); A. Griffin and E. Talbot, Phys. Rev. B 24, 5075 (1981).
63. A. Griffin (private communication).
64. E. C. Svensson in: "The Neutron and Its Application", Conference Proceedings, Cambridge, UK (1982).
65. E. C. Svensson, P. Martel, V. F. Sears and A. D. B. Woods, Can. J. Phys. 54, 2178 (1976).
66. A. Zawadowski, J. Ruvalds, and J. Solana, Phys. Rev. A. 5, 399 (1972); J. Ruvalds and A. Zawadowski, Phys. Rev. Lett. 5, 333 (1970).

67. F. Mezei, Phys. Rev. Lett. 44, 1601 (1980).
68. F. Mezei and W. G. Stirling in: "75th Jubilee Conference on Helium-4"; St. Andrews, UK, (August 1983).
69. T. J. Greytak and J. Yan, Phys. Rev. Lett. 22, 897 (1969).
70. J. A. Tarvin and L. Passell, Phys. Rev. B 19, 1458 (1979).
71. H. J. Maris, Rev. Mod. Phys. 49, 341 (1977).
72. H. R. Glyde and F. C. Khanna, Can. J. Phys. 58, 343 (1980).
73. K. Bedell, D. Pines and A. Zawadowski, Phys. Rev. B 29, 102 (1984).
74. V. F. Sears, Phys. Rev. B 28, 5109 (1983).
75. P. C. Hohenberg and P. M. Platzman, Phys. Rev. 152, 198 (1966).
76. V. F. Sears, Phys. Rev. 185, 200 (1969).
77. P. Martel, E. C. Svensson, A. D. B. Woods, V. F. Sears and R. A. Cowley, J. Low Temp. Phys. 23, 285 (1976).
78. V. F. Sears, Phys. Rev. B 29, (in press).
79. A. D. B. Woods and V. F. Sears, Phys. Rev. Lett. 39, 415 (1977).
80. V. F. Sears, E. C. Svensson, P. Martel, and A. D. B. Woods, Phys. Lett. 49, 279 (1982); E. C. Svensson, in Workshop on High Energy Excitations in Condensed Matter, LANL, Feb. 1984.
81. G. J. Hyland, G. Rowlands, and F. W. Cummings, Phys. Lett. 31A, 465 (1970); F. W. Cummings, G. J. Hyland, and G. Rowlands, Phys. Kondens. Mater. 12, 90 (1970).
82. V. F. Sears and E. C. Svensson, Phys. Rev. Lett. 43, 2009 (1979); V. F. Sears, E. C. Svensson, and A. F. Murray, (unpublished).
83. V. F. Sears and E. C. Svensson, Int. J. Quantum Chem. Symp. 14, 715 (1980).
84. H. N. Robkoff, D. A. Ewen, and R. B. Hallock, Phys. Rev. Lett. 43, 2006 (1979).
85. L. J. Campbell, Phys. Rev. B 27, 1913 (1983).
86. M. H. Kalos, M. A. Lee, P. A. Whitlock, and G. V. Chester, Phys. Rev. B 24, 115 (1981).
87. P. A. Whitlock, D. M. Ceperly, G. V. Chester and M. H. Kalos, Phys. Rev. B 19, 5598 (1979).
88. H. A. Mook, Phys. Rev. Lett. 51, 1454 (1983).
89. P. M. Lam and M. L. Ristig, Phys. Rev. B 20, 1960 (1979).
90. W. G. Stirling, R. Scherm, P. A. Hilton, and R. A. Cowley, J. Phys. C 9, 1643 (1976); P. A. Hilton, R. A. Cowley, W. G. Stirling, and R. Scherm, Z. Phys. B30, 107 (1978); P. A. Hilton, R. A. Cowley, R. Scherm, and W. G. Stirling, J. Phys. C 13, L295 (1980).
91. K. Sköld, C. A. Pelizzari, R. Kleb, and G. E. Ostrowski, Phys. Rev. Lett. 37, 842 (1976); K. Sköld and C. A. Pelizzari in: "Quantum Fluids and Solids", ed. by S. B. Trickey, E. D. Adams and J. W. Duffy, Plenum, New York, (1977); J. Phys. C 11, L589 (1978).

92. D. Pines in: "Quantum Fluids", ed. by D. F. Brewer, North Holland, Amsterdam, (1966).
93. A. D. B. Woods, E. C. Svensson, and P. Martel, Phys. Lett. 57A, 439 (1976).
94. C. H. Aldrich, III and D. Pines, J. Low. Temp. Phys. 32, 689 (1978), and earlier references therein.
95. A. D. B. Woods, (private communication).
96. L. van Hove, Phys. Rev. 95, 249 (1954).
97. R. Jullien, M. T. Béal-Monod and B. Coqblin, Phys. Rev. B 9, 1441 (1977); F. C. Khanna and H. R. Glyde, Can. J. Phys. 54, 648 (1976).
98. P. Schuck, J. Low Temp. Phys. 7, 459 (1972).
99. S. Takeno and F. Yoshida, Prog. Theor. Phys. 60, 1585 (1978); F. Yoshida and S. Takeno, Prog. Theor. Phys. 62, 37 (1979).
100. G. Baym and C. J. Pethick, in Ref. 14, Vol. II (1978).
101. A. I. Akhiezer, A. I. Akhiezer and I. Pomeranchuk, Sov. Phys. JETP, 14, 343 (1962).
102. S. W. Lovesey, J. Phys. C 8, 1649 (1975).
103. M. T. Béal-Monod and O. T. Valls, J. Low Temp. Phys. (submitted).
104. A. Widom and J. L. Sigel, Phys. Rev. Lett. 37, 1692 (1967); S. Babu and G. E. Brown, Ann. Phys. 78, 1 (1973).
105. O. T. Valls, G. F. Mazenko, and H. Gould, Phys. Rev. B 18, 263 (1978); O. T. Valls, H. Gould and G. F. Mazenko, Phys. Rev. B 25, 1663 (1982); see also K. N. Pathak and M. Lücke, Phys. Rev. B 28, 246B, (1983).
106. K. Bedell and D. Pines, Phys. Rev. Lett. 45, 39 (1980).
107. M. T. Béal-Monod, J. Low Temp. Phys. 37, 123 (1979); J. Magn. Mater. 14, 283 (1979).
108. H. R. Glyde and S. I. Hernadi, Phys. Rev. B (in press) (1984).
109. H. R. Glyde and S. I. Hernadi, Phys. Rev. B 28, 141 (1983).

# DYNAMICAL PROPERTIES OF CLASSICAL LIQUIDS AND LIQUID MIXTURES

G. Jacucci*, M. Ronchetti*, and W. Schirmacher**

* Dipartimento di Fisica
Universitá di Trento, Povo, Trento, Italy

**Physik-Department E13
Technische Universität München
D 8064 Garching, FRG

## INTRODUCTION

### a) Density Fluctuations in Single-Component Simple Classical Liquids

A simple liquid is defined as a system of N particles in which the structure dependent part of the potential energy can be represented as a sum over pairwise potentials:

$$E = \sum_{i<j} \phi(r_{ij}) \tag{1}$$

where i and j run over all particles of the system and $r_{ij}$ is the distance of a pair of particles. Once the pair potential $\phi(r)$ is specified a number of physical properties can be calculated by statistical physical methods or by computer simulation techniques. If one is interested in the dynamics of the system the central quantities one usually looks at are the coherent and incoherent neutron scattering laws $S(\vec{Q},\omega)$ and $S_S(\vec{Q},\omega)$. They can be directly measured by means of neutron inelastic scattering:

$$\begin{aligned} \frac{d^2\sigma}{d\Omega dE} &= \frac{k}{k_o} \{ b^2_{coh} S(\vec{Q},\omega) + b^2_{incoh} S_S(\vec{Q},\omega) \} \\ &= \left(\frac{d^2\sigma}{d\Omega dE}\right)_{coh} + \left(\frac{d^2\sigma}{d\Omega dE}\right)_{incoh} \end{aligned} \tag{2}$$

where $\frac{d^2\sigma}{d\Omega dE}$ is the partial differential cross section for inelastic scattering of neutrons of initial momentum $\hbar\vec{k}_o$ into the solid angle $\Omega$. The final momentum is $\hbar\vec{k}$, and $\hbar\vec{Q} = \hbar(\vec{k}-\vec{k}_o)$ is the momentum transfer. $\hbar\omega = \frac{\hbar^2}{2M}(k_o^2-k^2)$ is the energy transfer (M is the particle mass) and $b^2_{coh}$ and $b^2_{incoh}$ are coherent and incoherent averages over the squares of the neutron scattering lengths of the isotopes(se ref. 1). In isotropic liquids $d^2\sigma/d\Omega dE$ depends only on $Q = |\vec{Q}|$.

The double Fourier transforms of $S_S(Q,\omega)$ and $S(Q,\omega)$ which are called Van-Hove correlation functions have a simple intuitive meaning concerning the motion of the N classical particles:

$$\begin{aligned} G_S(\vec{r},t) &= \left(\frac{1}{2\pi}\right)^3 \int d^3\vec{Q} \int d\omega\, S_S(Q,\omega) \exp\{i\omega t - i\vec{Q}\vec{r}\} \\ &= \left(\frac{1}{2\pi}\right)^3 \int d^3\vec{Q}\, F_S(Q,t) \exp\{i\vec{Q}\vec{r}\} \end{aligned} \tag{3}$$

gives the probability for a specific particle appearing at $\vec{r}$ at time t if it started initially at the origin, i.e. $G_S(\vec{r},t)$ describes the individual motion of the particles.

On the other hand

$$\begin{aligned} G(\vec{r},t) &= \left(\frac{1}{2\pi}\right)^3 \int d^3\vec{Q} \int d\omega\, S(Q,\omega) \exp\{i\omega t - i\vec{Q}\vec{r}\} \\ &= \left(\frac{1}{2\pi}\right)^3 \int d^3\vec{Q}\, F(Q,t) \exp\{i\vec{Q}\vec{r}\} \end{aligned} \tag{4}$$

gives the probability for any particles being at time t given that the same or another had been at the origin at t=0, i. e. $G(\vec{r},t)$ describes the collective motion of the particles. $F_S(Q,t)$ and $F(Q,t)$ are called intermediate scattering functions.

$G(\vec{r},t)$ and $G_S(\vec{r},t)$ can be formally represented as the correlation functions of the number density

$$N(\vec{r},t) = \sum_{i=1}^{N} \delta(\vec{r} - \vec{r}_i(t)) \tag{5}$$

and the probability density of a tagged particle

$$N_o(\vec{r},t) = \delta(\vec{r} - \vec{r}_o(t))$$

in the following way:

$$G(\vec{r},t) = (1/N)\ \langle N(\vec{r},t)N(\vec{0},0)\rangle$$
$$G_S(\vec{r},t) = \langle N_o(\vec{r},t)N_o(\vec{0},0)\rangle \tag{7}$$

where <...> denotes a thermal average at temperature T.

The above correlation functions have the following initial values (sum rules for $S(Q,\omega)$ and $S_S(Q,\omega)$ ):

$$F_S(Q,t=0) = 1$$
$$F(Q,t=0) \equiv S(Q) = 1 + \rho_0 \int d^3r(g(r)-1)\exp\{i\vec{Q}\vec{r}\} \tag{8}$$

$\rho_0$ is the number of particles per volume, $S(Q)$ the static structure factor and $g(r)$ the radial pair distribution function of the liquid. In the hydrodynamic limit we have

$$\lim_{Q\to 0} S(Q) = \rho_0 k_B T\ \chi_T \tag{9}$$

where $\chi_T$ is the isothermal compressibility.

b) Individual motions

If one looks at the individual motions of a particle one is usually interested in the diffusion coefficient of the particle defined by

$$D = (k_BT/M) \int_o^\infty Z(t)dt \tag{10}$$

where

$$Z(t) = (M/3k_BT)\ \langle \vec{v}(t')\cdot\vec{v}(t+t')\rangle \tag{11}$$

is the normalised velocity autocorrelation function. Here, $\vec{v}(t)$ is the velocity of the particle at time t. In the hydrodynamic regime (ie. in the Q-regime where the wavelength $\lambda = 2\pi/Q$ is very large compared with a mean interatomic spacing a) particle number conservation yields the following connection between $S_S(Q,\omega)$ and D

$$S_S(Q,\omega) = (1/\pi)DQ^2\{\omega^2 + (DQ^2)^2\}^{-1} \tag{12}$$

i. e. $G_S(\vec{r},t)$ is simply the well known solution of the diffusion equation.

In the high-Q regime (collisionless regime $Q \gg 2\pi/a$) $S_S(Q,\omega) = S(Q,\omega) = S_{id}(Q,\omega)$ which is the scattering law of a non-interacting fluid (ideal gas):

$$S_{id}(Q,\omega) = (M/2\pi k_B T Q^2)^{1/2} \exp\{-M\omega^2/2k_B T Q^2\} \qquad (13)$$

Interpolation schemes between the limits (12) and (13) based on plausible physical ideas have been given in the literature[2,3] and might be useful for interpretation of measured $S_S(Q,\omega)$ data.

## c) Collective motions

The most interesting question concerning the collective motions of a liquid is, up to which values of Q do non-overdamped collective excitations exist. Such excitations manifest themselves in $S(Q,\omega)$ as peaks at finite frequency.

In the hydrodynamic regime momentum conservation assures that the sound damping constant is proportional to $Q^2$ which leads to long lived compressional sound waves of dispersion $\omega(Q)=uQ$, where u is the adiabatic sound velocity. At higher momentum transfer the corresponding Brillouin peak will broaden, accompanied with a non-trivial dispersion until the point where the collective excitation becomes overdamped. In analogy to the solid one usually defines the "liquid phonon dispersion" $\omega(Q)$ as the locus of the maxima in $S(Q,\omega)$.

First evidence for the existence of distinct phonon-like collective excitations in a simple liquid up to $Q=1.2\text{Å}^{-1}$ has been given by Copley and Rowe[4] who reported on neutron inelastic scattering (NIS) in liquid Rb. These data could be very nicely reproduced by a computer simulation of Rahman[5] using a pair potential derived from pseudopotential theory and the molecular dynamics (MD) technique[6,7]. In fig. 1 we show their $\omega(Q)$ curve. Very recently similar collective excitations have also been observed experimentally in liquid K[8]. The observed dispersion compares well with our simulation data on K to be presented in the next section.

It is interesting to note that such high-Q collective modes which are present in the alkalis are not observed in liquid Ar[6]. Bosse, Götze and Lücke[9] who developed a successful analytic theory of density and current fluctuations in simple liquids attribute this difference to the different compressibilities of the two types of liquids: the metallic bonding involves interatomic forces which are much more rigid that those of the rare gas liquids leading to a smaller compressibility and, in turn, to more pronounced solid-like behaviour of the metals. For a review of the conditions required for the propagation of a short wavelength collective mode see, for example, Section 6.6 of reference 2.

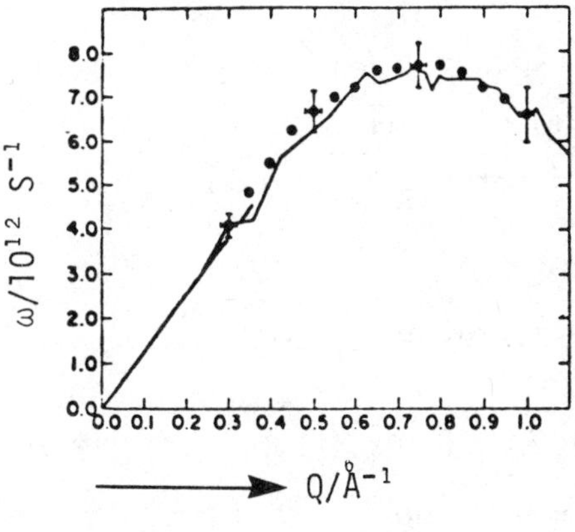


Fig. 1:

$\omega(Q)$ for liquid Rb obtained from NIS by Copey and Rowe[4] (dots) and from a MD calculation by Rahman[5] (continuous line).

One should mention that in the hydrodynamic regime entropy fluctuations lead to a peak at $\omega=0$ ("quasielastic peak" or "Rayleigh peak") of the form (12) with width $D_q Q^2$, where $D_q$ is the thermal diffusivity. In the Q range accessible to neutron scattering and molecular dynamics calculations this peak is usually quite broad and not very pronounced.

d) Extension to two components

Traditionally, in studying the dynamic fluctuations of a two-component liquid with $N_1=c_1N$ and $N_2=c_2N$ particles one considers the partial density fluctuations

$$N_\nu(\vec{r},t) = \sum_{i=1}^{N_\nu} \delta(r-r_i^\nu(t)) \tag{14}$$

where $r_i^\nu$ is the position of a particle of type $\nu$ ($\nu$ can be 1 or 2). One then defines partial dynamic structure factors by

$$S_{\mu\nu}(Q,\omega) = \frac{1}{4\pi N_\mu^{1/2} N_\nu^{1/2}} \int d^3\vec{r} \int d\omega \langle N_\mu(\vec{r},t)N_\nu(\vec{0},0) + N_\nu(\vec{r},t)N_\mu(\vec{0},0)\rangle \times \exp\{ i\vec{Q}\vec{r} - i\omega t\}$$

$d^2\sigma/d\Omega dE$ is then a weighted sum of these three functions.

It has, however, been demonstrated[10] that a very physical picture of what happens in the mixture arises if one considers overall number fluctuations

$$N(\vec{r},t) = N_1(\vec{r},t) + N_2(\vec{r},t) \tag{15}$$

and deviations from random mixing, i. e. concentration fluctuations defined by

$$c(\vec{r},t) = (1/N)\{N_1(\vec{r},t) - c_1N(\vec{r},t)\}$$
$$= (1/N)\{c_2N_1(\vec{r},t) - c_1N_2(\vec{r},t)\}. \tag{16}$$

One now defines scattering laws(and intermediate scattering functions in an analogous way) as follows:

$$S_{NN}(Q,\omega) = \frac{1}{2\pi N}\int d^3\vec{r}\int d\omega \; \langle N(\vec{r},t)N(\vec{0},0)\rangle \;\; \exp\{\, i\vec{Q}\vec{r} - i\omega t\} \tag{17}$$

$$S_{cc}(Q,\omega) = \frac{N}{2\pi}\int d^3\vec{r}\int d\omega \; \langle c(\vec{r},t)c(\vec{0},0)\rangle \;\; \exp\{\, i\vec{Q}\vec{r} - i\omega t\} \tag{18}$$

$$S_{Nc}(Q,\omega) = \frac{1}{2\pi}\int d^3\vec{r}\int d\omega \; \langle N(\vec{r},t)c(\vec{0},0) + c(\vec{r},t)N(\vec{0},0)\rangle \times \exp\{\, i\vec{Q}\vec{r} - i\omega t\} \tag{19}$$

and we have for the coherent NIS cross section

$$\frac{d^2\sigma}{d\Omega dE} \propto (\bar{b})^2 S_{NN}(Q,\omega) + (b_1-b_2)^2 S_{cc}(Q,\omega) + 2\bar{b}(b_1-b_2)S_{Nc}(Q,\omega) \tag{20}$$

where $b_1$, $b_2$ are the partial scattering lengths and $\bar{b} = c_1b_1+c_2b_2$.

We have the following useful thermodynamic sum rules:

$$\lim_{Q\to 0} F_{NN}(Q,t=0) \equiv S_{NN}(Q=0) = \rho_0 k_B T\chi_T + S_{NC}^2(0)/S_{cc}(0) \tag{21}$$

$$\lim_{Q\to 0} F_{cc}(Q,t=0) \equiv S_{cc}(Q=0) = \rho_0 k_B T/\left(\frac{\partial^2 G}{\partial c_1^2}\right)_{p,T,N} \tag{22}$$

$$\lim_{Q\to 0} F_{Nc}(Q,t=0) \equiv S_{Nc}(Q=0) = \rho_0(v_1 - v_2)S_{cc}(0) \tag{23}$$

Here G is the Gibbs free energy and $\partial^2 G/\partial c_1^2$ is the so-called stability of the alloy. $v_1$ and $v_2$ are the partial molar volumes per atom. We see that the cross-correlation functions between number and concentration fluctuations are related to size effects. To separate these size effects from other phenomena one is interested in one can study quantities which are uncoupled in the static limit. This can be achieved by isolating that part of $N(\vec{Q},t)$ which is orthogonal to $c(\vec{Q},t)$ considering $S_{NN}(Q)=F_{NN}(Q,t=0)$, $S_{cc}(Q)=F_{cc}(Q,t=0)$, and $F_{Nc}(Q,t=0)=S_{Nc}(Q)$ as scalar products of these two quantities, viz.

$$\tilde{N}(Q,t) = N(Q,t) - (S_{Nc}(Q)/S_{cc}(Q))c(Q,t) \tag{24}$$

yielding

$$S_{NN}(Q,\omega) = S_{\tilde{N}\tilde{N}}(Q,\omega) + (S_{Nc}(Q)/S_{cc}(Q))^2 S_{cc}(Q,\omega) \tag{25}$$

For $F_{\tilde{N}\tilde{N}}$ we have now the sum rule

$$F_{\tilde{N}\tilde{N}}(Q,t=0) = \rho_0 k_B T \chi_T \tag{26}$$

suggesting that the quantity $\tilde{N}$ behaves like the number density of a one-component liquid.

## 2. RESULTS OF COMPUTER SIMULATIONS OF LIQUID ALKALI METALS AND ALLOYS

As mentioned already above, computer experiments using the MD technique can provide a great deal of insight into the behaviour of collective excitations at high momentum transfer in classical liquids. Recently, a study of liquid Na and K, and of the alloy Na-K (at equal concentration) has been carried out[11] to investigate the interplay between concentration and number fluctuations.

The density dependent effective two-body potentials of Dagens, Rasolt and Taylor[12], which are known to give a good description of the static and dynamic properties of the alkali metals, were employed for the simulation. The main computational effort was directed at the calculation of the three partial dynamical structure factors and of the number and concentration fluctuation spectra $S_{NN}(Q,\omega)$ and $S_{cc}(Q,\omega)$

Results on $S_{NN}(Q,\omega)$ for the three systems are shown in fig. 2 for the smallest value of Q compatible with the periodic boundary conditions imposed for the simulation. In all three cases the spectrum has a hydodynamic-like structure with a well resolved Brillouin peak, and a central Rayleigh peak.

However, a dramatic difference is seen between the density fluctuation spectra of the alloy and the pure systems. While in pure Na and K the Rayleigh and Brillouin peak have similar height and width, in the alloy the central peak is narrower and much more intense. The conclusion is that the intermediate scattering function $F_{NN}(Q,t)$ has a component which decays very slowly with time. As Q increases, the Brillouin peak of the alloy shifts and broadens in the manner shown in fig. 3. The dispersion of the propagating mode is plotted for all three systems in fig.4.

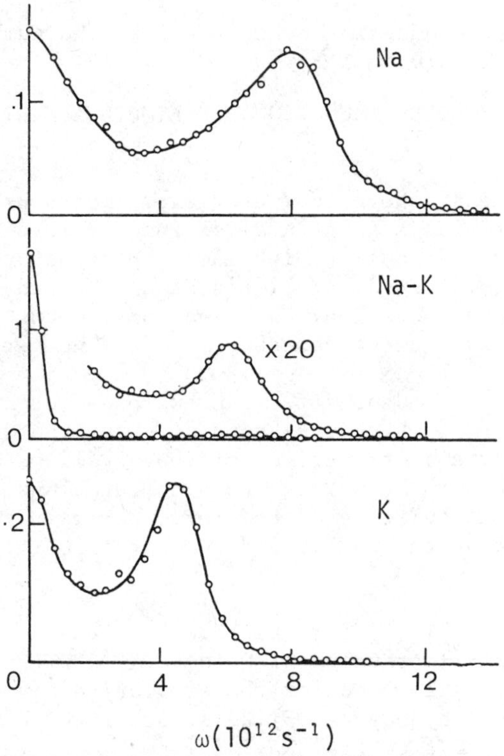


Fig.2: The function $S_{NN}(Q,\omega)$ (in units of $10^{-12}$s and normalized to unit area) for the three systems Na, K and NaK. The wavenumbers are thesmallest consistent with the periodic boundary condition: Q=0.289 $Å^{-1}$ (Na), 0.258 $Å^{-1}$ (NaK) and 0.232 $Å^{-1}$(K). The high frequency region for NaK is also shown on a scale twenty times larger.

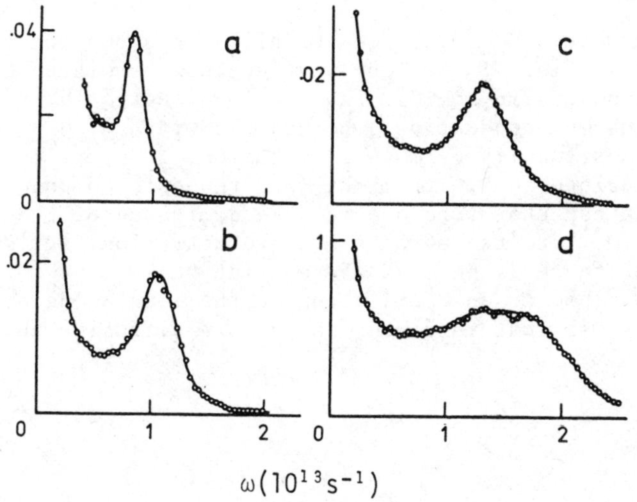


Fig. 3: The function $S_{NN}(Q,\omega)$ (same units as in fig. 2) for NaK at $Q=0.258$ Å$^{-1}$ (a), 0.447 Å$^{-1}$ (b), 0.577 Å$^{-1}$ (c), and 1.064 Å$^{-1}$ (d).

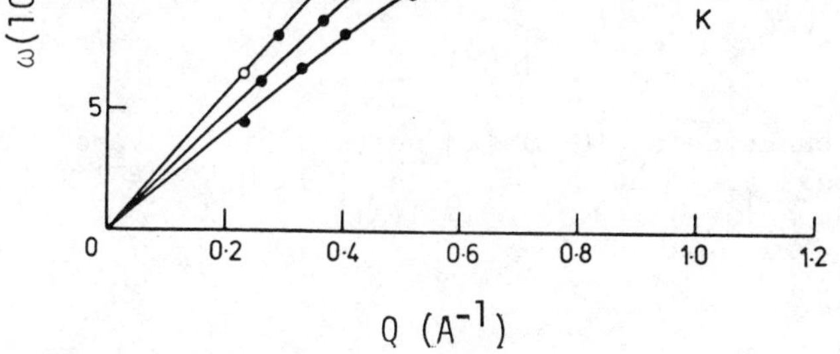

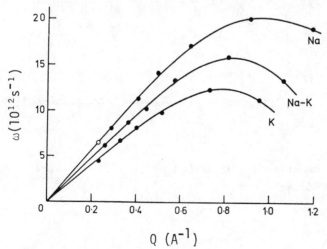

Fig. 4: Dispersion of the sound wave peak for the three model systems. Dots: ref. 11. circles: ref. 13. The curves are guidelines for the eye.

The figure shows that a well-defined propagating density fluctuation persists down to wavelengths which are comparable with the nearest neighbour separation in the liquid, i. e. $\lambda = 5.0$ Å (Na), 5.7Å (NaK) and 6.3 Å (K).

The function $S_{cc}(Q,\omega)$ for the alloy is shown for several values of Q in fig. 5. At long wavelengths the width of the Rayleigh component in $S_{NN}(Q,\omega)$ should in principle be controlled partly by thermal conduction, whereas the width of $S_{cc}(Q,\omega)$ is determined almost wholly by mutual diffusion. In practice $S_{CC}(Q,\omega)$ is almost indistinguishable from the low frequency component of $S_{NN}(Q,\omega)$ One can therefore ascribe the dominance of the central peak of $S_{NN}(Q,\omega)$ to the effects of interdiffusion. It is interesting to note that there is no sign of any side peak in $S_{cc}(Q,\omega)$ which could be ascribed to an oscillation in the local concentration fluctuations, so that this mode has a pure diffusive character.

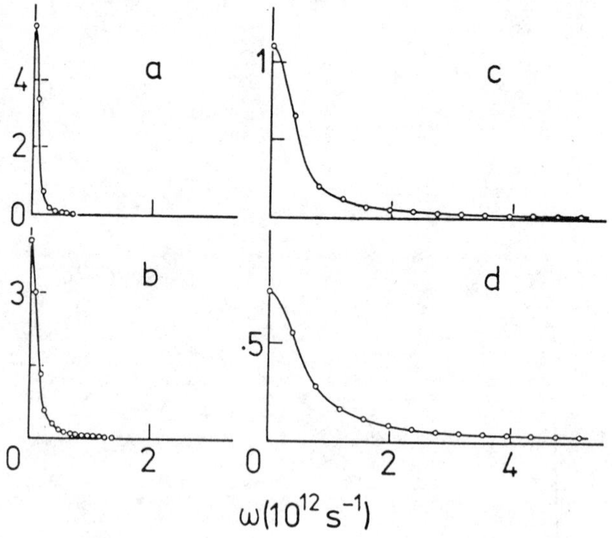


Fig. 5: The function $S_{cc}(Q, )$ (in units of $10^{-12}$s and normalized to unit area) at Q = 0.258 Å$^{-1}$ (a), 0.577 Å$^{-1}$ (b), 1.064 Å$^{-1}$ (c) and 1.569 Å$^{-1}$ (d).

It should be noted that the diffusion process is not usually visible in $S(Q,\omega)$ in one component liquids which corresponds to the function $S_{NN}(Q,\omega)$ of the alloy (for example, there is no sharp peak at zero frequency in pure K or Na), although $S_S(Q,\omega)$ is contained in $S(Q,\omega)$. The reason is that inhomogeneities in number density at wave vector Q die out without the need to transport individual atoms over a distance $2\pi/Q$, but with a simple rearrangement of the atomic positions in which each atom is redisplaced over only a very short distance. This implies a compensation of low frequency contributions from $S_S(Q,\omega)$ and $S_D(Q,\omega) = S(Q,\omega) - S_S(Q,\omega)$. This compensation no longer occurs precisely for $S_{NN}(Q,\omega)$ in the alloy. Inhomogeneities in the number density at wave vector Q can arise because of concentration fluctuations due to the different size of the two types of atoms. To restore homogeneity in the number density the excess Na- atoms must interdiffuse against the excess K-atoms over a distance of order $2\pi/Q$. Hence the appearance of the low frequency peak in the $S_{NN}(Q,\omega)$ of NaK.

If we use the more formal language introduced in part d) of the introduction the cross coupling between number and concentration fluctuations leads to a contribution to $S_{NN}(Q,\omega)$ as expressed by eq. (25). Comparing the prefactor (which is proportional to $S_{Nc}(Q)^2$) of this contribution with the static simulation data[14] shows that eq. (25) correctly describes the effect discussed above.

No such cross coupling between number and concentration fluctuations is present at low frequency in molten salts for the very good reason that, at low Q, concentration fluctuations are largely suppressed by the strong requirement of charge neutrality.

A process similar to the one observed in the Na-K alloy is to be expected in rare gas mixtures, although the coupling between number and concentration fluctuations will be smaller due to the much more similar sizes of the types of atoms.

## 3. SIMULATION RESULTS FOR THE STATIC AND DYNAMIC STRUCTURE FACTORS OF LIQUID $Li_4Pb$.

Another interesting class of liquid alloys is the one formed by metallic constituents which show strong deviations from ideal mixing behaviour at compositions determined by chemical stoichiometry. One example of such a system is the well investigated system $Li_4Pb$, which shows a tendency towards ionic bonding, although its conductivity is still that of a (poor) metal.

Recently a model in which the constituents interact (in addition to the usual short-range repulsion) via screened Coulomb potentials, has been employed for calculations of the partial structure factors of liquid $Li_4Pb$ using the mean spherical and hypernetted chain approximations[15]. These calculations show that the wave vector dependence of the measured concentration structure factor[16] $S_{CC}(Q)$ can be reasonably well accounted for by such a model.

Copestake et al.[15] used both hard-core and soft-core potentials for the short-range repulsion. They obtained best agreement with experiment using the hypernetted chain approximation and a soft-core repulsion. Their potential has the form

$$\phi_{ij}(r) = k_B Ta \exp\{(d-r)/p\} + Q_i Q_j e^2 \exp\{-\lambda r\}/r \qquad (27)$$

The parameters a and p determine the "softness" of the repulsive core which is assumed to be the same for all ij pairs. The authors arbitrarily chose a=2.0 and p=0.33 Å which are values similar to those employed for molten alkali halides. d is the hard core diameter which had been given the value 2.65 Å. The second part of the potential describes the chemical interaction of the alloy constituents. It is assumed that Pb (i=2) carries a negative charge $Q_2e$ as a result of charge transfer leaving a positive charge $Q_1e$ on the Li site.

Because the system is metallic electronic screening is present which in the Thomas Fermi approximation leads to the exponential factor in the second term of the pair potential with $\lambda = 4\pi e^2 N(E_F)$ ($N(E_F)$ is the electronic density of states at the Fermi level).

The choice of this potential was motivated by the fact that the ordering potential $W(r) = \phi_{12}(r) - (1/2)\{\phi_{11}(r) + \phi_{22}(r)\}$ could be directly evaluated from experiment in an approximate way for $3Å \leqq r \leqq 5Å$ which yielded $W(r) \propto \exp\{-\lambda r\}/r$ with $\lambda = 1.1$ Å$^{-1}$. This value is smaller than the free electron screening parameter indicating a pseudogap in the density of states.

Assuming $c_1Q_1 + c_2Q_2 = 0$ ("local charge neutrality") they obtained from the prefactor of the ordering potential $Q_1$=0.5 and $Q_2 = -2.0$. Recalculating $S_{cc}(Q)$, best results were obtained by choosing $Q_1 = -0.533$ and $Q_2 = -2.134$. We also took the latter values for our simulation.

We have undertaken a MD calculation using the model potential given by eq. (27) with the parameters listed above. The aim of this calculation is twofold:

(i) checking the integral equation results for the static correlation functions against "exact" computer simulation data and

(ii) extending the calculation to the dynamical structure factors to compare with recent neutron inelastic scattering experiments on liquid $Li_4Pb$ obtained by Soltwisch, Quitmann, Ruppersberg and Suck[16]. Since in this system the average neutron scattering length (see eq. (20) ) $\bar{b}=0$ ("zero alloy") these authors were able to measure directry the function $S_{cc}(Q,\omega)$.

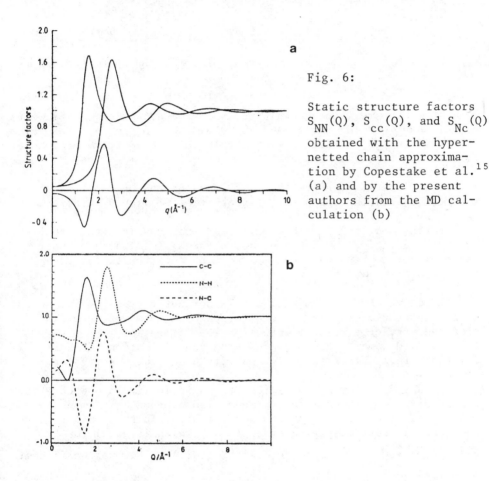

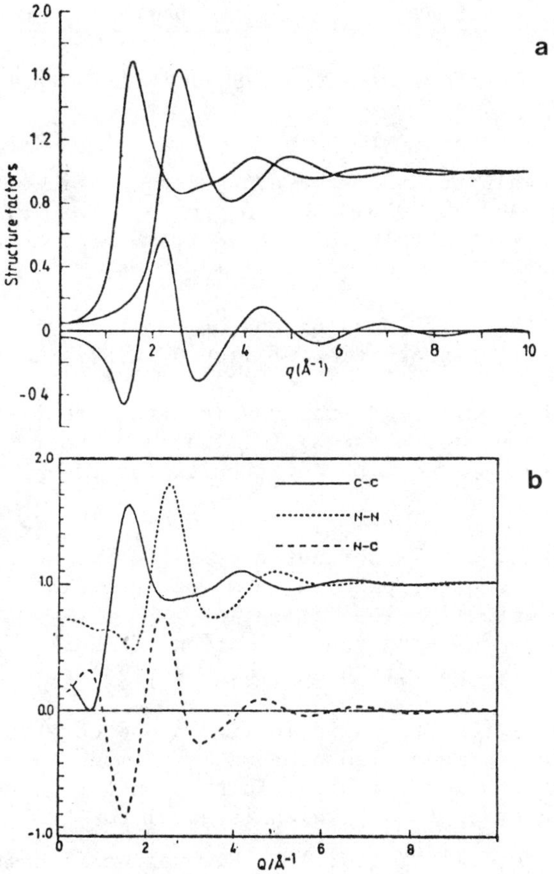


Fig. 6:

Static structure factors $S_{NN}(Q)$, $S_{cc}(Q)$, and $S_{Nc}(Q)$ obtained with the hypernetted chain approximation by Copestake et al.[15] (a) and by the present authors from the MD calculation (b)

With respect to the static structure factors, the MD results compare very favourably with the results obtained by Copestake and coworkers using the hypernetted chain approximation as can be seen from fig. 6.

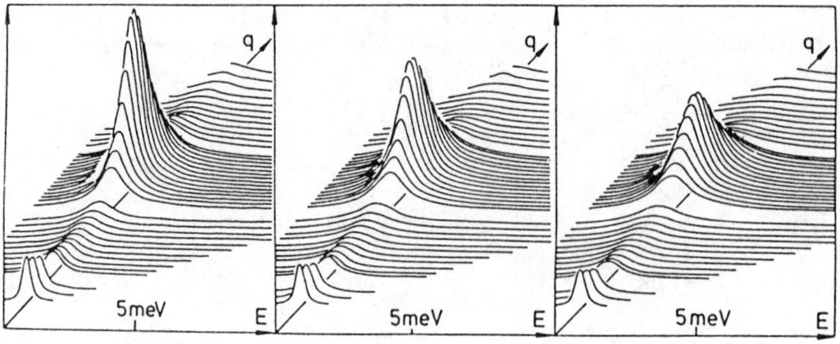


Fig. 7 $S_{cc}(q,E)$ measured at T = 1023 K, 1098K and 1273K by Soltwisch et al.[17]

Furthermore we have evaluated the self diffusion coefficients for Li ($D = 20 \times 10^{-5} cm^2/s$) and Pb ($D = 3.5 \times 10^{-5} cm^2/s$). The value for Li is in good agreement with a value extracted from the incoherent Li scattering intensity[17].

These results confirm the usefulness of the pair potential (27) for describing the physical properties of liquid $Li_4Pb$.

Let us now look at the results for the dynamic structure factors. In fig. 7 we have reproduced the $S_{cc}(Q,\omega)$ data of Soltwisch et al.[17] which can be compared with our results for the temperature T = 1085K shown in fig. 8.

As in the NaK system $S_{cc}(Q,\omega)$ is dominated by diffusive modes. Soltwisch et al. discuss the central peak in terms of interdiffusion and extract a Pb self diffusion coefficient of $D = 2.9 \times 10^{-5} cm^2/s$ at 1023 K using a generalization of Darken's relation. This number again agrees very nicely with our value.

Looking at our $S_{cc}(Q,\omega)$ results shown in fig. 8 one observes that besides the interdiffusive mode also a propagating mode is clearly visible for values of Q not too high. This is an ultrasound mode that is best displayed by $S_{NN}(Q,\omega)$ (which is equal to $c_c S_{11}(Q,\omega) + c_2 S_{22}(Q,\omega) + 2(c_1 c_2)^{1/2} S_{12}(Q,\omega)$ ). However we find it more spectacular to show the partial structure factors $S_{LiLi}(Q,\omega)$, $S_{PbPb}(Q,\omega)$ and $S_{LiPb}(Q,\omega)$ in figs. 9, 10 and 11, respectively.

The striking feature of this very peculiar alloy is that the lithium atoms alone participate in the sound propagation at short wavelengths, while the low frequency interdiffusion mode is dis-

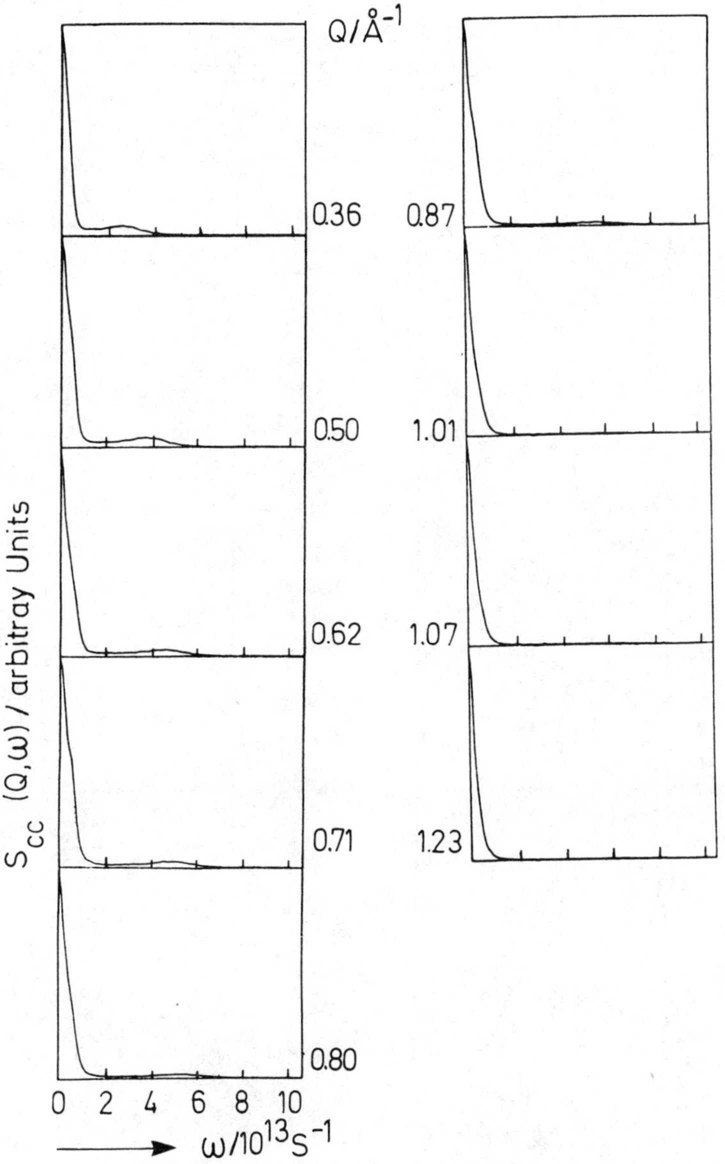


Fig. 8: $S_{cc}(Q,\omega)$ for $Li_4Pb$ at T=1085 K

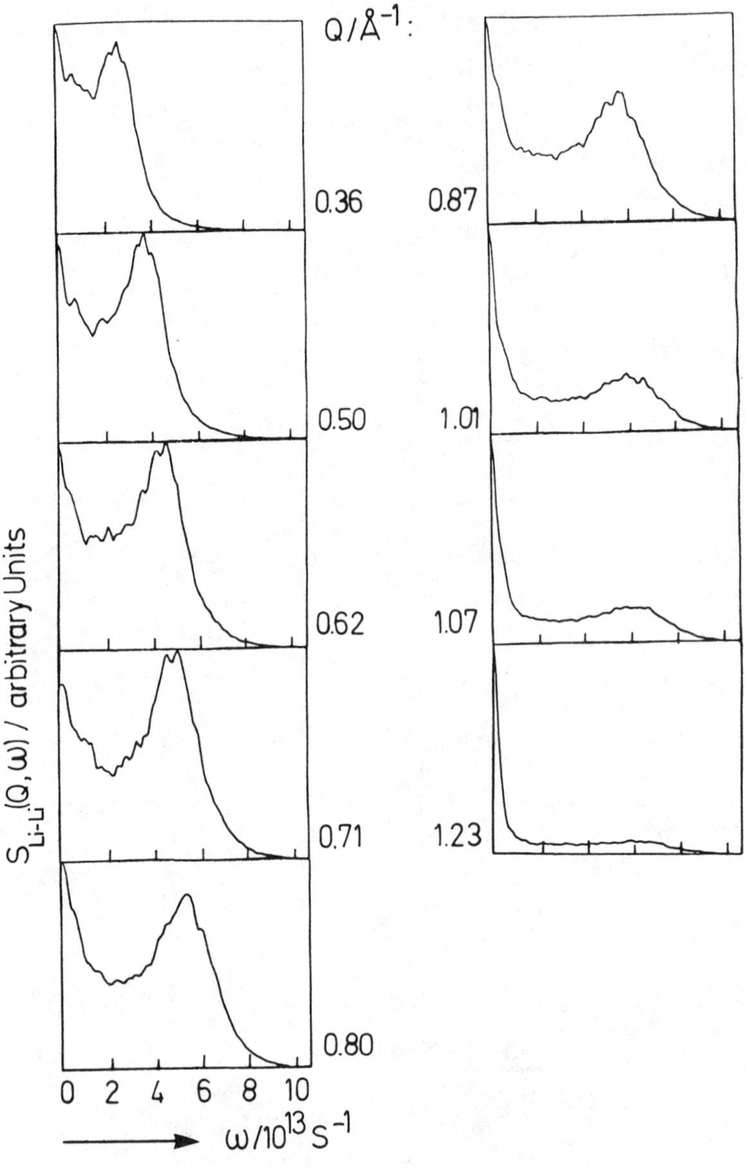


Fig. 9: $S_{LiLi}(Q,\omega)$ for $Li_4Pb$ at T=1085 K

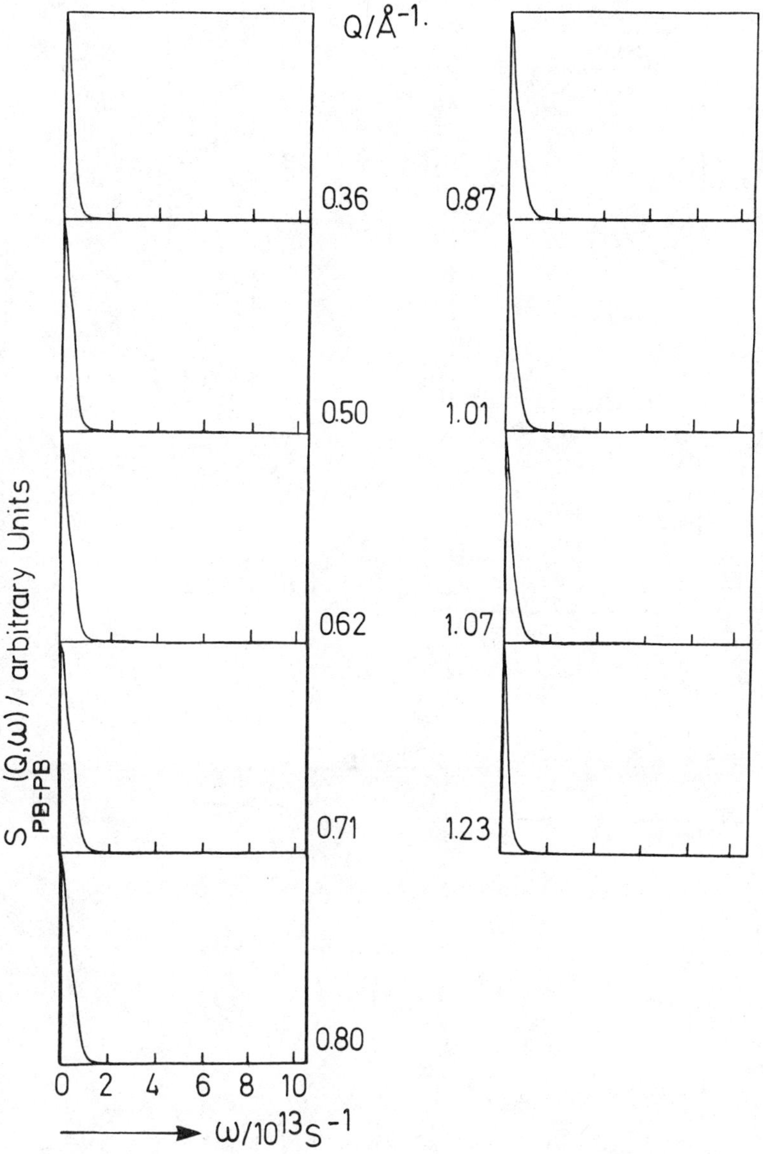


Fig. 10: $S_{PbPb}(Q,\omega)$ for $Li_4Pb$ at T=1085 K

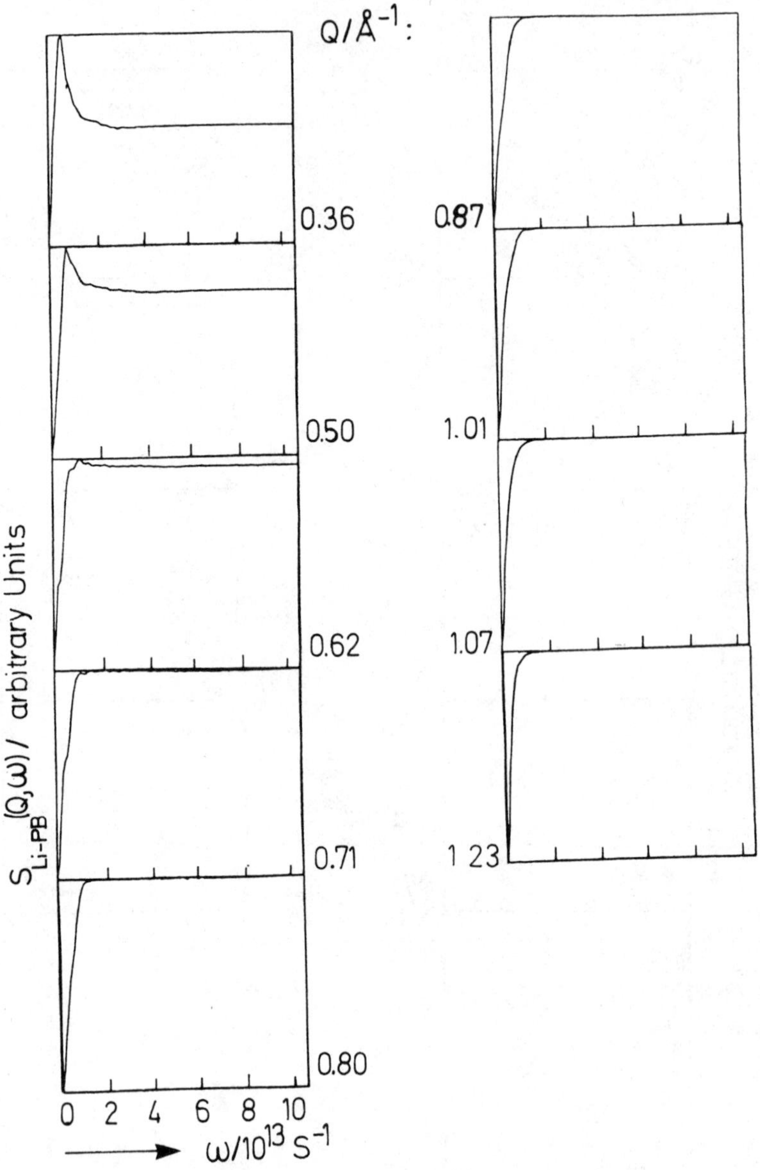


Fig. 11: $S_{LiPb}(Q,\omega)$ for $Li_4Pb$ at T=1085 K

played by both types of atoms for all Q-values.

This unforeseen fact originates from nothing else than the vastly different values of the two masses. The time scales of atomic motion for the two elements are separated in a Born-Oppenheimer-like fashion: Lithium atoms alone support the propagation of short period-short wavelength density waves while both types of atoms follow the slower interdiffusive motion.

The collective oscillations of the density of Li ions is also partly visible in our $S_{cc}(Q,\omega)$ data although it occurs with the Pb ions remaining essentially still.

The $S_{NN}(Q,\omega)$ shown in fig. 12 is intermediate between $S_{LiLi}$ and $S_{PbPb}$ at the lower Q values. At higher values of Q the contribution of the interdiffusion mode is reduced and is almost completely absent at our highest value of Q = 1.23 $\AA^{-1}$.

According to equation (25) the coupling between $S_{cc}(Q,\omega)$ and $S_{NN}(Q,\omega)$ is governed by $S_{NC}(q)$ (see fig. 6b). This means that the value of this function, divided by the corresponding value of $S_{cc}(Q)$, determines the relevance of the contribution of the interdiffusion mode in $S_{NN}(Q,\omega)$ as a function of Q. $S_{NC}(Q,\omega)$ in turn is also affected by the Q-dependence of the size difference effects.

Another peculiar observation is the presence of a shoulder in the interdiffusion spectrum shown by $S_{cc}(Q,\omega)$ and $S_{NN}(Q,\omega)$, below $Q \cong 1\ \AA^{-1}$. Its presence is systematic and represents a significant deviation from a single Lorentzian spectrum, and so inveighs against the presence of some kind of restoring force. This point is not clear and deserves further investigation (see also the discussion in ref. 17 in terms of "slow and fast relaxation channels").

In fig. 13 we plot the position and width of the Brillouin peak, as well as the positions of the shoulder. The first observation is that the peak is well visible at Q = 1.2 $\AA^{-1}$, and it should be possible to follow it still to higher values of Q.

If we extract a velocity (corresponding to the straight line in fig. 13) from our dispersion we obtain $u=\omega/Q=7500$ m/s. The experimental value for the velocity of sound for $Li_4Pb$, however, is only about 2000 m/s at similar conditions[18]. This discrepancy indicates that the compressibility of the model system is way too low, or that the simulation probes the $\omega$-Q region in which the ultrasound propagates.

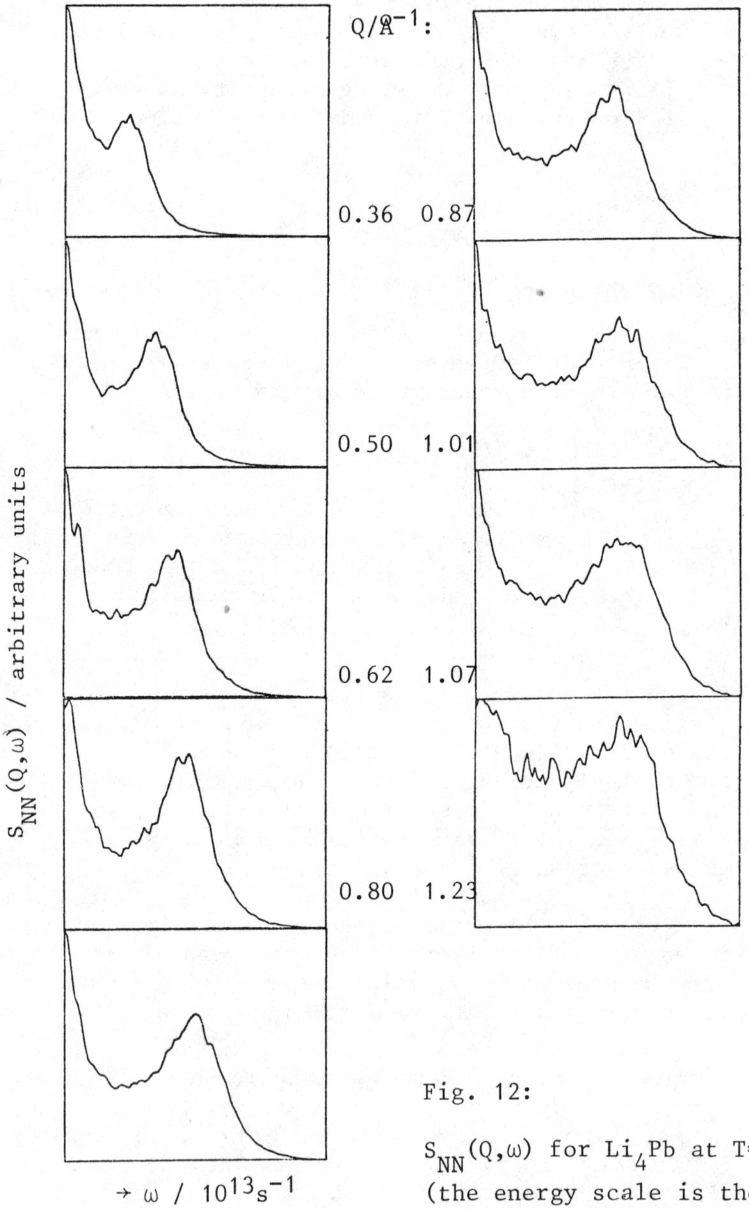


Fig. 12:

$S_{NN}(Q,\omega)$ for $Li_4Pb$ at T=1085 K (the energy scale is the same as in the preceding figures)

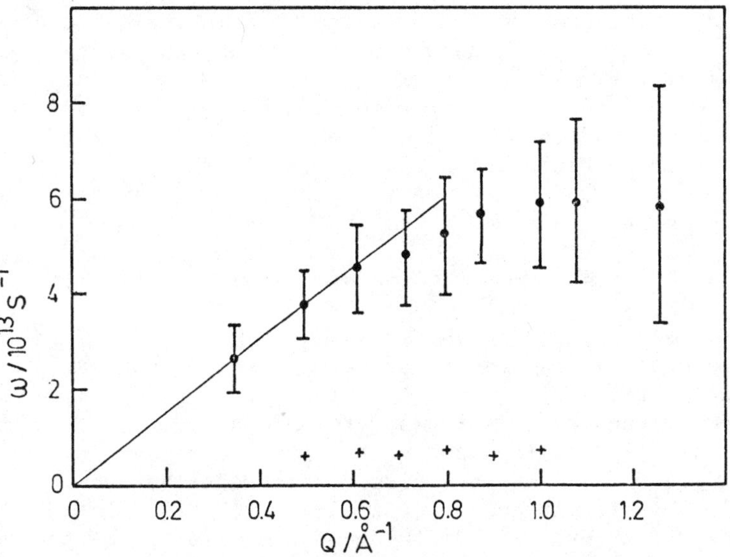


Fig.13: Sound dispersion for $Li_4Pb$ (dots) together with the widths of the peaks (bars). The crosses denote the positions of the shoulders in the interdiffusion peak.

Going back to the chosen pair potential it is of interest to investigate a slightly different model, i. e. to modify the values of the parameters describing the short range repulsion of the ionic core into the direction on increasing softness with the aim of increasing the compressibility.

It should be noted in passing that the separation of time scales in the two types of collective motion, i.e. compressional waves and interfiffusion, resulting in the lack of participation of the Pb ions in the sound mode, was supported by the artificially high value of the sound velocity. It amy be that the situation in the real system is not so clear cut. This is indicated by the absence of a Brillouin peak in the experimental data for $S_{cc}(Q,\omega)$.

## 4.SUMMARY AND CONCLUSION

It should be noted in passing that the separation of time scales in the two types of collective motion, ie. compressional waves and interdiffusion, resulting in the lack of participation of the Pb ions in the sound mode, was supported by the, seemingly, high value of the sound velocity. It may be that the situation in the real system is not so clear cut. This is indicated by the absence of a Brillouin peak in the experimental data for $S_{cc}(Q,\omega)$.

The interpretation of the data produced in inelastic neutron experiments on simple liquids and mixtures has been aided a great deal by computer simulation studies of model systems. Particularly so for mixtures, where the extraction of the partial structure factors from $d^2\sigma/d\Omega dE$ seen by the neutrons is not possible in general, the computer simulation can add invaluable insight into the respective role played by the constituent atoms in determining the dynamical properties of the system.

Recent findings we recalled or presented here include the following:

The interdiffusion mode in dense mixtures of non-strongly interacting particles is generally a sharp low-frequency feature. If the size of the two constituent atoms is appreciably different, this mode is well visible also in the number density dynamical structure factor. However, the size of its contributions to the spectrum is Q-dependent and is controlled by $S_{Nc}(Q)$. In rare gas mixtures (e. g. Ar-Kr) size differences are never large and this coupling should be much smaller than e.g. in Na-K.

In the monovalent alkali metals the interdiffusion mode has a Lorentzian shape. However, in the screened charge-transfer alloy $Li_4Pb$ it displays a shoulder for $Q \leqq 1\ \text{Å}^{-1}$,

A striking feature observed in the computer investigation of a system modelling $Li_4Pb$ which has the right structure factors and diffusion constants, but a propagation velocity four times larger than the experimental sound velocity is related to the vast differences of atomic masses of Li and Pb. It consists in a separation of time scales that occurs in the dynamics of atomic motion.

In particular, the Pb atoms are not able to follow the fast short wavelength oscillation of the sound mode, and only the lighter Li atoms contribute significantly to the propagation of density waves in the short wavelength regime.

REFERENCES


1. e. g. R. Scherm, Ann. Phys. (Paris) 7,349 (1972)
2. J. P. Hansen and I. R. McDonald, Theory of Simple Liquids (Academic Press, London, 1976)
3. J. P. Boon and S. Yip, Molecular Hydrodynamics (McGraw-Hill, New York, 1980)
4. J. R. D. Copley and J. M. Rowe, Phys. Rev. Lett. 32,49(1974)
5. A. Rahman, Phys. Rev. Lett. 32,52(1974)
6. D. Levesque, L. Verlet, and J. Kurkijärvi, Phys. Rev. A7,1690 (1973)

7. A. Rahman in "Neutron inelastic scattering",Vol. 1 (IAEA, Vienna, 1968)
8. G. Bucher and W. Gläser, Verhandl.DPG(VI)19,452(1984)
9. J. Bosse, W. Götze, and M. Lücke Phys. Rev. A17,434(1978) and Phys. Rev. A18,1176(1978)
10. A. B. Bhatia and D. E. Thornton, Phys. Rev. B1,3004(1970)
11. G. Jacucci and I. R. MacDonald in "Liquid and Amorphous Metals" ed. E. Lüscher and H. Coufal, Sijthoff and Noordhoff, Alphen, The Nederlands, 1980, p. 143
12. L. Dagens, M. Rasolt, and R. Taylor, Phys. Rev. B11,2726(1975)
13. A. Rahman in "Statistical Mechanics: New Concepts, New Problems, New Applications", ed. S. A. Rice, K. F. Freed, and J. C. Light, University press, Chicago, 1972
14. G. Jacucci, I. R. MacDonald, and R. Taylor, J. Phys. F8,L121 (1978)
15. A. P. Copestake, R. Evans, H. Ruppersberg, and W. Schirmacher, J. Phys. F13,1993(1983)
16. H. Ruppersberg and H. Reiter, J. Phys. F12,1311(1982)
17. M. Soltwisch, D. Quitmann, H. Ruppersberg, and J. B. Suck, Phys. Rev. B28,5583 (1983)
18. H. Ruppersberg and W. Speicher, Z. Naturforsch, 31A,47(1976)

# MAGNETIC EXCITATIONS IN INSULATORS AND METALS

Per-Anker Lindgård

Physics Department
Risø National Laboratory
DK-4000 Roskilde, Denmark

## ABSTRACT

Recently, we have seen a rapid development, using the dynamics to cast light on very fundamental problems in disordered or unknown ground state cases. Aspects to be discussed of this interesting development of both theory and experiment include the following topics. Spin waves in ordered structures depending on lattice dimensionality and frustration effects, the sinusoidal ordered structure and crystal field or anisotropy effects. The disordered ground state problems involve excitations in low dimensions, in the paramagnetic phase and in singlet ground state systems, with soft mode and central peak phenomena. The itinerant magnets are similar to the singlet ground state because in both cases a moment must be induced before spin wave excitations can propagate. The recent situation will be commented on. The task of the theory has been to include correlation effects and to treat non-linear phenomena. In some cases these are so prominent that new non-linear excitation modes can be identified, such as solitons and the Villain mode. The future holds a number of further problems for the study of magnetic excitations. For example in two dimensions (or surface physics) where so far mainly ground state properties have been investigated.

## INTRODUCTION

The topic spinwaves and magnetic excitations is vast and diversified. Alone during the last five years more than 2000 articles have been published, which are indexed under "spinwaves" - and even more including other excitations. A magnetic structure per se does not give much information about the fundamental interactions, the Hamiltonian, of which it is a consequence. However, by sending perturbative waves through the structure and studying the wavevector q

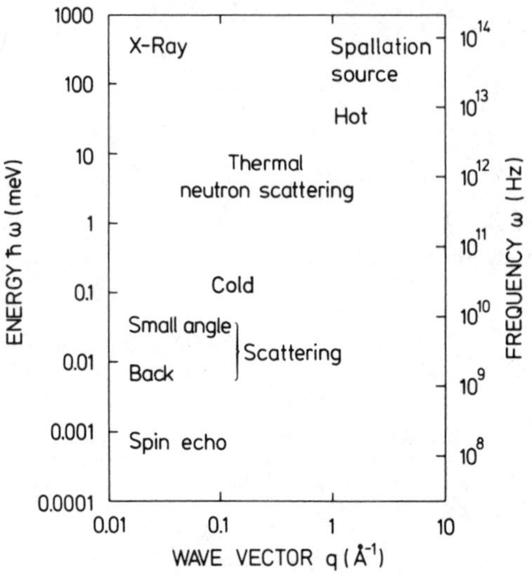


Fig. 1. $\chi^{\alpha\beta}(q,\omega)$ can be measured in this window of q and ω by various neutron scattering techniques.

and frequency ω dependence and localization and damping properties of the waves, one can in the most direct way available determine the physics and strength of the parameters of the Hamiltonian. This explains the importance of the subject and the numerous papers. The information about the fate of the test waves is contained in the response function tensor $\chi^{\alpha\beta}(q,\omega)$. The neutrons, ranging from cold, thermal to hot neutrons, covers an ideal range of velocities and energies to study the q and ω scale of relevance for the microscopic interactions on the interatomic level in condensed matter. The scattering of waves of the magnetic neutrons is a resonance technique, which directly measures the components of $\chi^{\alpha\beta}(q,\omega)$ It is the only technique which covers the whole window $10^{-2}<q<10$ $Å^{-1}$ and $10^{-4}<\omega<10^{3}$ meV with a direct magnetic crossection. It, therefore, is and will remain so, a technique of central importance in the present field, and any effort to extend the window even further to higher or lower frequencies will be important, see Fig. 1. Other techniques may cover special corners of this window. The use of X-rays in the form of synchrotron radiation has undergone a substantial development recently. Although this radiation only has an indirect magnetic crossection, the scattering of X-rays could, because of the strong intensity available, possibly be used as a supplementary technique, in particular for the high frequency range.

Since this paper should not only indicate the status of the field of magnetic excitations and neutron scattering today, but also show avenues for future developments, an emphasis will be made on where the theoretical understanding at present is not yet fully clear and tested experimentally. In systems with perfect long range order the theory is perfectly clear and exact, and experiments can be used to determine the Hamiltonian parameters[2]. This aspect, which is extremely important, will not be discussed here, although comments will be made on the uniqueness of the deduced parameters. The main discussion will be centered on situations where the theoretical situation is difficult. This is in short in the presence of disorder, or when only shortrange correlations play a role. In this case the deduction of Hamiltonian parameters is definitely not unique and depends even on the accuracy of theory. In this case one may also encounter new nonlinear excitations, which are of fundamental interest on their own. The correlation effects are particularily large in low, especially 1-dimensional systems[3]. The study of these may be regarded as test ground for more generally applicable theories dealing with correlation effects. The space here will not permit a discussion of quenched disordered systems like chemical disorder[4], spin glass[5], amorphous[6] systems etc.[4] in which the study of correlation effects in the dynamic behaviour also is of importance. Another whole field which will not be covered, is where magnetic excitations are used as a probe on the medium in which they propagate[2] through interactions with other modes like phonons, electronic or nuclear levels etc. The uniqueness of the interpretation of such studies again depends on how well correlation effects are understood. The correlation effects will be the main theme of the following.

## EXACT THEORIES OF MAGNETIC EXCITATIONS

### Exactly known groundstate

When the magnetic groundstate of an infinite system is known it is easy to find the exact excitation spectrum by linearizing the equation of motion for the spin operators around this ground state. There are several methods for doing this using transformations of the spin operators to the simpler operators. For example to bosons by the Dyson-Maleev[7](DM), the Holstein Primakoff[8](HP) transformations and the matching of matrix element method[9](MME), or the more general standard basis operator transformation[10]. If only the bilinear terms of the bose Hamiltonian obtained by the (DM), (HP) or (MME) transformations is considered we will call it simple or linear spin wave theory.

The effect of the anharmonic terms can be treated by perturbation theory[11], but this will not be discussed here. Another approach is to linearize the equation of motion for the spin operators by a selfconsistent decoupling of the non-linear terms. This method is not exact, but very convenient and usually reproduces the exact limiting cases; it will be discussed in the next section.

Unfortunately there is only one exactly known ground state, namely the fully aligned ferromagnetic state. This is the ground state for Hamiltonians of the form

$$\mathcal{H} = -\sum_{ij} \{J_{ij}\vec{S}_i\cdot\vec{S}_j + K_{ij}(\vec{S}_i\cdot\vec{S}_j)^2\} - \sum_i (HS_i^z + DS_i^{z2}) \qquad (1)$$

if the interactions are not too strongly competing. That is when $J_q$ and $K_q$ have a maximum for q=0 and H,D > 0. The Fourier transformed interaction constants are defined as $F_q = \Sigma_n z_n F_n \gamma_q^{(n)}$, where $F_n$ is the interaction constant to the n'th neighbour group with $z_n$ members and $\gamma_q^{(n)} = \Sigma\, \exp(iq\rho_n)/z_n$ is the phase factor for this neighbour group. For T=0 the exact dispersion relation is given by the simple spin wave theory[12]

$$\omega_q = H+D(2S-1) + 2S\sum_n[J_n + 2(S-1)K_n]z_n(1-\gamma_q^{(n)}) \quad (2)$$

When the axial anisotropy terms H and D are zero the fully aligned ground state is destroyed by domain walls for lattice dimensions less than 3, and (2) is no longer valid. However, when we have long range order it is clear from (2) that a measurement of $\omega_q$ for different q and H yields a most direct determination of the interaction constants. It is also clear that one cannot separate the bilinear and biquadratic exchange terms $J_n$ and $K_n$[12]. This is the first indication that the parameters deduced from spin wave measurements are effective and model dependent even in this very simple exactly solvable example. The reason is that $\omega_q$ only represents the propagation energy for a single spin flip. A study of the temperature dependence of the spin wave energy and damping (which involves several spin flips) would make a separation possible[12]. But this can be done much simpler by measuring the energy spectrum directly for two isolated spins. This spectrum consists of several energy levels, which can be measured by neutron scattering. This method has in particular been advocated by Furrer _et al_[13] and it was demonstrated that $K_q$ was indeed not zero as usually tacidly assumed.

For isolated clusters of spins not only the ground state, but all states are exactly known. Since neutron scattering can determine several transition energies and the type of cluster can be identified by the scattering formfactor, this cluster method should be recognized as an important supplement to spin wave scattering for separating different interaction mechanisms. This is particularly true for the many possible anisotropic interactions. The disadvantage with the method is that one is not studying the pure substance and the intensity is small because one is not studying a collective excitation.

## Approximately known ground states

If we consider the planar Heisenberg Hamiltonian with $D<0$ and $J>0$

$$\mathcal{H} = -J\sum_{ij} S_i \cdot S_j - D\sum_i {S_i^z}^2 \quad (3)$$

the exact ground state is no longer the fully aligned ferromagnetic state, the anisotropy introduces a singlet component into it. However, if the anisotropy is relatively small i.e. $d = D(S-1/2)/2J_oS \ll 1$ one can find the exact spin wave dispersion relation by perturbation theory[14] which gives to second order in d and first order in 1/S

$$\omega_q = 2J_oS\{(1-\gamma_q)(2d+1-\gamma_q + \frac{d^2}{2S}[(1+\gamma_q)+(3+\gamma_q)c])\}^{1/2} \quad (4)$$

where $c = \Sigma_q\gamma_q/(1-\gamma_q)$ is proportional to the nearest neighbour correlation function $\langle S_o^x S_1^x\rangle$. The quantum nature of the problem shows up in the kinematical factor (S-1/2) in d, and was verified experimentally[15] for S=1, see Fig. 2. The last term in (4) is due

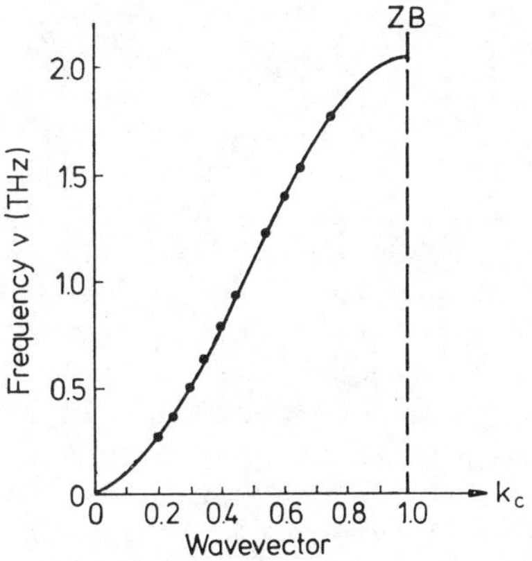


Fig. 2. The spin wave dispersion[15] for the 1-dimensional $CsNiF_3$ at T = 4.9 K. The curve is the calculated using MME transformation and simple spin wave theory. The result is in fact quite non-trivial because there is no long range order. The spin waves are well defined due to the extended shorth range order present in 1-dimension (after Steiner et al[15]).

to corrections to the ferromagnetic ground states; notice it makes the excitation energies higher than those obtained by simple spin wave theory. In order to calculate similar corrections for a general crystal field anisotropy the MME bose operator expansion[9] of the spin operators was derived. When the resulting bilinear bose Hamiltonian is diagonalized the crystal field effects are exactly included in a perturbation sense. Thus we obtain for (3) the exact (4) with the exception of the spin correlation term c. The MME transformation bridges the gap between the simple expansions[7,8] and the standard basis operator method[10], which requires a complete (numerical) diagonalization of the local crystal field Hamiltonian. The advantage of the MME method is that it requires no a priori knowledge on the crystal field. It is clear from (4) that the ground state corrections can be absorbed in effective parameters for the simple spin wave theory. However, if the dispersion relation $\omega_q(H)$ is studied as a function of an external perturbation, such as a magnetic field H or pressure, the ground state properties changes and it is important to master the ground state correction for a correct interpretation of the results. The correlation effects have not yet been experimentally studied for a simple system like (3) for $S \geq 3/2$. This would be of interest as a test on the interpretation, based on simple spin wave theory, of the field dependence of the spin wave data for Tb[16].

Another example of a nearly known ground state is the simple antiferromagnet[11] (eq. 3 with $J<0$ and $D=0$). Simple spin wave theory based on the Néel ↑↓↑↓ state gives the familiar result (5a)

$$a)\ \omega_q = 2J_oS\sqrt{1-\gamma_q^2} \qquad b)\ \omega_q = \pi J_oS\sqrt{1-\gamma_q^2} \qquad (5)$$

The problem is again that the ground state has a component of quantum mechanical singlets i.e. a random pairing like (↑↓)...(↓↑). The corrections are exclusively of the correlation type, as the term proportional to c in (4). This is particularly large for 1-dimension and for S=1/2. In this case $\omega_q$ was found exactly by des Cloizeaux and Pearson[17] to be (5b). Again the exact excitation energies are higher than given by simple spin wave theory, by as much as 60%; this was verified experimentally[18]. It is clear from (5) that even this large correction can be absorbed in effective parameters for the simple spin wave theory. So are there no qualitative new effects of the zero point motion? We could expect a linewidth of the spin wave spectrum. On this there are no exact results for S=1/2, but a 1/S expansion theory by Mikeska[19] indicates that the response at T=0 is not like the usual $\delta(\omega-\omega_q)$, but approximately of the form $1/(\omega^2-\omega_q^2)^{1/2}$ for $\omega \geq \omega_q$, where (5b) is the lower diverging edge of a spectrum, which has a large high frequency tail. Indications of this behaviour has been verified experimentally[18]. However, the most striking difference comes in the field dependence $\omega_q(H)$, which according to Shiba[20] for d=1 and S=1/2 is radically different from the simple spin wave results, which is identical to the classical S = ∞ result, see Fig. 3. The experiments on the field dependence gave unexpected results and did not confirm any of the theories[21]. This is in fact more the rule than the exception and has been found in other model systems viz. TMMC[22], Tb[16] etc. The conclusion is either that the theories for the field dependence is much more complicated than expected or that the success of

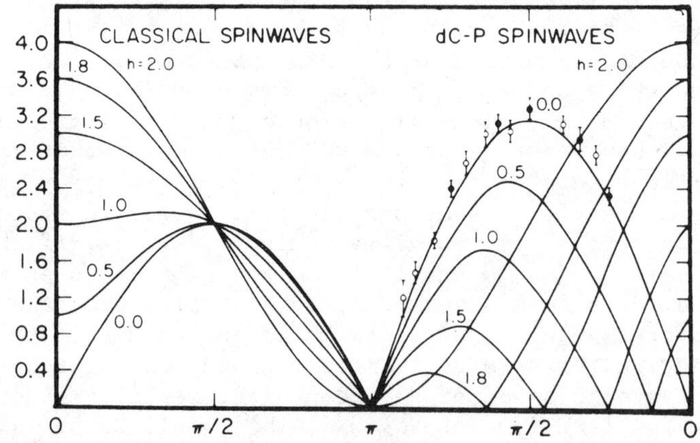


Fig. 3. Magnetic field dependence of spin wave energies in a 1-dimensional antiferromagnetic nearest-neighbour Heisenberg chain where $h = g\mu_B H/2J$. The left-hand side shows the results obtained in classical linear spin wave theory. The right-hand side shows the field dependence of the des Cloizeaux Pearson states[20]. Experimental points for h = 0 are shown with open circles (after Heilmann et al[18]).

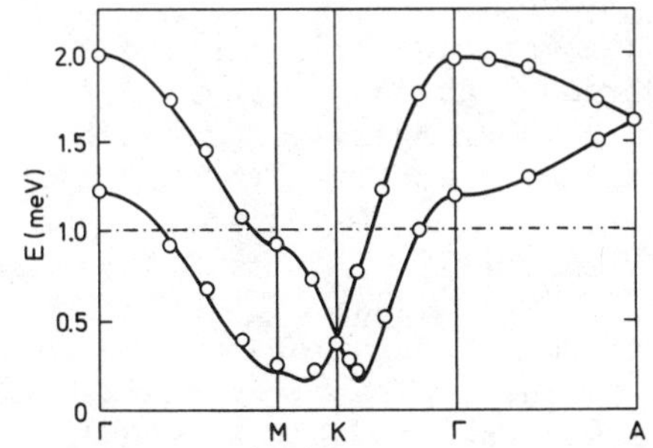


Fig. 4. Spin wave dispersion in the singlet-triplet cluster magnet[23] $Cs_3Cr_2Br_9$ The curve represents the RPA theory. The potentially soft modes occur as a ring around the K-point (after Leuenberger et al[23]).

the simple spin wave theory hides a wealth of physical phenomena in effective parameters, or both. We will return to the discussion of the antiferromagnet in a field.

One can more consciously study the effect of correlated singlets in the antiferromagnetic ground state. The $Cs_3Cr_2Br_9$ compound[23] represent systems in which there is a chemical pair clustering (↑↓)-(↑↓)-(↑↓) which gives pair singlet-triplet clusters which may be coupled with a variable strength according to the compounds[24]. The dispersion relation for a weakly coupled pair system is identical to that for the planar single ion anisotropy system (4) in a simple spin wave theory, when the pair states are first exactly diagonalized[23], see Fig. 4. It would be interesting to see the field and coupling strength dependence for these systems.

Another case for which it would be interesting to study correlation effects is for the 2-dimensional triangular antiferromagnet. Because one cannot satisfy that all neighbour spins are antiparallel for a triangle, see Fig. 5a, one calls the system frustrated[25].

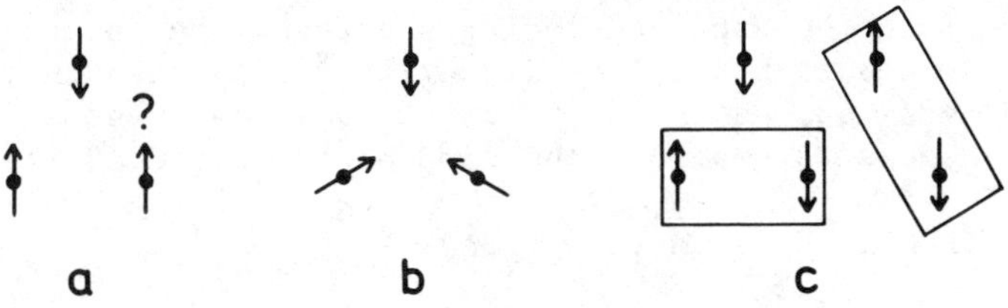


Fig. 5. Triangular antiferromagnet. a) frustrated structure. b) Néel 120° spiral structure, classical ground state. c) Anderson singlet ground state.

The Néel state is the compromise spiral state where all spins are canted, Fig. 5b. Anderson[26] has shown that for S=1/2 one can find ground states with much lower energies than the Néel state by distributing singlet pairs in different ways, Fig. 5c. Experimental realization of such system have so far only shown antiferromagnetically coupled ferromagnetic chains[27]. They may more closely represent S=∞, for which the classical Néel state, Fig. 5b, should be good. The 2-dimensional frustration problem seems to have analogies in many other fields such as 2-dimensional melting, surface physics and lattice gauge theories for elementary particles. It may therefore be worthwhile to study the physics in the experimentally rather clean magnetic systems.

## RESPONSE FUNCTION THEORIES

A neutron scattering determination of a magnetic structure gives the average long range order and not the true ground state. However, for the Green's function theory in combination with decouplings such as the random phase approximation (RPA) this information on the average of the spin components is sufficient for the calculation of an approximate energy spectrum for the excited states, and thereby an approximate determination of the Hamiltonian parameters from a spin wave measurement.

The theory[1] considers the generalized susceptibility tensor (or Green's function[28]) $\chi^{\alpha\beta}(q,\omega) = \langle\langle S^{\alpha}_{q}, S^{\beta}_{-q}\rangle\rangle_{\omega}$. The poles of $\chi^{\alpha\alpha}(q,\omega)(\alpha=x,y)$ give the so-called transverse excitations and the poles of $\chi^{zz}(q,\omega)$ give the longitudinal excitations, when z is the moment direction. These excitations can be directly measured by neutron scattering. For simple magnets the longitudinal response is centered around $\omega=0$, a so-called central peak. The cross section is

$$\frac{d^2\sigma}{d\Omega\, dE} = \frac{K}{(1-e^{-\omega/kT})} \sum_{\alpha\beta} \left(\delta_{\alpha\beta} - \frac{Q_\alpha Q_\beta}{Q^2}\right) \mathrm{Im}\chi^{\alpha\beta}(q,\omega) \qquad (6)$$

where K is essentially constant, Q and $\omega$ are the momentum and energy transfer of the neutrons and $q = Q-\tau$ ($\tau$ is a reciprocal-lattice vector). The residues of the poles of $\chi^{\alpha\beta}(q,\omega)$ give the intensities of the peaks in the neutron scattering spectrum. The factor $[\delta_{\alpha\beta} - (Q_\alpha Q_\beta/Q^2)]$ makes it possible to identify the longitudinal and transverse excitations. A polarization analyses gives an additional important handle to separate the components of the excitations.

## THE RANDOM PHASE APPROXIMATION (RPA)

The theory of magnetic excitations in systems strongly influenced by a crystal field is very intriguing, because the ground state is basically unknown. In fact, the problem is very similar to that we encountered in the spin wave theory for antiferromagnets.

However, as usual the mean-field concept allows us to surmount all difficulties and to give an approximate answer for the regions of interest. Let us generalize the usual formula for the enhanced susceptibility $\chi(q) = \chi_o[1 + 2J(q)\chi(q)]$ by including the frequency dependence. In general we have

$$\underline{\underline{\chi}}(q,\omega)=\underline{\underline{\chi}}_o(\omega)[1+2\underline{\underline{J}}(q)\underline{\underline{\chi}}(q,\omega)]=[1+2\underline{\underline{\chi}}_o(\omega)\underline{\underline{J}}(q)]^{-1}\underline{\underline{\chi}}_o(\omega) \tag{7}$$

J(q) is the Fourier-transformed bilinear exchange interaction tensor. This expression holds for any crystal field $V_c$. $\chi_o(\omega)$ is the local susceptibility tensor which is calculated in terms of the states $|p\rangle$ and energies $E_p$ of the single site Hamiltonian which includes crystal field $V_c$ and the average, molecular field from the neighbour interactions. The strong on-site-correlations are therefore exactly included. The fundamental operators are the so-called standard basis operators (SBO) for the transitions between the local states $|p\rangle$ and $\langle r|$ although the spin operators still are the physically relevant linear combinations $S_i^\alpha = \Sigma\langle p|S_i^\alpha|r\rangle|p\rangle\langle r|$. The $\alpha\beta$ component of the single-site susceptibility tensor is

$$\chi_o^{\alpha\beta}(\omega) = \sum_{pr} \frac{\langle p|S^\alpha|r\rangle\langle r|S^\beta|p\rangle}{E_r-E_p-\omega}(n_p-n_r) \tag{8}$$

where $n_p \sim \exp(-E_p/T)$ is the population factor of the state $|p\rangle$ with energy $E_p$, and S denotes total angular momentum. The Greek indices denote the cartesian components x, y and z. The mean-field equation (7) was derived independently by many authors for the crystal field problem (Buyers et al[29], Haley and Erdös[30] using the standard basis operator technique, and by Peschel et al[31] using field-theoretical methods). It is convenient (but not self-consistent) to replace $n_p$ by the Boltzmann factor. Self-consistency can be obtained for S = 1, but not for a general value of S.

The physics of the random phase approximation can be described as follows. We assume a spin $S_R$ and its neighbours at R+$\delta$ are participating in a certain spin wave mode with wave vector and frequency q and $\omega_q$. If the neighbour spins participate in other spin waves one assumes that these have an arbitrary phase, wave vector k and frequency $\omega_{k\neq q}$ which average out in the ensemble average for $S_R$. To a first approximation $S_R$ therefore only sees the local average of its neighbour spins $\langle S_{R+\delta}\rangle = \langle S^z\rangle_{local}$. In the RPA the local average is furthermore replaced by the global average $\langle S^z\rangle$. In the equation of motions, pairs of spin operators are therefore approximated as $S_i^\alpha S_j^\beta \sim \langle S_i^\alpha\rangle S_j^\beta + S_i^\alpha\langle S_j^\beta\rangle$. In the mean-field or RPA approximation the poles of equation (7) are $\delta$ functions in $\omega$. In short, the RPA theory considers one mode in an effective medium. If there are couplings between different modes the picture is modified and we obtain also a finite lifetime or a width of the peaks.

When applied to the Hamiltonian (3) the SBO-RPA theory gives the exact dispersion (4) with the exception of the last pair correlation term proportional to c, the same result as obtained by the MME transformation[9,14] and simple spin wave theory. We showed in the previous section that neglecting pair correlation effects could give effective parameters which were 60% off the correct values. Perhaps this may be considered as an upper limit for the uncertainty in parameters deduced from a SBO-RPA analyses? The effect is most serious when analysing differential measurements like $d\omega_q(H)/dH$.

The relative brevity of the discussion of the SBO-RPA theory here is inversely proportional to the number of studies of this kind during the past decade - and to the success and the importance for obtaining the (effective) interaction parameters. A brief review on this aspect was given previously[32]. Here we shall mainly discuss the features which go beyond the RPA-theory.

## THE CORRELATED EFFECTIVE FIELD (CEF) THEORY

In an attempt to, in a simple way, improve the RPA theory Lines[33] proposed a phenomenological method for calculating the local average for the neighbour spins, which may differ from the global ensemble average. Suzuki[34] generalized the method slightly and called it a dynamical correlated effective field (DCEF) theory. The idea is simply that the average for $\langle S_{R+\delta}\rangle$ may depend on the actual state of $S_R$. Therefore a nearest neighbour spin product is decoupled as

$$\begin{aligned} S_i^\alpha S_j^\beta \sim & \left[\langle S_i^\alpha\rangle - c^{\alpha\beta}(S_j^\beta - \langle S_j^\beta\rangle)\right]S_j^\beta \\ & + S_i^\alpha\left[\langle S_j^\beta\rangle - c^{\alpha\beta}(S_i^\alpha - \langle S_i^\alpha\rangle)\right], \end{aligned} \tag{9}$$

where the phenomenological correlation parameter $c^{\alpha\beta}$ is determined[34] by calculating $\langle S_i^\alpha\rangle$ and $\langle S_i^\alpha S_i^\beta\rangle$ selfconsistently using $\chi^{\alpha\beta}(q,\omega)$. For the isotropic antiferromagnet $c^{\alpha\beta}$ is exactly[34] the nearest neighbour correlation function $\langle S_o^\alpha S_1^\beta\rangle$. We notice from (9) that the CEF decoupling introduces an effective single ion anisotropy $-c^{\alpha\beta}(S_j^{\beta 2} + S_i^{\alpha 2})$. This method therefore gives a temperature and field dependent renormalization of the crystal field, which cannot be obtained by the RPA theory. However, the (D)CEF theory again gives undamped spin waves, and a $\delta(\omega-\omega q)$ response, because the neighbour spins are replaced by static quantities (9).

The DCEF theory was applied to $RbFeCl_3$ with a good result, see Fig. 6. The $XFeCl_3$, (X = Rb,Cs,Tl,$NH_4$ etc.), compounds[27] are 2-dimensional triangular antiferromagnets of ferromagnetic chains, the basic spin is S=1 and a large planar anisotropy term $DS^{z2}$ provides additional singlet character in the ground state. The Néel ground state is the triangular spiral state, fig. 5b, with the ordering vector Q = q at the K-point in the Brillouin zone. $RbFeCl_3$ orders below 1.95 K and spin wave measurements show dispersion curves

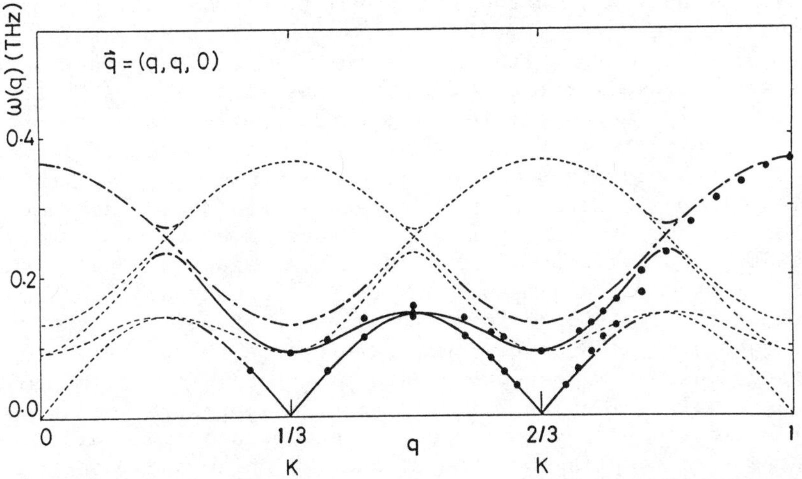


Fig. 6. Spin wave dispersion for the possibly frustrated $RbFeCl_3$ system (experimental points by Petitgrand et al[36]). The DCEF theory by Zuzuki[35] assumes a spiral structure, the 120° canted Néel ground state. The thickness of the lines indicates the intensity. The crystal field effects introduce splittings of the modes (after Zuzuki[35]).

which cannot be explained by the simple spin wave spiral dispersion relation[12]

$$\omega_q = 2S\{[J_Q - \tfrac{1}{2}(J_{q+Q} + J_{q-Q})][J_Q - J_q + D]\}^{1/2}.$$

The crystal field levels introduce large splittings in the dispersion, see Fig. 6. Similar splittings were actually observed also for the cone phase of Er[37] and have not been explained. They may be of the same origin. The $CsFeCl_3$ compound has a larger D and does not order spontaneously. However, by applying a magnetic field one can make the singlet system order. This gives an excellent opportunity for studying the phase transitions for a singlet system. A fully selfconsistent SBO-RPA theory[38] explains the field dependence of the exitations at high frequencies, but fails at low frequencies near the transition point.

## CORRELATION THEORY

When a system is far from the ground state and has small or no long-range order (LRO) the previously discussed methods fails. But there will of course still be some excitations determined by the local properties and the short-range order (SRO). This situation is more common than rare, for example a) any magnet at finite temperatures in particular near $T_c$ and in the paramagnetic phase, b) the singlet ground state magnets, c) itinerant magnets and d) disordered compunds or alloys etc. One needs a theory which can treat (LRO) and (SRO) on the same level of accuracy. The response function theories[39] are ideal for this because it is easy to build in a physical picture (although the theories are not exact in a perturbative sense). First let us reemphasize that the RPA theory is tremendously successful in explaining experimental data when correlation effects are small or when (LRO) is dominant. It would be good to preserve this feature as a basis for more advanced theories. (LRO) is here taken in a generalized sense, as when all spins on average are in a particular state, i.e. not only a maximum spin state $|S\rangle$, but for example also a non-magnetic singlet state $|0\rangle$. A diminished (LRO) means that there are many spin wave modes excited in the system and in our previous picture this means that the neighbour spins $S_{R+\delta}$ commonly participate in spin wave modes $\omega_k$, other than the $\omega_q$ we were considering for $S_R$ (in each member of the ensemble). The RPA effective medium approximation is no longer valid because $S_R$ now sees a time dependent molecular field, and there is a coupling between $\omega_k$ and $\omega_q$, which does not vanish in the ensamble average.

The first step beyond the RPA is therefore to consider the effect of coupling two modes selfconsistently. The selfconsistent modes are not harmonic oscillation modes, but to a first approximation damped or overdampled oscillations and diffusion modes. This approximation we call the correlation theory[40]. It has the virtues of the RPA theory of being quite simple and applicable to many systems including disordered systems, because (SRO) effects are taken into account by including selfconsistency of both static and dynamic properties. In the traditional application of the mode-mode coupling theory, as advocated in particular by Kawasaki[41], Lovesey et al[42] and discussed by de Raedt et al[43], the dynamical variables are usually taken to be the operators (and derivatives thereof) which in the hydrodynamic limit satisfy conservation laws. The rationale is that these modes should be slowly varying with time and therefore dominate the dynamical behaviour. For the antiferromagnet these variables are the magnetization density $M_q$ and the energy density $E_q$. A coupling between these should occur for finite external fields. Clearly this basis is designed to treat the long wavelength limit $q \to 0$ and $\omega \to 0$. However, this limit is not easily observable by either neutron scattering or in numerical simulation studies. In the correlation theory[40] an alternative set of dynamical variables is chosen; namely dynamical variables, which allows the local or short range properties to be calculated exactly. This provides a description of the normal modes at high q and $\omega$. It is therefore a theory designed to obtain properties which can be tested by the above mentioned measurements. This basis turns out to be the same as that convenient for the description of the ordered phase. Consequently all temperatures can be treated in the same framework. The correlation theory does not use the hydrodynamic concept energy modes, but gives a rather simple picture of the dynamics in terms of correlation functions. For the antiferromagnet the interplay between these on different sublattices is important. If no approximations were made the choice of dynamical variables is only a matter of taste and convenience. However, since approximations are necessary the most physical basis (i.e. the one giving the best non-interacting normal modes) should be the most adequate.

A number of facts are known about the dynamics in terms of the static properties[39,44]. The frequency moments[1] of the response function

$$\langle\omega_q^n\rangle \equiv \int_{-\infty}^{\infty} \omega^n S(q,\omega)/\chi_q[1-e^{-\omega/T}]/\omega\, d\omega$$

give the simplest, but not very detailed, information. Let us in order to be specific consider a dynamical variable vector $\underline{A} = (a,b)$ with only two components a and b (for example $S_q^x$ and $S_q^y$) for which the static susceptibility matrix is diagonal i.e.

$$\underline{\underline{\chi}}_q = \begin{pmatrix} \chi_{aa} & 0 \\ 0 & \chi_{bb} \end{pmatrix} \text{ and } 1/\underline{\underline{\chi}}_q = \begin{pmatrix} 1/\chi_{aa} & 0 \\ 0 & 1/\chi_{bb} \end{pmatrix}.$$

Then the first moment matrix is

$$\langle\underline{\underline{\omega}}_q\rangle \equiv \langle[\underline{A},\underline{A}^+]\rangle/\underline{\underline{\chi}}_q = iM\begin{pmatrix} 0 & 1 \\ -1 & 0 \end{pmatrix}/\underline{\underline{\chi}}_q$$

where $M = -i\langle[a,b]\rangle$. If the response is of the type $\delta(\omega-\omega_q)$, as found in the RPA theory, $\omega_q$ can be obtained by diagonalizing $\langle\underline{\underline{\omega}}_q\rangle$ one finds the exact result

$$\omega_q = \pm M/\sqrt{\chi_{aa}\chi_{bb}}, \tag{10}$$

in terms of the exact static susceptibility components and M. In our example $M = \langle S^z\rangle$ is the (LRO) parameter, which is often the case. The first moment frequency $\omega_q$ is then a soft mode, which vanishes at the transition temperature $T_c$ like $(T_c-T)^{\beta+(\gamma_a+\gamma_b)/2}$, where $\beta$ and $\gamma_a,\gamma_b$ are the critical (LRO) and susceptibility exponents for $\chi_{aa}$ and $\chi_{bb}$. However, near $T_c$ we expect a line broadening and in general $\omega_q$ no longer even represents the peak position of the response. The simplest information on the line width is given by the second moment which has the exact form

$$\langle\underline{\underline{\omega}}_q^2\rangle \equiv \langle[\dot{\underline{A}},\underline{A}^+]\rangle/\underline{\underline{\chi}}_q = \langle\underline{\underline{\omega}}_q\rangle^2 + \text{terms involving } \{\langle aa\rangle \text{ and } \langle bb\rangle\}.$$

The line broadening is proportional to $\langle\omega_q^2\rangle-\langle\omega_q\rangle^2$ and is therefore proportional to the short range correlation functions $\langle aa\rangle$ and $\langle bb\rangle$. In fact if we know the lineshape we can from $\langle\omega_q\rangle$ and $\langle\omega_q^2\rangle$ calculate both the peak position and the linewidth exactly in terms of static quantities[45]. However, the simple Breit-Wigner formula, which is often used to represent a spinwave life time and which replaces $\delta(\omega-\omega_q)$ by $\Gamma_q/[(\omega-\omega_q)^2+\Gamma_q^2]$, is not adequate because it gives an infinite $\langle\omega_q^2\rangle$. It is much better to generalize the harmonic oscillator response

$$S(q,\omega) = \chi_q \frac{1}{2}\{n(\omega_q)\delta(\omega+\omega_q) + [n(\omega_q)+1]\delta(\omega-\omega_q)\}\omega_q$$

where $n(\omega) = (e^{\omega/T}-1)^{-1}$ to the damped harmonic oscillator lineshape[40]

$$S(q,\omega) = \frac{1}{\pi}\chi_q[n(\omega)+1] \quad \frac{2\beta_q\langle\omega_q^2\rangle\omega}{(\omega^2-\langle\omega_q^2\rangle)^2+4\beta_q^2\omega^2} \tag{11}$$

This is also called the two-pole-approximation because it represents two complex poles at $\omega = \pm\alpha_q + i\beta_q$ such that $\alpha_q^2 + \beta_q^2 = \langle\omega_q^2\rangle$. When the damping $\beta_q$ reduces to zero $\alpha_q = \omega_q$ the first moment frequency; for finite $\beta_q$ one has $\alpha_q \neq \omega_q$. The peak positions for (11) is not even at $\alpha_q$ for finite $\beta_q$; it goes to zero already for $\alpha_q = \beta_q$. This is due to overdamping of the oscillations. Knowledge about $\langle\omega_q\rangle$ and $\langle\omega_q^2\rangle$ is not sufficient to determine $\alpha_q$ and $\beta_q$ separately, unless their ratio is assumed fixed, which may be quite reasonable in some cases[45]. In general the damping $\beta_q$ is determined by the mode-mode decoupling using the Mori theory[39] for the relaxation function

$$\underline{\underline{\chi}}(q,z) = (\underline{A}|\underline{A}^+)_{q,z} = \underline{\underline{\chi}}_q \frac{1}{\underline{\underline{z}}+\underline{\underline{\langle\omega_q\rangle}}+(\underline{X}_1|\underline{X}_1^+)_{q,z}} \tag{12}$$

where $z = i\omega$ and the random force $\underline{X}_1 = \dot{\underline{A}}(t) - i\langle\omega_q\rangle\underline{A}(t)$ is the force acting in a rotating coordinate system following the spin wave frequency. For the Heisenberg ferromagnet $(X_1|X_1^+)$ involves terms of the kind

$$(S_k^x(t)S_{q-k}^y(t)|S_{-k}^x(0)S_{k-q}^y(0)) \sim (S_k^x(t)|S_{-k}^x(0))(S_{q-k}^y(t)|S_{k-q}^y(0))$$

which is decoupled in the mode-mode coupling approximation[41]. Here each relaxation function $(S^\alpha(t)|S^\alpha)$ is expressed in the two-pole approximations and the equations are closed by inserting back into (12).

Analysing damping of spin waves should at least be done with the two-pole-approximation (11), but there may be further effects as well. Firstly the higher moments $\langle\omega_q^n\rangle$ for $n>2$ are infinite. To make them finite a simple cut-off at high frequencies may be used, which ensures that $\langle\omega_q^4\rangle$ is correct[40]. Secondly the line-shape (11) may be too simple and involve also a central peak or the lineshape is more skew, for example the $\{\omega^2-\omega_q^2\}^{1/2}$ type, as suggested[19] for the S=1/2 linear antiferromagnet. These last effects may only occur in extremely non-linear systems, whereas (11) plus a cut-off should cover the weakly non-linear cases.

The damping of spin waves was measured by neutron scattering in the ideal Heisenberg ferromagnets EuO[46] and EuS[47]. In EuO one

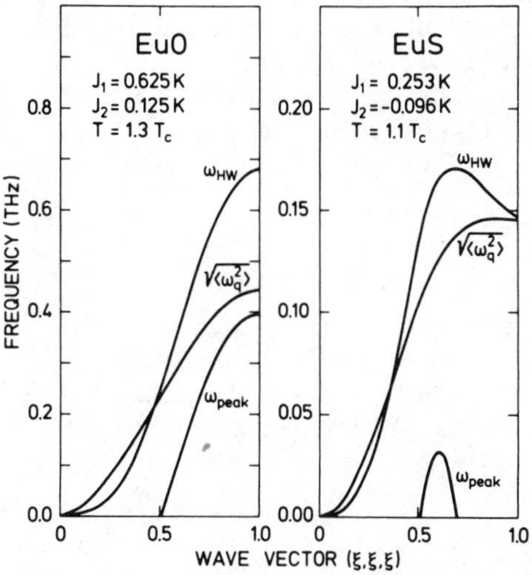


Fig. 7a. Calculated[40] peak position $\omega_{peak}$ and half width at half maximum $\omega_{HW}$ for EuO and EuS at T = 1.3 $T_c$ and 1.1 $T_c$.

found that "sloppy" spin waves persisted near the zone boundary[46] to at least 2 $T_c$, see Fig. 7. It was also found that (11) gave the best representation of the data. Excellent agreement was found between all available both static and dynamic data for $T > T_c$ and the selfconsistent correlation theory with no adjustable parameters[40]. In this case $\langle\omega_q\rangle = 0$ and there are no simple spin waves. Notice the cut-off becomes important near $T_c$ at high q.

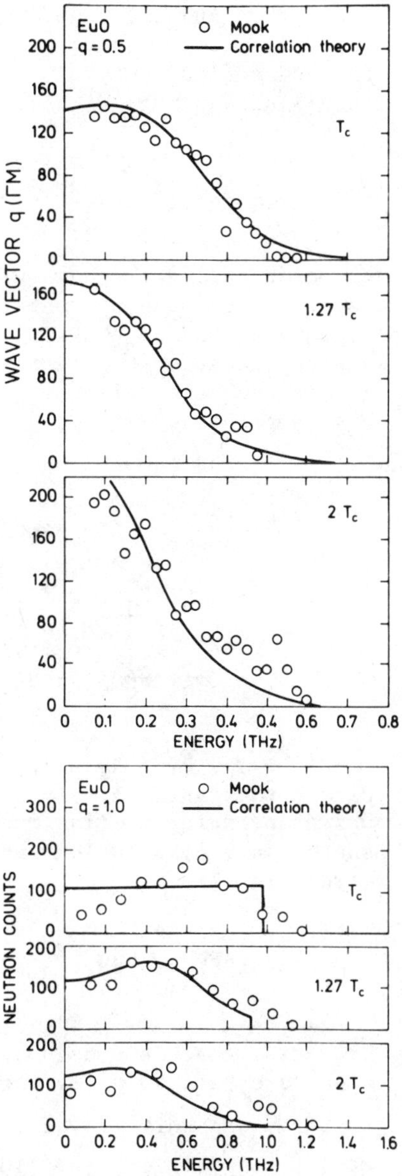


Fig. 7b. Measured (Mook[46]) and calculated[40] lineshapes for EuO at $q = 0.5\ q_{zone}$ and $q = q_{zone}$ for $T = T_c$, 1.3 $T_c$ and 2$T_c$ (after Lindgård[40] and Mook[46]).

Any use of the Mori response function theory[39] requires an a priori knowledge of the static properties. In the correlation theory the same level of accuracy for the static and dynamic approximations is ensured, by using the second moment for the determination of $\chi_q$ and a decoupling corresponding to the mode-mode coupling. Since $\langle\omega_q^2\rangle$ only involves static correlation functions this decoupling very closely corresponds to the familiar Hartree Foch[11] decoupling. We have three exact expressions for

$$\langle\underline{\omega}_q^2\rangle\underline{\chi}_q = \langle[\underline{\dot{A}},\underline{A}^+]\rangle = (\underline{\dot{A}}|\underline{\dot{A}}^+) = -(\underline{\ddot{A}}|\underline{A}^+) \tag{13}$$

to be decoupled. We have used the first two in the dynamic theory[40]. For the Heisenberg magnet one obtains the same $\chi_q$ by decoupling any of the last two terms, but the last is most convenient because it directly gives terms proportional to $\chi_q = (\underline{A}|\underline{A}^+)$. For the Heisenberg ferromagnet for example one finds for any temperature[40,44]

$$\chi_q^{xx}=\chi_q^{yy} = \frac{1}{J_o}\,\frac{1}{R_\perp-\gamma_q} \quad \text{and} \quad \chi_q^{zz} = \frac{1}{J_o}\,\frac{1}{R_\parallel-\gamma_q}$$

$$R_\perp = \sum_q \gamma_q^2(\langle S_q^x S_{-q}^x\rangle+\langle S_q^z S_{-q}^z\rangle)/\sum_q \gamma_q(\langle S_q^x S_{-q}^x\rangle+\langle S_q^z S_{-q}^z\rangle) \tag{14}$$

$$R_\parallel = \sum_q \gamma_q^2(\langle S_q^x S_{-q}^x\rangle+\langle S_q^y S_{-q}^y\rangle)/\sum_q \gamma_q(\langle S_q^x S_{-q}^x\rangle+\langle S_q^y S_{-q}^y\rangle)$$

For $T \gg T_c$, $R_\perp = R_\parallel = 1/\chi_o = 3T/[J_oS(S+1)]$. Below $T_c$, $R_\perp$ and $R_\parallel$ differ substantially since $\langle S^z\rangle > 0$. $T_c$ is given by $R_\perp = R_\parallel = 1$, which according to (14) happens when a certain relation between the short range correlation functions is fulfilled. For a simple cubic structure this is

$$\tfrac{1}{2}(\langle S_o^\alpha S_o^\alpha\rangle+\langle S_o^\alpha S_2^\alpha\rangle+\langle S_o^\alpha S_3^\alpha\rangle)_{T_c}=\langle S_o^\alpha S_1^\alpha\rangle$$

We notice that it does not mean that any of the correlation functions are particularily large. Thus for EuO we found[40] at $T_c$ that $\langle S_o^z S_1^z\rangle_{T_c}$ was only 12% of that for a fully aligned system and $\langle S_o^z S_2^z\rangle_{T_c}$ only 8%. Using (10) and (14) we can now calculate the first moment frequency for $T<T_c$ for example for the variables $S_q^x$ and $S_q^y$. This is much better for $T>0$ than the RPA result, which is obtained when the (SRO) correlation functions are neglected i.e. when $\langle S_q^z S_{-q}^z\rangle \sim M^2\delta_{qo}$ and $R_\perp = 1$. As discussed above the peak positions are further renormalized due to line broadening. Now the method (13) can be used for more complicated system (crystal field systems for example[45]) to include correlation effects in the susceptibilities and consequently in the first moment frequencies (10). Already at this level the correlation theory[40] goes beyond the correlated effective field[33,34] method by including the correlation functions in a more systematic way. Additionally it of course offers the possibility of calculating the lineshape and linewidths.

THE CENTRAL PEAK PHENOMENA

A central peak shows up as a resonance near $\omega = 0$ with some width, but usually no dispersion, which means maximum response at $\omega = 0$. We can distinguish between two cases; one in which linear theory can account for it and the second where it must be attributed to non-linear phenomena. But this is not a principal classification because it depends on how sophisticated the linear theory is. It is usually the central peak which drives a phase transition.

An example of a linear central peak is found in crystal field systems in which there are degenerate levels $E_r = E_p$ which are coupled by the spin operators. It is clear from (7) and (8) that in the RPA theory this gives a $\delta(\omega)$ response even if the degeneracy does not occur in the ground state. It is the Curie Weiss part of the general Van Vleck susceptibility which gives the $\delta(\omega)$ response. This was found in the singlet-triplet system $Pr_3Tl$[48]. It is in fact the same phenomena as the paramagnetic response for the Heisenberg ferromagnet, which only contributes to the Curie Weiss susceptibility. However, using the non-linear correlation theory[40] we saw that the central peak could acquire both a width and a dispersion, the sloppy spin waves. An interesting observation[48] was that the correlation length $1/\kappa_1$ is longer in a singlet ground state system than in a pure Heisenberg magnet at $T_c$. This is generally true for disordered systems: the LRO is replaced by SRO.

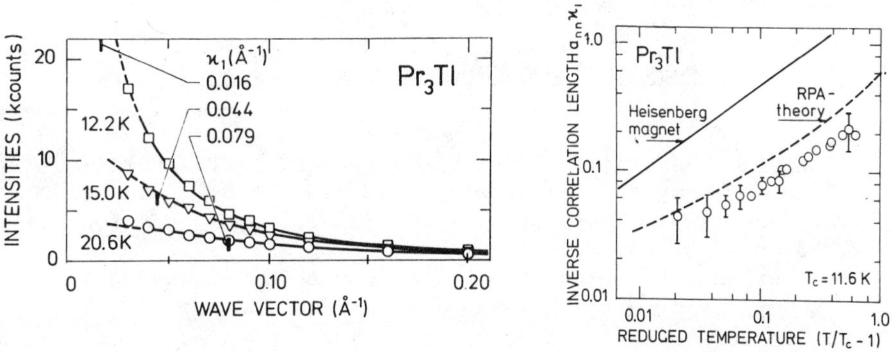


Fig. 8. a) Intensity of the Central peak versus wave vector. The full lines are least squares fitted, resolutionsfolded Lorentzians with the indicated widths. b) The inverse correlation length $\kappa_1$ is smaller than for a pure Heisenberg magnet (after Als-Nielsen et al[48]).

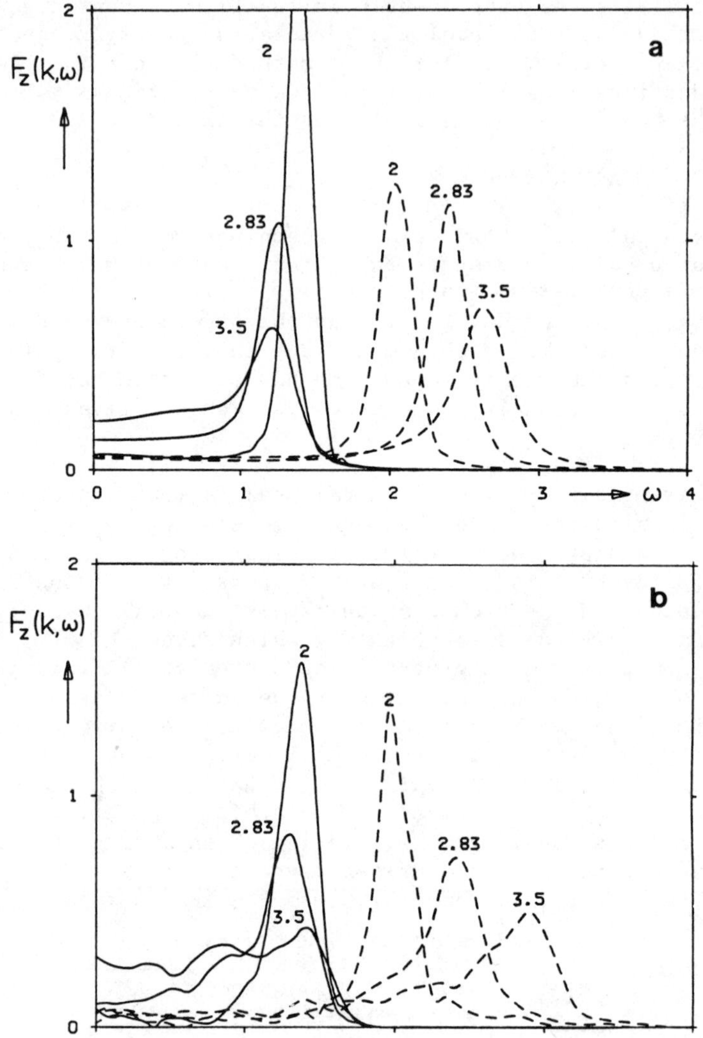


Fig. 9. Longitudinal response for the classical antiferromagnetic chain in a magnetic field $h = g\mu_B/2J$ at $T = 0.1$ J and at two wavevectors —— $k=0.3\ q_{zone}$ and --- $k=0.7\ q_{zone}$ and for h = 2, 2.83 and 3.5.
a) continued fraction theory (Lovesey *et al*[42]) and
b) computer simulation (Balcar *et al*[50]).
Notice intensity is building up at small ω and a second resonar (energy-mode?) might be showing up at h = 3.5.
(After Balcar *et al*[50]).

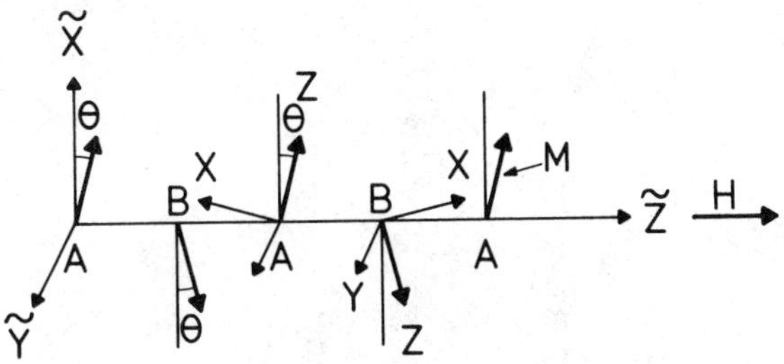


Fig. 10. An antiferromagnetic chain in a field having this local structure shows in the total longitudinal response $\tilde{\chi}^{zz}(q,\omega)$ both the local transverse and the local longitudinal response - and therefore both a spin wave and a central peak resonance.

For an antiferromagnet in a magnetic field the coupling between "magnetization" and "energy" modes could perhaps give rise to a central peak[49]. For the 1-dimensional classical antiferromagnet the Mori theory by Lovesey et al[49] in fact gave two different $\delta$-function resonances, but no central peak. It was surprising that the "energy" mode was propagating and undamped. A more refined calculation by de Raedt et al[43] gave basicly the same result with some broadening of the $\delta$-functions in particular of the "energy" mode. Computer simulations support this picture[50], see Fig. 9. The physical picture is much simpler (and equivalent!) if we use the correlation theory[44]. First we calculate the local situation, which as shown on Fig. 10 is an antiferromagnetic spin structure perpendicular to the field apart from a small canting angle $\theta$. The total $\tilde{\chi}^{\alpha\beta}(q,\omega)$ therefore is composed of both the local transverse uniform and staggered susceptibility as well as the local longitudinal. Using only the first moment frequency, the former gives a spin wave response $\delta(\omega-\omega_q^{uniform})$ and $\delta(\omega-\omega_q^{staggered})$ and the last a central $\delta(\omega)$ response. We found $\omega_q^{uniform} = J_oM\{(R-\gamma_q)(R+\cos\theta\gamma_q)\}^{1/2}$ and $\omega_q^{staggered} = J_oM\{(R+\gamma_q)(R-\cos\theta\gamma_q)\}^{1/2}$ where R depends on the field and M is the local average moment. The response will be broadened by the correlation effects. This result also agrees with the computer simulations and gives in addition in a simple way a "linear" central peak. We need not invoke an "energy" mode concept.

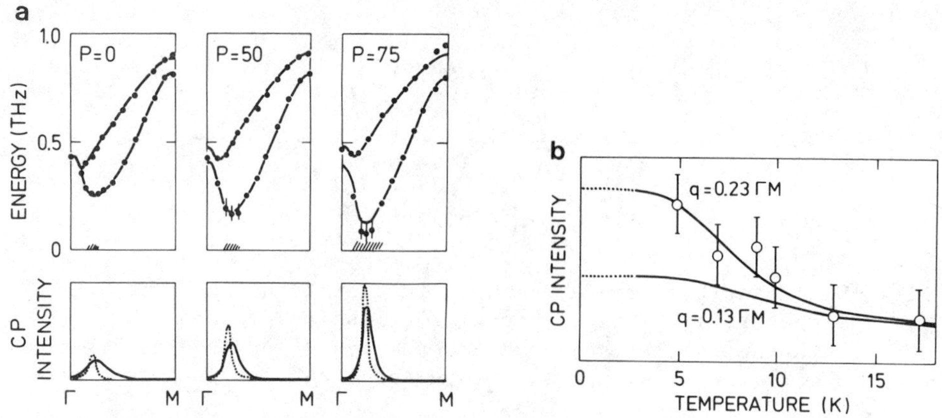


Fig. 11. a) Pr under uniaxial pressure shows a softmode behaviour and the growth of a central peak a low frequencies. The experimental data by McEwen *et al*[72] are denoted by •, ////// and ...... The full curve is the result of the correlation theory.
b) The temperature dependence of the central peak in Pr at zero pressure measured by polarised neutron scattering o (Burke *et al*[51]) compared with the correlation theory result (after Lindgård[52]).

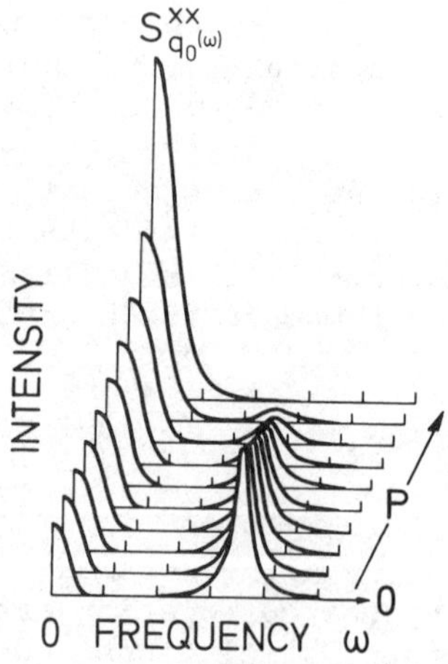


Fig. 12. Pressure dependence of $S^{xx}(q_o,\omega)$, for a singlet-doublet model, showing a soft mode and a central peak. The foremost curve is calculated using parameters relevant for Pr at $P = 0$, $T = 5$ K and $q_o = 0.2$ ΓM. The curves are normalized to the same area (after Lindgård[52]).

The planar S=1 magnet is a singlet-doublet system and has no dipole transition between the degenerate $|\pm 1\rangle$ states - and therefore no linear central peak[48]. Nevertheless a central peak was observed[51] in the transverse susceptibility $\chi_\perp(q_o,\omega)$ in Pr upon approaching the ordering temperature, see Fig. 11. The phenomenon can be accounted for by the correlation theory[52] which for $\chi_\perp(q_o,\omega)$ gives a resonance both at the first moment frequency $\omega \sim \pm\, \omega_q$, and at $\omega = 0$, the central peak, see fig. 12. It is convenient to describe the width of these peaks by two damped harmonic oscillators and we can write the response function

$$S_\perp(q,\omega)=\frac{2}{\pi}\chi_\perp(q)\left[n(\omega)+1\right]\left\{(1-P)\frac{\omega\Gamma_q(\omega_q^2+\Gamma_q^2)}{(\omega^2-\omega_q^2-\Gamma_q^2)^2+4\omega^2\Gamma_q^2}+P\frac{\omega\delta_q(\varepsilon_q^2+\delta_q^2)}{(\omega^2-\varepsilon_q^2-\delta_q^2)+4\omega^2\delta_q^2}\right\}$$

The weight P of the central peak ($\varepsilon_q < \delta_q \ll \omega_q$) is given by the ratio between nearest neighbour correlation functions and $\omega_q^2$. The central peak, therefore, is predicted to absorb all the spectral weight at the soft mode transition $\omega_q \to 0$ and also to show critical narrowing, see Fig. 12. The central peak is also important, but broad, at high temperatures when $\langle S_x^2\rangle \approx \langle S_z^2\rangle$. The physical picture of the narrow central peak is critical fluctuations in regions with induced magnetic short range order. If the (SRO) is sufficiently large sloppy spin waves could emerge for large q as in EuO. The situation resembles that of itinerant magnets considerably. The Stoner excitations between electron levels corresponds to the local $|0\rangle \to |\pm 1\rangle$ transitions between crystal field levels.

The presently most intensively investigated central peak is that found in $CsNiF_3$. This substance is an ideal planar magnet with S=1 described by the Hamiltonian (3). Furthermore it is ideally 1-dimensional, which makes correlation effects particularily large. Pr is a 3-dimensional, planar, effective S=1 model. A very thorough review of the results was published recently[53] and we need here only make a few comments. The temperature and field dependence was measured of a large central peak (not too surprising) in the longitudinal $\chi_\parallel(q,\omega,H)$ ($\parallel$ to the magnetic field which is in the easy plane). Theoretically this was attributed to 1) soliton effects and 2) two spin wave effects (which are quite pronounced in a 1-dimensional system because the density of states has two sharp peaks) and 3) breathers and other non-linear effects. The soliton is the solution to the classical sine-Gordon equation, which corresponds to a $S=\infty$ continuum, spin chain. Experimentally[53] the predictions of this model for $\chi_\parallel(q,\omega,H)$ seems to be valid for a range of field and temperatures, but one found that both mechanism 1) and 2) are playing a role. A soliton is a $2\pi$-domain wall, which in 1-dimension can move quite easily and thereby contribute to the

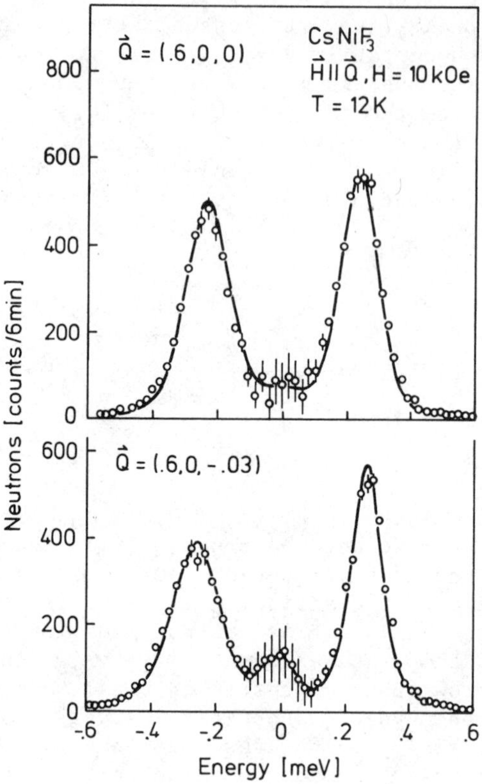


Fig. 13. Spin waves and central peak in the transverse response function for $CsNiF_3$ in a magnetic field (after Steiner et al[53]).

dynamics. A recent polarized neutron experiment[69] showed that a large part of $\chi_{\parallel}(q,\omega,H)$ came from out-of-plane fluctuations which are neglected in the S =∞ soliton theory. It is likely that quantum effects do play a significant role in $CsNiF_3$. It was much harder[53] experimentally to see a small central peak in the transverse susceptibility $\chi_{\perp}(q,\omega,H)$ in addition to the usual spin wave peaks, see Fig. 13. There are no contributions from two spin wave processes to this, but there are both from the soliton effect and, as we have seen for Pr, from the quantum effect of the singlet ground state. The influence of quantum effects have not yet been fully investigated for the S=1 $CsNiF_3$. The correlation theory can accurately treat the quantum effects, but is is not yet clear how well the very non-linear soliton is described by the mode-mode coupling approximation.

The phenomena that the motion of a domain wall can be considered as an elementary excitation in 1-dimension, was nicely demonstrated by Villain[54], who considered an antiferromagnetic Ising chain with a small transverse interaction ε. The ground state

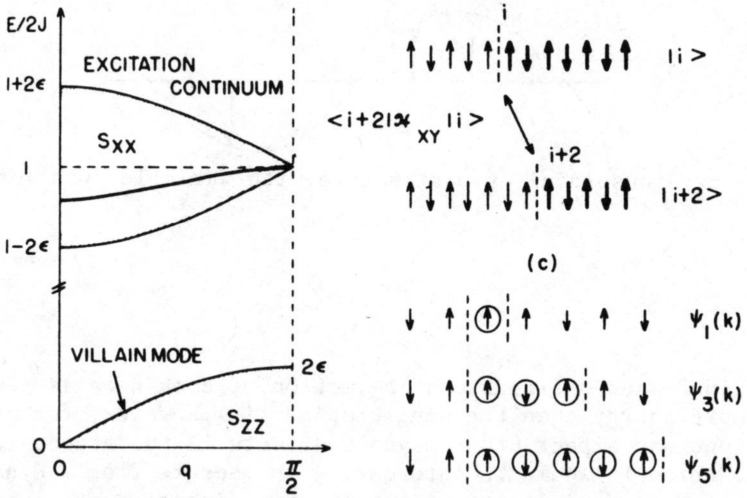


Fig. 14. (a) 1-D antiferromagnetic domain walls are shown. The domain walls can move from i to i+2 due to the transverse matrix element ε $(i+2|H_{xy}|i)$. (b) The spin wave excitation continuum and the Villain mode. The thick line in the continuum corresponds to the peak position of the spectral weight. The domain-wall motion causes the excitation continuum. (c) shows spin wave excitations (after Yoshizawa et al[55]).

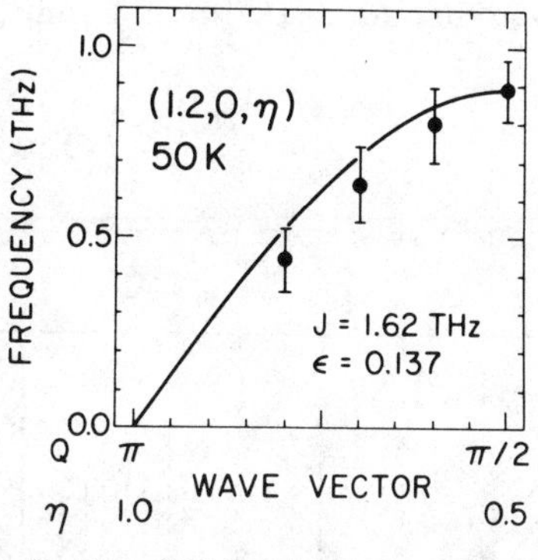


Fig. 15. Dispersion of the Villain mode. The solid line shows the Villain prediction, $\omega_Q = 4\varepsilon J \sin(Q)$, for the ε and J which describe the spin wave response in $CsCoBr_3$ (after Nagler et al[56]).

involves states of the type, see Fig. 14a, where the dotted lines indicate the domain wall. Two simultanous spin flips move the wall and cost less energy than the single spin flip spin wave process, Fig. 14c. This extremely non-linear "central peak" in $\chi_{\parallel}(q,\omega)$ has a dispersion and small width like a linear spin wave! This simple soliton, or Villain mode, was seen experimentally[55] in $CsCoCl_3$ and[56] $CsCoBr_3$, see Fig. 15.

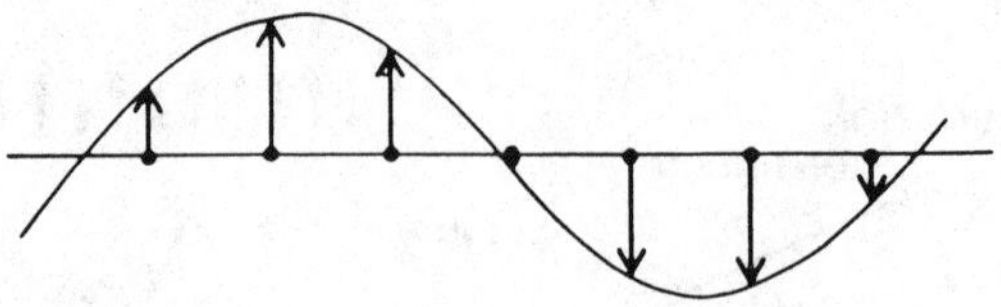

Fig. 16. Sinusoidial spin structure with variable spin length.

In 2- and 3-dimensions the motion of domain walls requires much more energy than the single spin flip. However, for 2-dimensions one may expect other movable defects in the structure, vortices, which it would be interesting to measure from a dynamical point of view[56]. A first attempt in this direction has been done by Rossat-Mignod et al[70]. Since such phenomena are almost macroscopic they are expected to occur in the lower left corner of Fig. 1.

For a system with an incommensurate sinusoidial order, see Fig. 16, there is a zero frequency Goldstone mode corresponding to a uniform moving of the node points. This is called a phason mode, but has not yet been seen experimentally. The spin waves in such a system are strongly broadened because of the variable length of the average moment[57]. The first experimental observation in such a structure for Nd[58] showed broad spin wave bands, see fig. 17.

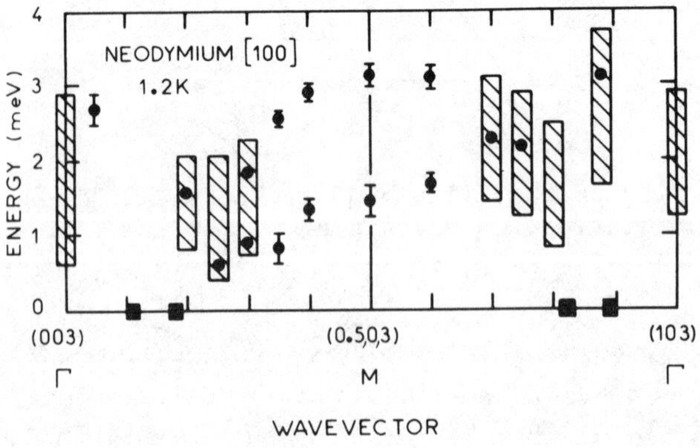


Fig. 17. Magnetic excitations in the sinusoidally ordered Nd at T = 1.2 K showing broadened spin waves and zero frequency response (after McEwen et al[51]).

## ITINERANT MAGNETS

Finally we shall briefly discuss the current understanding of excitations in the itinerant magnets. At low temperatures where the induced magnetization is large the RPA theory[59] is excellent and gives exactly (6) and (7), with the non-interacting susceptibility (8) replaced by the metallic electronic susceptibility $\chi_o(q,\omega)$. This can be computed from the electronic energy bands and the resulting calculated spin wave dispersion and Stoner band edge agrees excellently with the measurements[60]. The problem comes when the induced moment is small (either as in weak itinerant systems or at high temperatures). We recognize the similarity with the singlet ground state crystal field system. For a ferromagnet the transition temperature is reached when the directional disorder among the effective spins gets large, and not when the magnitude of the moments disappear (the Stoner picture). A local moment can be formed in an itinerant magnet in the presence of (SRO) only. The problem for itinerant magnet theories is therefore first to calculate the induced local moment and then the directional disorder. The first part already requires a mode-mode coupling theory and this was developed by Moriya[61] and Kawabata. It covers both the localized and weak itinerant limits. The result can be expressed as an effective Hamiltonian with non-integer spin variables and where the exchange integral depends on the local moment average $M_o^2 = \Sigma_q \langle S_q^z S_{-q}^z \rangle$. For $T>T_c$ the effective Hamiltonian is simply[62]

$$\mathcal{H} = -\sum_{ij} J_{ij} M_o^2 \vec{S}_i \cdot \vec{S}_j \qquad (16)$$

The local moment is calculated selfconsistently and varies of course slowly with temperature, with a small dip at $T_c$. Using (16) one can go on and calculate spin dynamics below and above Tc using the correlation theory for example. One would not expect a result very different from that found in insulators except that the Stoner excitations in some cases may wipe out possible sloppy spin waves at high q.

A lot of discussion[63] has arisen because neutron scattering using constant energy scans[64] observed peaks both for Ni and Fe above $T_c$ at quite small wave vectors $q > 0.25$ Å$^{-1}$. If these were sloppy spin wave resonances, one argued[65] that one needed an unusual and unexpected giant (SRO). Recently a thorough polarized neutron study of $\chi(q,\omega)$ using the conventional constant q scans was performed[66] for $T>T_c$ in $^{60}$Ni and Fe(4%-Si). The data could be described as broad Lorentzian lineshape with no sign of any sharp "sloppy spin waves", thus not supporting the picture of strong (SRO). However, both observations are correct. By plotting the $\chi(q,\omega)$ data in a Lowde-Winsor diagram[67] one finds[66] that $^{60}$Ni and EuO are very similar at $T = 1.25\ T_c$ showing a broad ridge of response peaking roughly at the spin wave frequency $\omega_q$ below Tc, see fig. 18.

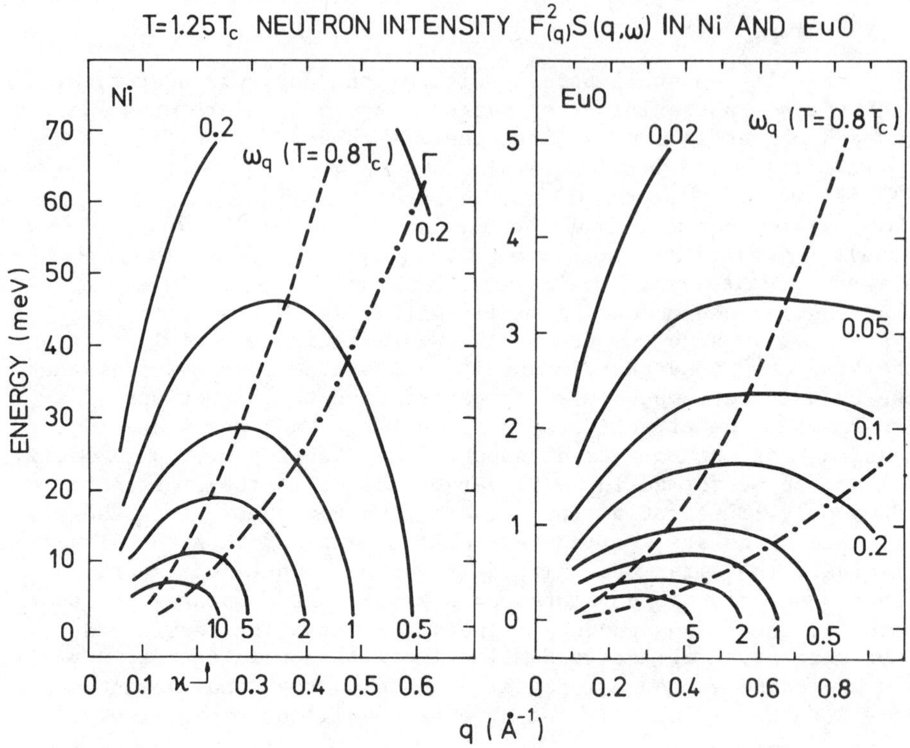


Fig. 18. Comparison between the paramagnetic response at T = 1.25 $T_c$ for the itinerant magnet Ni and the Heisenberg magnet EuO, represented by an intensity contour map. --- shows the spin wave dispersion $\omega_q$ at T = 0.8 $T_c$ - • - • - indicates the half width frequency Γ (after Wicksted et al[56]).

The explanation must be that the criterion for finding sloppy spin waves is not the static condition $q < \kappa$, the inverse correlation length, but it is a dynamic one: When the spin wave has a sufficiently high frequency it sees a stronger, instant (SRO) than the long time average. This is consistent with the calculations for EuO which shows sloppy spin waves even when $\langle S_0^z S_1^z \rangle = 0$ at $T = \infty$. In the weak itinerant magnet MnSi the spin wave excitations observed below the Stoner band for $q < 0.3$ Å$^{-1}$ were damped out above $T_c$, but the ridge in the Stoner band[68] which seems to extrapolate along the simple spin wave $\omega_q$, was not found to be temperature dependent. The ridge in the Stoner band has not yet been understood theoretically, see fig. 19.

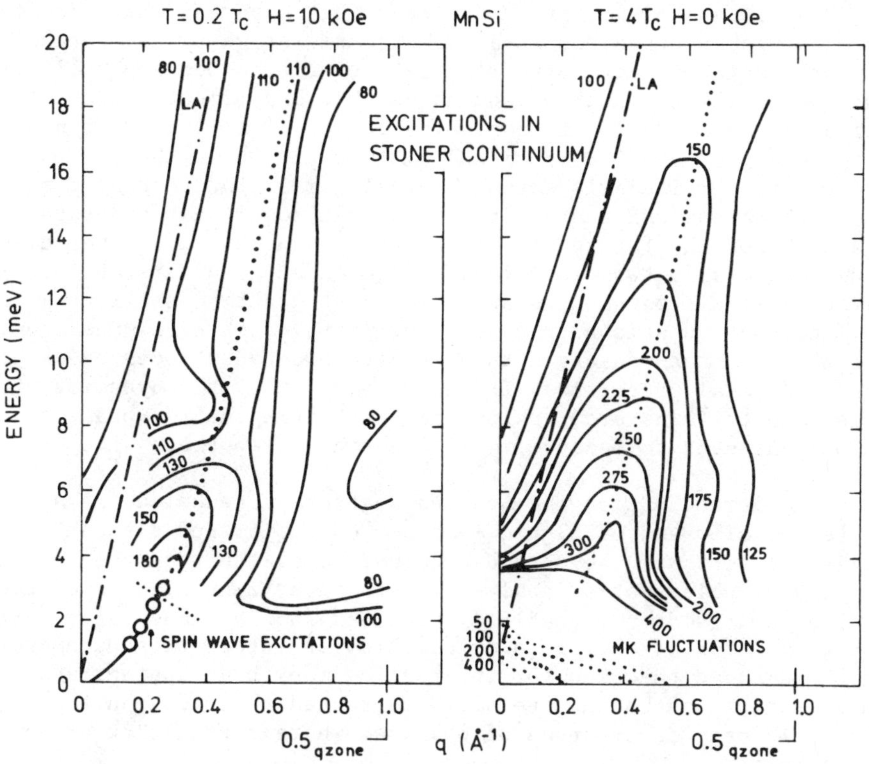


Fig. 19. The magnetic response for the weak itinerant magnet MnSi at T = 0.2 $T_c$ and T = 4 $T_c$ showing the Stoner band and at low q and T spin waves, which reduce to paramagnetic (MK) fluctuations above $T_c$ (after Ishikawa[68]).

## CONCLUDING REMARKS

Magnetic excitations in well ordered structures are by now well understood and experiments in this situation are of the outmost importance for obtaining the physical interaction parameters, exchange interactions, crystal field etc. The linear excitations give, however, only somewhat phenomenological, model dependent parameters, which may contain several contributions from different basic physical mechanisms. A task for the future is to develop methods which can distinguish between various such mechanisms, by for example applying external perturbations or by investigating small clusters.

Magnetic excitations in disordered structures, where one only has short range order, is in a rather early stage of understanding. The correlation effects give rise to interaction between excitation modes and new non-linear phenomena occurs. Experimentally and theoretically the situation is beginning to be clarified. Most notable, is the study of the 1-dimensional systems, in which the correlation effects are particularly strong, revealed many interesting non-linear phenomena which are perhaps not yet completely understood. It is a task for the future to follow these phenomena in higher dimensions, especially two-dimensions, where a deeper understanding of the quite simple magnetic systems may cast light on interesting and important questions in many other parts of physics. One should bear in mind that measurements of finite frequency modes essentially probe the local properties. If one wish to study the dynamics of large objects (clusters), which are fluctuating near a phase transitions one need to measure the low q and $\omega$ response.

The demand put on the neutron scattering technique is a further development of high q - and $\omega$-resolution and also an ability to separate the various components of the response function $\chi^{\alpha\beta}(q,\omega)$ by polarization analyses. With experimental tools of this calibre the field of magnetic excitations should remain an important test ground for a) ab initio calculation of interaction parameters and b) advanced theories for interacting many body systems and non-linear phenomena in general. Such tested theories can then be applied in many other areas of physics, where correlation effects play a role.

REFERENCES


1. W. Marshall and S.W. Lovesey, Theory of Thermal Neutron Scattering (1970), (Oxford University Press).
2. See reviews by S.K. Sinha and P. Fulde (1979), Handbook on the Physics and Chemistry of Rare Earths, Ed. K.A. Gschneidner and L. Eyring (North Holland).
3. See "Physics in 1-Dimension" (1981), Ed. J. Bernasconi and T. Schneider (Springer Press).
4. See "Excitations in Disordered Systems" (1982) Ed. M.F. Thorpe (Plenum Press) and R.A. Cowley (1976) AIP Conf. Proc. 29, 243. R.A. Cowley, G. Shirane, R.J. Birgeneau and H.J. Guggenheim (1977) Phys. Rev. B15, 4292, and B.D. Rainford (1979) J. Mag. Magn. Mat., 14, 197.
5. A.P. Murani (1978), J de Phys. C6, 1517, F. Mezei (1983), J. Mag. Magn. Mat. 31-34, 1327, J. Villain (1980), J. Mag. Magn. Mat. 15-18, 105.
6. R.A. Cowley, D.McK. Paul W.C. Stirling, N. Cowlam (1983), Physica 120B, 373, G. Shirane, J.D. Axe, C. Majkarak and T. Mizoguchi (1982), Phys. Rev. B26, 2575.
7. F. Dyson (1956), Phys. Rev. 102, 1217, S.V. Maleev (1958), Sov. Phys.-JETP 6, 776.
8. T. Holstein, H. Primakoff (1940), Phys. Rev. 58, 1098.

9. P.-A. Lindgård and A. Kowalska (1976), J. Phys. C 9, 2081.
10. A. Messiah (1961), "Quantum Mechanics", p. 266.
11. K. Keffer (1966), "Handbook der Physik" XVII, p. 1 (Springer Press).
12. P.-A. Lindgård, A. Kowalska and P. Laut (1969), J. Phys. Chem. Sol. 28, 1357. See also the comprehensive paper by B.R. Cooper, R.J. Elliott, S.J. Nettel, and H. Suhl (1962) Phys. Rev. 127, 57.
13. A. Furrer and H.U. Düdel (1979), J. Magn. Magn. Mat. 14, 256.
14. E. Rastelli and P.-A. Lindgård (1979), J. Phys. C. 12, 1899.
15. M. Steiner, J. Villain and C.G. Winsor (1976), Adv. in Phys. 25, 87.
16. J. Jensen, J.G. Houmann and H.J. Bjerrum Møller (1975), Phys. Rev. B12, 303.
17. J. des Cloizeaux and J.J. Pearson (1962), Phys. Rev. 128, 213.
18. I.U. Heilmann, G. Shirane, Y. Endoh, R.J. Birgeneau and S.L. Holt (1978), Phys. Rev. 18B, 3530, and M.T. Hutchings, H. Ikeda and J.M. Milne (1979), J. Phys. C. 12, L793.
19. H.J. Mikeska (1975), Phys. Rev. B12, 7, 2794.
20. N. Ishimura and H. Shiba (1977), Progr. Theor. Phys. 57, 1862, (1980) ibid 64, 479.
21. G. Müller, H. Beck and J.C. Bonner (1979) Phys. Rev. Lett. 43 75, and M. Mohan and G. Müller (1983) Phys. Rev. 27, 1776.
22. I.U. Heilmann, J.K. Kjems, Y. Endoh, G. Reiter, G. Shirane and R.J. Birgeneau (1981), Phys. Rev. B24, 3939.
23. B. Leuenberger, H.U. Güdel, R. Feile and J.K. Kjems (1983), Phys. Rev. B28, 5368.
24. H. Kadowaki, K. Hirakawa and K. Ubukoshi (1983). J. Phys. Soc. Japan 52, 1799.
25. G. Toulouse (1977), Commun. Phys. 2, 115.
26. P.W. Anderson (1973), Mat. Res. Bull. 8, 153.
27. K. Hirakawa, H. Kadowaki and R. Ubukoshi (1983), Physica 120B, 187 and (1983), J. Phys. Soc. Japn. 52, 1814.
28. D.N. Zubarev (1960), Sov. Phys. Usp. 3, 320.
29. W.J.L. Buyers, T. Holding, T.M. Svensson, R.A. Cowley and H.T. Hutchings (1971), J. Phys. C. 4, 2139.
30. S.B. Hayley and P. Erdös (1972), Phys. Rev. B5, 1106.
31. I. Pechel, M. Klenin and P. Fulde (1972), J. Phys. C. 5, L194.
32. P.-A. Lindgård (1978), "Durham conference on Rare Earth", Inst. Phys. Conf. Ser. 37, 97.
33. M.E. Lines (1974), Phys. Rev. B9, 3927 and (1975), Phys. Rev. B11, 1134.
34. N. Suzuki (1978), J. Phys. Soc. Japan 45, 1791.
35. N. Zuzuki (1983), J. Phys. Soc. Japan 52, 3907.
36. D. Petitgrand, B. Hennion, P. Radhakrisna, C. Escribe and S. Legrand (1982), "Recent developments in Condensed Matter Physics" Ed. J.T. Devreese et al 4, 205 (Plenum Press) and H. Yoshizawa, M. Kozukua and K. Hirakawa (1980), J. Phys. Soc. Japan 49, 144.
37. R.M. Nicklow, N. Wakabayashi, M.K. Wilkinson and R.E. Reed (1971), Phys. Rev. Lett. 27, 334.

38. W. Knop and P.-A. Lindgård (to be published).
39. H. Mori (1965), Progr. Theor. Phys. 33, 432 and 34, 399 and K. Kawasaki (1968), Progr. Theor. Phys. 39, 1133, and 40, 399.
40. P.-A. Lindgård (1982), J. Appl. Phys. 53, 1861 and (1983), Phys. Rev. B27, 2980.
41. K. Kawasaki (1976) "Phase Transitions and Critical Phenomena" Ed. C. Domb and M.S. Green (Academic Press), p. 166.
42. S.W. Lovesey and J.M. Loveluck (1979), J. Phys. C12, 4015.
43. H. de Raedt and B. de Raedt (1980), Phys. Rev. B21, 304 and (1981), Phys. Rev. B23, 4597.
44. P.-A. Lindgård (to be published)
45. P.-A. Lindgård (1979), Proc. Conf. on Rare Earth, Philadelphia Ed. J.E. Crow, R.P. Guerten and T.W. Mihalisin (Plenum Press), 153.
46. H.A. Mook (1981), Phys. Rev. Lett. 46, 508 and L. Passell, J. Als-Nielsen and O. Dietrich (1976), Phys. Rev. B14, 4897-4923.
47. H.G. Bohn, W. Zinn, B. Dorner and A. Kollmar (1980), Phys. Rev. B22, 5447.
48. J. Als-Nielsen, J.K. Kjems, W.J.L. Buyers and R.J. Birgeneau (1977) Physica 86-88B, 1162 see also W.J.L. Buyers (1975), AIP Conf. Proc. 24, 27, and T.M. Holden, W.J.L. Buyers, E.C. Svensson and H.G. Purvins (1976) "Crystal Field Effects in Metals and Alloys" Ed. A. Furrer (Plenum Press) p. 189.
49. S.W. Lovesey (1980), Z. Physik B37, 307 and "Condensed Matter Physics, Dynamic Correlations (1980) (Benjamin Press).
50. E. Balcar, S.W. Lovesey and J.M. Loveluck (1984), Z. Phys. B54, 195.
51. B. Lebech, K.A. McEwen and P.-A. Lindgård (1975), J. Phys. C8, 1684, J.C.G. Houmann, B. Lebech, A.R. Macintosh, W.J.L. Buyers, O.D. McMasters and K.A. Gschneider, Jr. (1977), Physica 86-88B, 1156, S.K. Burke, W.G. Stirling and K.A. McEwen (1981), J. Phys. C. L976.
52. P.-A. Lindgård (1983), Phys. Rev. Lett. 50, 690.
53. M. Steiner, K. Kakurai and J.K. Kjems (1983), Z. Phys. B53, 117.
54. J. Villain (1975), Physica 79B, 1.
55. H. Yoshizawa, K. Hirakawa, S.K. Satija and G. Shirane (1981), Phys. Rev. B23, 2298 and S.E. Nageler, W.J.L. Buyers, R.L. Armstrong and B. Briat (1982), J. Mag. Magn. Mat. 31-34, 1213 and (1983) Phys. Rev. B27, 1784.
56. S.E. Nagler, W.J.L. Buyers and R.L. Armstrong (1982), Phys. Rev. Lett. 49, 590.
57. S.H. Liu (1980), J. Mag. Magn. Mat. 22, 93 and P.-A. Lindgård (1983), J. Mag. Magn. Mat. 31-34, 603, see also A.D. Bruce and R.A. Cowley (1978), J. Phys. C, 11, 3609.
58. K.A. KcEwen and W.G. Stirling (1982), J. Mag. Magn. Mat. 30, 99.
59. T. Izuyama, O. Kim and R. Kubo (1963), J. Phys. Soc. Japan 50, 47.
60. J.F. Cooke (1976), Proc. Conf. Neutron Scatt. (Gatlinburg) 723, J.F. Cooke, J.W. Lynn and H.L. Davis (1980) B21, 4118.
61. T. Moriya and A. Kawabata (1973), J. Phys. Soc. Japan 34, 639.
62. T. Moriya (1983), J. Mag. Magn. Mat. 31-34, 11.

63. Panel discussion at ICM 82 (1983), J. Mag. Magn. Mat. 31-34, 313.
64. H.A. Mook, J.W. Lynn and R.M. Nicklow (1973), Phys. Rev. Lett. 30, 556, J.W. Lynn and H.A. Mook (1981), Phys. Rev. B23, 198, J.W. Lynn (1975), Phys. Rev. B11, 2624.
65. R.E. Prange and V. Korenman (1975), Phys. Rev. B19, 4691, H. Capellmann (1979), Z. Phys. B34, 29.
66. Y.J. Uemura, G. Shirane, O. Steinsvoll and J. Wicksted (1983), Phys. Rev. Lett. 51, 2322 and J.P. Wicksted, G. Shirane and O. Steinsvoll (1984), Phys. Rev. Lett.
67. R.D. Lowde and C.G. Winsor (1970) Adv. in Phys. 19, 813.
68. Y. Ishikawa (1979) J. Mag. Magn. Mat. 14, 123 and Y. Ishikawa, G. Shirane, J. Tarvin and M. Kohgi (1977), Phys. Rev. B16, 4956 and Y. Ishikawa, Y. Noda, G. Fincher and G. Shirane (1982), Phys. Rev. B25, 254.
69. K. Kakurai, R. Pynn, B. Dorner and M. Steiner (1984) J. Phys. C. 17, L123.
70. J. Rossat-Mignod, L.P. Reynault and R. Pynn (1982), ILL annual report (annex) 93. See also ref. 71.
71. L.P. Regnault, I. Rossat-Mignod, I.Y. Henry, R. Pynn and D. Petitgrand, Proc. of "Elementary Excitations and Fluctuations in Magnetic Systems", San Miniato (1984) (Springer Press).
72. K.A. McEwen, W.G. Sterling and C. Vettier, Phys. Rev. Lett. (1978) 41, 343 and Proc. Conf. CEF effects in f-Electron Systems (Plenum Press, New York, 1982).

# VIBRATIONAL AND TRANSPORT PROPERTIES OF MOLECULAR SOLIDS

K.H. Michel

Department of Physics, University of Antwerp
2610 Wilrijk (Belgium)

## INTRODUCTION

The study of crystals built up from molecules or molecular ions is a subject of increasing interest. Many interesting thermoelastic, electric and optic properties depend on the dynamics of the molecules. In addition to their center of mass motion ("translations") the molecules possess orientational degrees of freedom ("rotations"). Very often there exists a high temperature crystalline phase where the molecules have no preferential orientation, such a crystal is called orientationally disordered crystal (ODIC). In general the symmetry of the molecule is lower than the symmetry of the surrounding site (for instance a dumbbell in an octahedral potential[1]). The molecule then occupies with equal probability several sterically different positions. This situation is realized by frequent reorientations of the individual molecule. The average orientational probability distribution function then has the symmetry of the crystalline site. At lower temperature, the high symmetry distribution breaks down and there occurs a transition to an orientationally ordered state of lower symmetry. In general the orientational ordering of the molecules is accompanied by a change in lattice structure. Such a situation is an indication of translation-rotation coupling.

Symmetry adapted rotator functions have to be used for the description of the statics [2] [3] [4] and dynamics[5] of orientational disorder in crystals. On the other hand, in presence of translation-rotation coupling due to sterical hindrance, large amplitude orientational motions imply large amplitude center of mass displacements of molecules or individual atoms. Therefore we conclude that we have to use concepts from the theory of strongly anharmonic crystals.

In the present article we will study the following problems. We start with a formulation of molecular interactions in terms of symmetry adapted functions, taking into account the symmetry of the molecule and of the crystal site (Sect. 2). Next (Sect. 3) we study in detail the bilinear translation-rotation coupling and its relation to molecular symmetry. As particular examples we consider KCN and $NaNO_2$. Using concepts from the theory of quantum crystal we propose in Sect. IV a renormalization of various coupling coefficients. Such a description is important near phase transitions where one observes a peculiar temperature dependence of interaction potentials. We treat the ferroelastic transition in KCN and the incommensurate transition in $NaNO_2$. The following section V is devoted to dynamic equations for translation-rotation coupled systems and to a discussion of the corresponding resonances. Finally in Sect. VI we consider frequency dependent transport coefficients related to orientational relaxation and phonon viscosity.

## SYMMETRY ADAPTED FUNCTIONS AND INTERACTIONS

The center of mass of the molecule is located at a crystalline site with point group symmetry S. The molecule itself is taken as a rigid body with symmetry M. We start with the situation where a rectangular system of axes (x',y',z') which is rigidly attached to the molecule coincides with the rectangular system (X, Y, Z) fixed in the crystal. The mass distribution on a spherical shell with fixed radius r and centered at the center of mass of the molecule is given by:

$$f(\Omega) = \sum_{\nu} \delta(\Omega-\Omega'_{\nu}) \quad . \tag{2.1}$$

Here the summation $\nu$ runs over a same type of atoms located on the spherical shell and with angular position $\Omega_\nu = (\Theta_\nu, \varphi_\nu)$. Orientational properties are often expressed by means of spherical harmonics $Y_\ell^m(\Omega)$. For fixed $\ell$ they form the basis of an irreducible representation of the full rotation group. At a crystal site of symmetry S, the reduction of symmetry causes a decomposition into irreducible representations $\Gamma$ of lower dimensionality. The corresponding basis functions are symmetry adapted functions.[6]

$$S^{\tau}_{\ell(P)}(\Omega) = \sum_{m=-\ell}^{\ell} Y^m_\ell(\Omega)\, \alpha^{m\tau}_{\ell(P)} \quad . \tag{2.2}$$

Here $\tau = (\Gamma,\beta)$ labels the representation $\Gamma$ and the components $\beta$. The index (P) indicates the point group. The coefficients $\alpha^{m\tau}_{\ell(P)}$, for fixed $\ell$, form a unitary matrix. They have been tabulated[7]. We use conventions and notations of Ref. 7. The symmetry adapted functions form a complete and orthonormal set. Expanding Eq. (2.1) by taking advantage of site symmetry, we write

$$f(\Omega) = \sum_{\nu} \sum_{\ell} \sum_{\tau} S^{\tau}_{\ell(S)}(\Omega'_{\nu})\, S^{\tau}_{\ell(S)}(\Omega) \quad . \tag{2.3}$$

A rotation of the molecule by the Euler angles $(\alpha,\beta,\gamma) \equiv \vec{\omega}$ brings the orientation $\Omega'_{\nu}$ into position $\Omega_{\nu}$. Using the transformation of spherical harmonics

$$Y^{m}_{\ell}(\Omega_{\nu}) \equiv R(\vec{\omega})\, Y^{m}_{\ell}(\Omega'_{\nu}) = \sum_{n=-\ell}^{\ell} Y^{n}_{\ell}(\Omega')\, D^{nm}_{\ell}(\vec{\omega}) \quad , \tag{2.4}$$

we obtain the mass distribution of the rotated molecule:

$$f(\Omega;\vec{\omega}) = \sum_{\ell}{}' \sum_{\tau} b^{\tau}_{\ell}(\vec{\omega})\, S^{\tau}_{\ell(S)}(\Omega) \quad , \tag{2.5}$$

where

$$b^{\tau}_{\ell}(\vec{\omega}) = \sum_{\nu} S^{\tau}_{\ell(S)}(\Omega_{\nu}) = \sum_{\lambda} g^{\lambda}_{\ell}\, U^{\lambda\tau}_{\ell(M,S)}(\vec{\omega}) \quad , \tag{2.6a}$$

$$g^{\lambda}_{\ell} = \sum_{\nu} S^{\lambda}_{\ell(M)}(\Omega'_{\nu}) \quad , \tag{2.6b}$$

$$U^{\lambda\tau}_{\ell(M,S)}(\vec{\omega}) = \sum_{n,m} (\alpha^{n}_{\ell(M)})^{\star}\, D^{nm}_{\ell}(\vec{\omega})\, \alpha^{m\tau}_{\ell(S)} \quad . \tag{2.6c}$$

The rotator function[2] $U^{\lambda\tau}_{\ell(M,S)}$ depends on the symmetry of the molecule and of the site. Only those values of $\ell$ occur in Eq. (2.5) for which the list of irreducible representations of the molecular group M contains the identity representation $\Gamma_{o}(M)$. The use of rotator functions for the analysis of orientationally disordered structures was shown in Ref. 3,4, on the basis of cubic (site) rotator functions[6]. Recently a group theoretical generalization has been given in Ref. 8.

The potential between two molecules at center of mass positions $\vec{R}(\vec{r})$ and $\vec{R}(\vec{r}\,')$ is a sum of atom-atom potentials:

$$V(\vec{r},\vec{r}\,') = \sum_{\nu,\nu'} V(\vec{R}(\vec{r})-\vec{R}(\vec{r}\,');\Omega_{\nu}(\vec{r}),\Omega_{\nu'}(\vec{r}\,')) \quad . \tag{2.7}$$

Here $\vec{r} = (\vec{n},\kappa)$ labels the lattice position $\vec{n}$ of the cell and the position $\kappa$ within the cell. Separating $\vec{R}$ in an equilibrium position $\vec{X}$ and a displacement $\vec{u}$

$$\vec{R}(\vec{n},\kappa) = \vec{X}(\vec{n},\kappa) + \vec{u}(\vec{n},\kappa) \quad , \tag{2.8}$$

we expand the r.h.s. of Eq. (2.7) and obtain

$$V(\vec{r},\vec{r}\,') = \sum_{p} \sum_{\nu,\nu'} \frac{1}{p!} V^{(p)}_{i_1 \ldots i_p}(\vec{r},\vec{r}\,';\nu,\nu')[u_{i_1}(\vec{r})-u_{i_1}(\vec{r}\,')] \times \ldots$$

$$\times [u_{i_p}(\vec{r}) - u_{i_p}(\vec{r}')] \,, \tag{2.9}$$

where $V^{(P)}$ stands for the p-th derivative (p = o, 1, 2...) taken at equilibrium positions. In Eq. (2.9), summation is understood over the repeated Cartesian indices $i_1 \ldots i_p$. Next the coefficients $V^{(P)}$ which depend on the orientation of the molecules are expanded in terms of site symmetry adapted functions. Using Eq. (2.6a), we then rewrite (2.9) as [9]

$$V(\vec{r},\vec{r}') = \sum_p \frac{1}{p!} \sum_{LL'} v^{(p)}_{i_1 \ldots i_p} \left(\begin{smallmatrix} L & L' \\ \vec{r} & \vec{r}' \end{smallmatrix}\right) b^{L}(\vec{r})\, b^{L'}(\vec{r}') \times$$

$$\times [u_{i_1}(\vec{r}) - u_{i_1}(\vec{r}')] \times \ldots [u_{i_p}(\vec{r}) - u_{i_p}(\vec{r}')] \,. \tag{2.10}$$

Here we have used the notation $L \equiv (\ell,\tau)$, $b^{L}(\vec{r})$ stands for $b^{\tau}_{\ell}(\omega(\vec{n},\kappa))$. The coefficients $v^{(p)}$ are given by

$$v^{(p)}_{i_1 \ldots i_p} \left(\begin{smallmatrix} L & L' \\ \vec{r} & \vec{r}' \end{smallmatrix}\right) = \int d\Omega_{\nu} \int d\Omega_{\nu'}\, V^{(p)}_{i_1 \ldots i_p}(\vec{r},\vec{r}';\nu,\nu') \times$$

$$\times S^{L}_{(S)}(\Omega_{\nu})\, S^{L'}_{(S)}(\Omega_{\nu'}) \,. \tag{2.11}$$

Of particular interest are special cases of the general expression (2.10). If p = o,

$$V^{RR}(\vec{r},\vec{r}') = \sum_{L\,L'} v^{(o)} \left(\begin{smallmatrix} L & L' \\ \vec{r} & \vec{r}' \end{smallmatrix}\right) b^{L}(\vec{r})\, b^{L}(\vec{r}') \,, \tag{2.12}$$

describes the orientational interaction of two molecules on a rigid lattice. If in addition one of the molecules is approximated by a sphere or happens to be a single atom, $\ell$ = o, and the corresponding function $S^{o}$ or $b^{o}$ is a constant. The crystal field potential $V^{R}$ on a molecule at site r is obtained by summation over the surrounding particles taken in spherical approximation:

$$V^{R}(\vec{r}) = \sum_{r'} \sum_{L_o} v^{(o)} \left(\begin{smallmatrix} L_o & \ell'=o \\ \vec{r} & \vec{r}' \end{smallmatrix}\right) b^{L_o}(\vec{r}) \,. \tag{2.13}$$

Only those values $L_o$ of $L \equiv (\ell,\Gamma,\beta)$ occur which belong to the unit representation of the point group of site: $L_o \equiv (\ell_o, \Gamma_o, 1)$. As an example we quote $CD_4$ in the plastic phase [2] [10] where the crystal field is expanded in terms of cubic rotator functions.

Of particular interest is the bilinear translation-rotation coupling term, which results from the case p = 1 if one of the partners has spherical symmetry ($\ell' = o$)

$$V^{TR}(\vec{r},\vec{r}') = \sum_L \sum_i v_i^{(1)}\left(\begin{smallmatrix} L & o \\ \vec{r} & \vec{r}' \end{smallmatrix}\right) b^L(\vec{r}) \quad [u_i(\vec{r}) - u_i(\vec{r}')] . \tag{2.14}$$

Finally, in case p = 2, we see that the term with $\ell = o$, $\ell' = o$ contributes to the harmonic translation-translation interaction:

$$V^{TT}(\vec{r},\vec{r}') = \sum_{i,j} v_{ij}^{(2)}\left(\begin{smallmatrix} o & o \\ \vec{r} & \vec{r}' \end{smallmatrix}\right) [u_i(\vec{r})-u_i(\vec{r}')][u_j(\vec{r})-u_j(\vec{r}')] . \tag{2.15}$$

On the other hand, if $\ell \neq o$, $\ell' = o$, we obtain a correction to (2.14) of the form

$$V^{TTR}(\vec{r},\vec{r}') = \sum_L \sum_{i,j} v_{ij}^{(2)}\left(\begin{smallmatrix} L & o \\ \vec{r} & \vec{r}' \end{smallmatrix}\right) b^L(\vec{r})[u_i(\vec{r})-u_i(\vec{r}')][u_j(\vec{r})-u_j(\vec{r}')]. \tag{2.16}$$

The interaction coefficients $v^{(p)}$ depend on the symmetry of the site and on the symmetry of the molecule. In particular this symmetry determines the nature of the bilinear translation-rotation coupling:

$$v_i^{(1)}\left(\begin{smallmatrix} L & o \\ \vec{r}, & \vec{r}' \end{smallmatrix}\right) = \int d\Omega_\nu \; V_i^{(1)}(\vec{r},\vec{r}';\Omega_\nu) \; S_{(S)}^L(\Omega_\nu) \quad . \tag{2.17}$$

Using relation (2.2) and the transformation property of $Y_\ell^m(\Omega)$ under spatial inversion assuming that the center of mass of the molecule is located at a center of inversion in the crystal, we find

$$v_i^{(1)}\left(\begin{smallmatrix} \ell\tau, & o \\ \vec{r}, & -\vec{r}' \end{smallmatrix}\right) = (-1)^{\ell+1} \; v_i^{(1)}\left(\begin{smallmatrix} \ell\tau, & o \\ \vec{r}, & \vec{r}' \end{smallmatrix}\right) \quad . \tag{2.18}$$

If the molecule is centrosymmetric, $\ell$ is even and $v_i^{(1)}$ changes sign under inversion. On the other hand, if the molecule is not centrosymmetric and odd values of $\ell$ are essential, $v_i^{(1)}$ is invariant under inversion. In the next chapter we shall study two real crystals, KCN and $NaNO_2$, and investigate the consequences of property (2.18).

The formulation of orientational properties in terms of symmetry adapted functions becomes particular simple in the case of linear molecules. Then the polar angles $(\Theta,\varphi) \equiv \Omega$ are sufficient to describe the molecular orientations in the crystal fixed frame. Another simplification occurs in strong crystalline fields where there is only one fixed axis for molecular reorientations.

TRANSLATION-ROTATION COUPLING

We want to describe two different systems, KCN and $NaNO_2$ within a same theoretical framework. KCN is a model system for the study of translation-rotation coupling[11]. In the high temperature phase above $T_c$ = 168K, the structure is cubic Fm3m, the CN "molecules" are orientationally disordered[12]. Below the transition, the structure is orthorhombic with a preferential orientation of the CN molecules but without electrical order. Below 83K, the system is antiferroelectric. Above the 168K transition, the structural change is announced by an anomalous softening of the elastic constants[13] or corresponding phonons[14]. The 168K phase transition can be explained by a model[15] based on translation-rotation coupling where the CN molecule is approximated by a symmetric dumbbell (symmetry $D_{\infty h}$). On the other hand, $NaNO_2$ has also an orientationally disordered crystal phase (structure Immm). The $NO_2$ "molecules" (symmetry $C_{2v}$) occupy with equal probability two sterically different positions. Reorientations occur essentially by rotations around the $\vec{c}$-axis[16] [17] [18]. At 165 °C, there occurs a transition to an incommensurate phase[19]. Below 163°C, the $NO_2$ molecules have a preferential orientation, the crystal is ferroelectric[19].

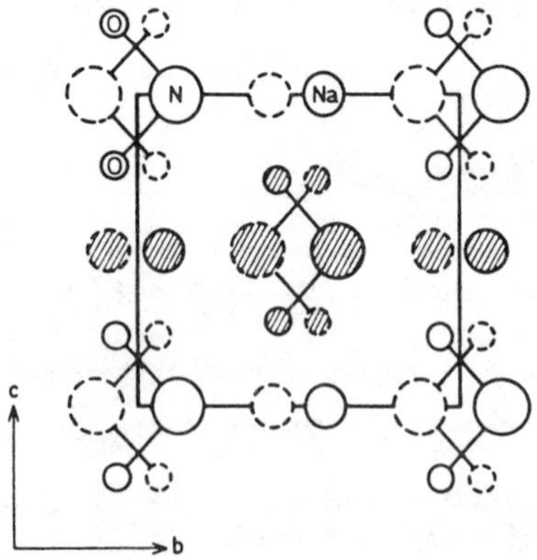


Fig. 1. Average structure of $NaNO_2$ in paraelectric phase projected on (bc) plane.

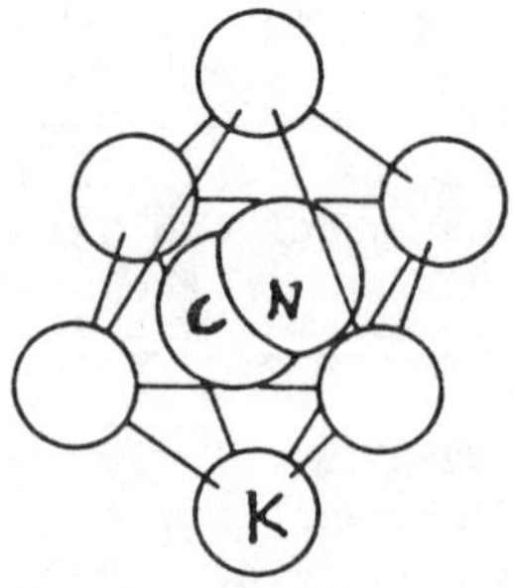


Fig. 2. CN molecular ion surrounded by six K atoms.

In both systems, the equilibrium positions $\vec{X}(\vec{n})$ of the molecular centers of mass are symmetry centers, surrounded by six nearest single atoms in position

$$\vec{X}(\vec{n},\kappa) = \vec{X}(\vec{n}) + \vec{\tau}(\kappa) \quad , \quad \kappa = 1 - 6 . \tag{3.1}$$

The vectors $\vec{\tau}(\kappa)$ depend on the lattice structure (see Ref. 15 for KCN and Ref. 16 for $NaNO_2$). Since the molecular site is a center of symmetry, we can choose $\vec{\tau}(1) = -\vec{\tau}(4)$, $\vec{\tau}(2) = -\vec{\tau}(5)$, $\vec{\tau}(3) = -\vec{\tau}(6)$. The actual position of the $\kappa$-th atom surrounding the $\vec{n}$-th molecule reads

$$\vec{R}^A(\vec{n},\kappa) = \vec{X}(\vec{n},\kappa) + \vec{u}^A(\vec{n},\kappa) \quad , \tag{3.2a}$$

where $\vec{u}^A(\vec{n},\kappa)$ is the atomic lattice displacement. The $\nu$-th atom belonging to the $\vec{n}$-th molecule (taken as a rigid body) has the actual position

$$\vec{R}^\nu(\vec{n}) = \vec{X}(\vec{n}) + \vec{d}^\nu(\Omega(\vec{n})) + \vec{u}^C(\vec{n}) \quad . \tag{3.2b}$$

Here $\vec{u}^C(\vec{n})$ is the center of mass displacement. The vector $\vec{d}^\nu$ specifies the position of the $\nu$-atom in the rigid molecule with respect to the molecular center of mass, $\Omega$ is the molecular orientation with respect to the crystal fixed frame. Since the CN molecule is linear, $\Omega = (\Theta,\varphi)$. On the other hand, in $NaNO_2$ we assume that reorientations about the $\vec{c}$ axis are dominant. The molecular orientation is then specified by $\Omega = (\Theta = \pi/2,\varphi)$ where $\varphi$ specifies the position of the N-atom in the (ab) plane. We take the system of axes XYZ in coincidence with the crystallographic axes $\vec{a}$, $\vec{b}$, $\vec{c}$. In KCN, (XYZ) coincide with the cubic axes. The interaction of the $\vec{n}$-th molecule with atom A in position $(\vec{n},\kappa)$ is given by

$$V(\vec{n},\kappa) = \sum_\nu V^\nu(\vec{n},\kappa) \quad , \tag{3.3a}$$

with the atom-atom potential

$$V^\nu(\vec{n},\kappa) = V^\nu(|\vec{R}^\nu(\vec{n}) - \vec{R}^A(\vec{n},\kappa)|) \quad . \tag{3.3b}$$

As an example we quote a sterical hindrance potential[15] [16] of Born-Mayer type or an electrostatic multipole potential[20]. Expanding the r.h.s. of Eq. (3.3b) in terms of lattice displacements, expanding subsequently in terms of symmetry adapted functions (see Eqs. (2.9) - (2.11)), and summing finally over all molecules $\vec{n}$ and surrounding atoms $(\vec{n},\kappa)$ of the crystal, we obtain

$$V = V^R + V^{TR} + \ldots \quad , \tag{3.4}$$

where $V^R$ is a sum of single particle potentials

$$V^R = \sum_{\vec{n}} \alpha_{\lambda_o} S_{\lambda_o}(\Omega(\vec{n})) , \tag{3.5}$$

and where $V^{TR}$ is the translation-rotation-coupling

$$V^{TR} = \sum_{\vec{n}} \sum_{\kappa} v_{\lambda i}^{(1)}(\kappa) \; S_\lambda(\Omega(\vec{n})) \; [u_i^C(\vec{n}) - u_i^A(\vec{n},\kappa)] . \tag{3.6}$$

Here

$$\alpha_{\lambda_o} = \sum_{\kappa} \sum_{\nu} v_{\lambda_o}^{(o)\nu}(\kappa) , \tag{3.7a}$$

$$v_{\lambda i}^{(1)}(\kappa) = \sum_{\nu} v_{\lambda i}^{(1)\nu}(\kappa) , \tag{3.8a}$$

with

$$v_{\lambda}^{(o)\nu}(\kappa) = \int d\Omega \; S_\lambda(\Omega) \; v^{(o)\nu}(\Omega,\kappa) , \tag{3.7b}$$

$$v_{\lambda i}^{(1)\nu}(\kappa) = \int d\Omega \; S_\lambda(\Omega) \; v_i^{(1)\nu}(\Omega,\kappa) . \tag{3.8b}$$

Here $V^{(o)\nu A}$ and $V_i^{(1)\nu A}$ correspond to the potential (3.3b) and its first derivative respectively, taken at equilibrium lattice positions. The single particle potential has full site symmetry (Sect. II). In KCN the functions $S_{\lambda_o}$ are the cubic surface harmonics $K_{41}$ etc. with $A_{1g}$ symmetry. In $NaNO_2$, $S_{\lambda_o}$ corresponds to $\cos m\varphi$, m even. In strong crystal field, $\varphi$ is restricted to $\pi/2$, or $3\pi/2$ and both directions along the $\vec{b}$-axis are equivalent. On the other hand, lattice displacements lead in general to a reduction of symmetry at the molecular site. In KCN the lowest value of $\ell$ in agreement with the molecular symmetry $D_{\infty h}$ is 2. The functions $S_2^\tau$ then fall in two irreducible representations of the cubic group, with $\Gamma = E_g$ corresponding to a doublet and $\Gamma = T_{2g}$ to a triplet. In case of $NaNO_2$, the corresponding functions $S_\lambda$ is $Y_1^{1,s} = -i(Y_1^1 - Y_1^{-1})/\sqrt{2}$ with $\Theta = \pi/2$, or equivalently $Y_1^{1,s}(\pi/2,\varphi) \propto \sin\varphi$. This function is uneven with respect to a two fold rotation about the $\vec{c}$-axis, it belongs to the $B_u$ representation and plays the role of an order parameter[16]. Applying the symmetry rule (2.18) we find

$$v_{\lambda i}^{(1)\nu}(\kappa) = \mp \, v_{\lambda i}^{(1)\nu}(\kappa+3) \; , \; \kappa = 1, 2, 3. \tag{3.9}$$

Here the - sign refers to KCN ($\ell$ even) and the + sign to $NaNO_2$ ($\ell$ odd). The same property holds for expression (3.8a). The relevance of property (3.9) becomes most evident if we study the coupling to long wavelength lattice displacements. Following Ref. 15, we find for the interaction of rotations with acoustic phonons:

$$V_{ac}^{TR} = \sum_{\vec{k}} \hat{v}_{\lambda i}(\vec{k}) \; S_{\lambda}(-\vec{k}) \; u_i(\vec{k}) \; . \tag{3.10}$$

Here we have defined Fourier transforms of rotator functions

$$S_{\lambda}(\vec{k}) = \frac{1}{\sqrt{N}} \sum_{\vec{n}} S_{\lambda}(\Omega(\vec{n})) \, e^{i\vec{k}.\vec{X}(\vec{n})} \tag{3.11}$$

and of acoustic displacements $u_i(\vec{k})$. The latter are obtained from a linear combination of $u^C$ and $u^A$ (two particles per primitive unit cell, mass $m = m_A + m_c$). The coupling $\hat{v}_{\lambda i}(\vec{k})$ reads

$$\hat{v}_{\lambda i}(\vec{k}) = \frac{1}{m} [ v_{\lambda i}^{(1)}(\vec{k} = o) - v_{\lambda i}^{(1)}(\vec{k})] \; , \tag{3.12a}$$

with

$$v_{\lambda i}^{(1)}(k) = \sum_{\kappa} v_{\lambda i}^{(1)}(\kappa) \, e^{i\vec{k}.\vec{\tau}(\kappa)} \; . \tag{3.12b}$$

Combining Eqs. (3.9), (3.8a) with (3.12b), (3.12a) we find

$$\hat{v}_{\lambda i}(\vec{k}) = \frac{-i}{m} \sum_{\kappa} v_{\lambda i}^{(1)}(\kappa) \, \sin(\vec{k}.\vec{\tau}(\kappa)), \; \ell \; \text{even} \tag{3.13a}$$

or

$$\hat{v}_{\lambda i}(\vec{k}) = \frac{2}{m} \sum_{\kappa} v_{\lambda i}^{(1)}(\kappa) \, \sin^2(\frac{\vec{k}.\vec{\tau}(\kappa)}{2}), \; \ell \; \text{odd.} \tag{3.13b}$$

Consequently for long wavelength, the coupling of centrosymmetric molecules to acoustic lattice displacements is linear[15] in $\vec{k}$ while for non centrosymmetric molecules (like $NO_2$), it is quadratic[21] in $\vec{k}$. Since a factor $\vec{k}$ in Fourier space corresponds to a spatial gradient $\vec{\nabla}$ in real space, and since $\nabla_i u_j = \varepsilon_{ij}$, we find that $V^{TR}$ reduces to a coupling of the orientational order to uniform lattice strain in the case of a symmetric molecule. On the other hand, in the case of a non centrosymmetric molecule, $V^{TR} \propto (\vec{\nabla}S)(\vec{\nabla}u)$, i.e. we obtain a coupling of lattice strain to the gradient of the orientational order parameter. In the first case ($\ell$ even), the coupling gives rise to thermoelastic anomalies[15] (softening of elastic constant), while in the second case ($\ell$ uneven), it is relevant for the occurence of an incommensurate phase transition[22][21].

Starting from Eq. (3.6), one can also study the coupling of rotational motion to optical phonons[16]. The coupling $v^{o}_{\lambda i}(k)$ is now given by

$$v^{o(1)}_{\lambda i}(\vec{k}) = \sqrt{\mu}\,[\frac{1}{m_c}\, v^{(1)}_{\lambda i}(\vec{k} = o) + \frac{1}{m_A}\, v^{(1)}_{\lambda i}(\vec{k})] \ . \qquad (3.14)$$

Here $\mu$ is the reduced mass $\mu = (m_A\, m_c)/m$.
From Eqs. (3.12b) and (3.9) it follows that for the case $\ell$ even, $v^{o(1)}(k) \propto \vec{k}$, i.e. vanishes at the zone center. On the other hand, for the case $\ell$ odd, $v^{(1)}_{\lambda i}(\vec{k} = o) \neq 0$ and there exists a non zero coupling to optical phonons at the zone center. In the case of non centrosymmetric molecules, optical measurements give direct information on the microscopic coupling coefficients $v^{(1)}_{\lambda i}(\kappa)$. On the other hand, neutron or X-ray experiments allow us to study the coupling to acoustic phonons at finite $\vec{k}$. From Eq. (3.13b) we see that these experimental methods give information about the same coefficient $v^{(1)}_{\lambda i}(\kappa)$. We mention that recent molecular dynamics computer calculations on[23] KCN and [18] $NaNO_2$ are to a large extent in agreement with the present theory.

ANHARMONICITIES AND HARD CORE INTERACTIONS

As a consequence of sterical hindrance, large amplitude flucuations, which are characteristic of ODIC phases, imply large translational displacements. There is ample evidence from experiment[24] and molecular dynamics[23]. These effects are most important near structural phase transitions. One then has to extend the series expansion (2.9a) or (3.4) to infinite order in the lattice displacements and to use methods that are known from the theory of quantum crystals. Here we follow Ref. 25.

From the theory of strongly anharmonic crystals[26], we know that a partial summation of the infinite series (2.9) results in a replacement of the coupling coefficients $V^{(p)}$ by the corresponding renormalized quantities. In particular $V^{(o)}$ and $V^{(1)}$ in Eqs. (3.7b) and (3.8b) become:

$$\bar{V}^{(p)\nu}_{i_p}(\vec{n},\kappa) = \int d\vec{r}\; g(\vec{n},\kappa;\vec{r})\; V^{(p)\nu}_{i_p}(\vec{n},\vec{r}) \ . \qquad (4.1)$$

where p = o or 1. The pair distribution function g refers to the center of mass position of the molecule and to the position $(\vec{n},\kappa)$ of the atom A. Explicitly it reads

$$g(\vec{n},\kappa;\vec{r}) = A(\vec{n},\kappa)\gamma(\vec{n},\vec{r})\; f(\kappa,\vec{r}) \ . \qquad (4.2)$$

Here $A(\vec{n},\kappa)$ is a normalization factor, given by

$$A(\vec{n},\kappa) = [\int d\vec{r}\ \gamma(\vec{n},\vec{r})\ f(\kappa,\vec{r})]^{-1} \quad . \tag{4.3a}$$

The thermal part is a Gaussian

$$f(\kappa,\vec{r}) = \exp\{-\frac{1}{2}[(\vec{\tau}(\kappa) - \vec{r}).\underline{\Delta}^{-1}.(\vec{\tau}(\kappa)-\vec{r})]\} , \tag{4.3b}$$

the width of which is determined by the displacement-displacement correlation

$$\Delta_{ij}(\kappa) = <[u_i^C(\vec{n}) - u_i^A(\vec{n},\kappa)][u_j^C(\vec{n})-u_j^A(\vec{n},\kappa)]>. \tag{4.4}$$

The thermal average is taken over the canonical ensemble of the crystal. The Jastrow factor $\gamma$ in Eq. (4.2) suppresses fluctuations near the hard core[27]. Since the sterical hindrance depends on the instantaneous molecular orientation and on the shape of the molecule, we choose

$$\gamma(\vec{n},\vec{r}) = \exp[-\beta \sum_\nu V^{(o)\nu}(\Omega(\vec{n}),\vec{r})] , \tag{4.3c}$$

where $\beta = (k_B T)^{-1}$.

For the moment we assume $\underline{\Delta}$ known (see later). Then Eq. (4.2) determines $g(\vec{n},\kappa;\vec{r})$ for a given molecular orientation $\Omega(\vec{n})$. Therefore the vertices $\overline{V}^{(p)}$, Eq. (4.1) depend on the molecular orientation. Expanding in terms of symmetry adapted functions, we calculate the coupling coefficients according to (3.7b, 8b) with $V^{(p)}$ replaced by $\overline{V}^{(p)}$. Finally we obtain all results of the previous section where now the coefficients $v_{\lambda i}^{(1)}(\kappa)$ and $v^{(o)\nu}(\kappa)$ are replaced by the corresponding renormalized quantities $\bar{v}^{(1)}$, $\bar{v}^{(o)\nu}$ or $\bar{\alpha}_{\lambda_o}$. The symmetry relations (3.9) remain valid for the renormalized coefficients. We quote the results

$$\bar{v}_{\lambda i_p}^{(p)\nu}(\kappa) = \int d\Omega S_\lambda(\Omega)\ \overline{V}_{i_p}^{(p)\nu}(\Omega,\kappa) , \tag{4.4}$$

with p = 0, 1;

$$V^R = \sum_{\vec{n}} \bar{\alpha}_{\lambda_o}\ S_{\lambda_o}(\Omega(\vec{n})) , \tag{4.5}$$

$$V^{TR} = \sum_{\vec{n}} \sum_{\kappa} \bar{v}_{\lambda_i}^{(1)}(\kappa)\ S_\lambda(\Omega(\vec{n}))\ [u_i^C(\vec{n}) - u_i^A(\vec{n},\kappa)] . \tag{4.6}$$

We now turn to the calculation of $\underline{\Delta}$, Eq. (4.4). Taking Fourier transforms of the lattice displacements, we obtain

$$\Delta_{ij}(\kappa) = \frac{1}{N}\sum_{\vec{k}} \left\{ \frac{\langle u_i^C(\vec{k})\, u_j^C(-\vec{k})}{m_C} + \frac{\langle u_i^A(\vec{k})\, u_j^A(-\vec{k})\rangle}{m_A} \right.$$

$$\left. - 2\cos(\vec{k}.\vec{\tau}(\kappa)) \frac{\langle u_i^C(\vec{k})\, u_j^A(-\vec{k})\rangle}{\sqrt{m_A\, m_C}} \right\} . \qquad (4.7)$$

The correlation functions on the r.h.s. are elements of the 6 x 6 matrix $\langle u_\rho(\vec{k})\, u_\sigma(-\vec{k})\rangle$, where $u_\rho(\vec{k})$ is the $\rho$-th component of the six dimensional vector

$$(u_\rho(\vec{k})) = (u_x^C(\vec{k}), u_y^C(\vec{k}), u_z^C(\vec{k}), u_x^A(\vec{k}), u_y^A(\vec{k}), u_z^A(\vec{k})) . \qquad (4.8)$$

The translation-rotation interaction (4.6) is rewritten as

$$V^{TR} = \sum_{\vec{k}} \bar{v}_{\lambda\rho}^{(1)}(\vec{k})\; S_\lambda(-\vec{k})\; u_\rho(\vec{k}) \qquad (4.9)$$

where

$$(\bar{v}_{\lambda\rho}^{(1)}(\vec{k})) = \left( \frac{\bar{v}_{\lambda x}^{(1)}(\vec{k}=o)}{\sqrt{mc}}, \ldots_{\lambda y}, \ldots_{\lambda z}, \frac{-\bar{v}_{\lambda x}^{(1)}(\vec{k})}{\sqrt{m_A}}, -\ldots_{\lambda y}, -\ldots_{\lambda z} \right) \qquad (4.10)$$

The correlation functions $\langle u_\rho(\vec{k})\, u_\sigma(-\vec{k})\rangle$ are calculated from an interaction potential

$$V = V^{TT} + V^{TR} + V^{RR} + V^{R} . \qquad (4.11)$$

Here $V^{TT}$ is the displacive (renormalized) harmonic interaction

$$V^{TT} = \frac{1}{2}\sum_{\vec{k}} u_\rho(-\vec{k})\; M_{\rho\sigma}(\vec{k})\, u_\sigma(\vec{k}) , \qquad (4.12)$$

where $M(\vec{k})$ is the bare dynamical matrix. $V^{RR}$ is a direct orientational interaction (e.g. multipole-multipole) of the form:

$$V^{RR} = \frac{1}{2}\sum_{\vec{k}} S_{\lambda_1}(\vec{k})\, J_{\lambda_1\lambda_2}(\vec{k})\; S_{\lambda_2}(-\vec{k}) . \qquad (4.13)$$

We obtain for the displacement-displacement correlation matrix the exact result[15]

$$\langle u_\rho(\vec{k})\, u_\sigma(-\vec{k})\rangle = k_B T\, (\underline{M}^{-1}(\vec{k}))_{\rho\nu} [\, \underline{1} + \underline{\bar{v}}^{(1)\tau}(-\vec{k})\, \underline{\chi}(\vec{k})\, \underline{\bar{v}}^{(1)}(\vec{k})\, \underline{M}^{-1}(\vec{k})]_{\nu\sigma} \,. \tag{4.14}$$

In the absence of coupling, $\langle uu\rangle_o = k_B T\, M^{-1}$. The second term within brackets on the r.h.s. becomes dominant at an orientational phase transition, where some elements of the orientational susceptibility $\underline{\chi}(\vec{k}) = \langle S(-k)S(k)\rangle\beta$ diverge at $T = T_o$ and $\vec{k} = \vec{k}_o$. Explicitly one finds

$$\underline{\chi}(\vec{k}) = \{\underline{1} - [\underline{K}(\vec{k}) - \underline{C}^S]\, \underline{\chi}^o\}^{-1}\, \underline{\chi}^o \ , \tag{4.15}$$

where

$$\underline{K}(\vec{k}) = \underline{C}(\vec{k}) + \underline{J}(\vec{k}) \ , \tag{4.16a}$$

$$\underline{C}^S = \frac{1}{N} \sum_{\vec{k}} \underline{C}(\vec{k}) \ , \tag{4.16b}$$

with

$$\underline{C}(\vec{k}) = \underline{\bar{v}}^{(1)\tau}(-\vec{k})\ \underline{M}^{-1}(\vec{k})\ \underline{\bar{v}}^{(1)}(\vec{k}) \ . \tag{4.16c}$$

Here $\underline{C}(\vec{k})$ stands for the lattice mediated interaction, $C^S$ represents a correction due to self interaction[28] and $\chi^o$ denotes the single particle orientational susceptibility:

$$\chi^o_{\lambda\lambda} = \beta y_{\lambda\lambda} = \beta Z_o^{-1} \int d\Omega\, S_\lambda^2(\Omega)\, \exp[-\beta V_s^R] \ , \tag{4.17a}$$

and

$$V_s^R = V^R - \sum_\lambda C^S_{\lambda\lambda} S_\lambda^2(\Omega) \ , \tag{4.17b}$$

and

$$Z_o = \int d\Omega\, \exp[-\beta V_s^R] \ . \tag{4.17c}$$

Here the second term on the r.h.s. of Eq. (4.17b) represents the self energy correction to the single particle potential. This correction is essential for a comparison of a theoretical model with experimental data of the single particle orientational distribution (J.M. Rowe).

Since $\underline{\Delta}(\kappa)$ is a quantity in real space, divergences in $\langle u(\vec{k})u(-\vec{k})\rangle$, Eq. (4.14), are smoothened out by the $\vec{k}$-sum on the r.h.s. of Eq. (4.7). This is well known from the study of local properties at phase transitions[30][31]. In general one expects an

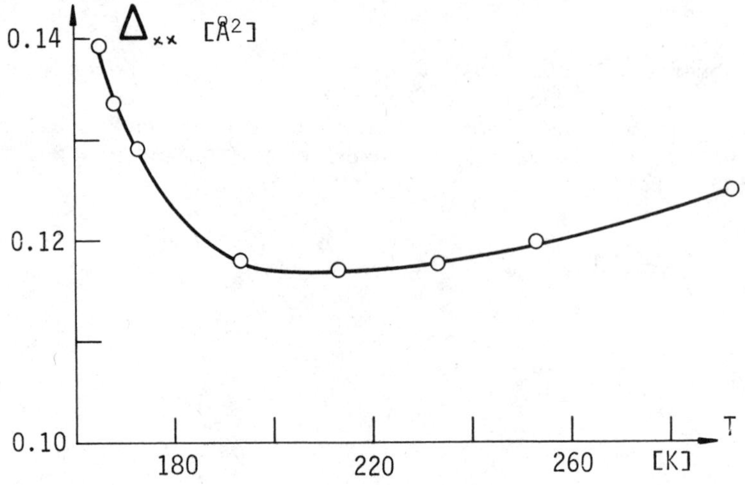


Fig. 3.

increase of $\underline{\Delta}$ of $T^+ \to T_o$ in the disordered phase[31]. J.M. Rowe and the author have calculated numerically $\underline{\Delta}$ on the basis of Eqs. (4.7)-(4.16c) for the ODIC phase of KCN:

The result is in qualitative agreement with experiment[24][12][32]. The temperature variation of $\Delta$ has important consequences. An increase in $\underline{\Delta}(T)$ with T approaching $T_o$ leads to a temperature dependence of the pair distribution function (4.3b). With increasing width, the domain of integration in Eq. (4.1) for the renormalized potentials is extended. On the other hand, the Jastrow factor $\gamma$ in Eq. (4.2) excludes the region close to the hard core, regions of space where the sterical hindrance is weaker, are favored in the integration. As a consequence, the strength of the single particle potential and of the translation rotation coupling diminishes in absolute value for $T^+ \to T_o$. Similarly the lattice mediated interaction decreases, if in Eq. (4.16c) we use the renormalized coupling, and in turn this effect influences the orientational susceptibility. As a result one obtains a set of coupled non linear equation[25] which should be solved selfconsistently by numerical methods.

Our conclusions are in qualitative agreement with several experimental results. We first consider the alkalicyanides. From the decrease of the strength of the single particle potential, it follows that the averaged orientational probability distribution function $\langle f(\Omega) \rangle$ becomes more isotropic near $T_o$. Such an effect has indeed been measured by neutron scattering[12][33].

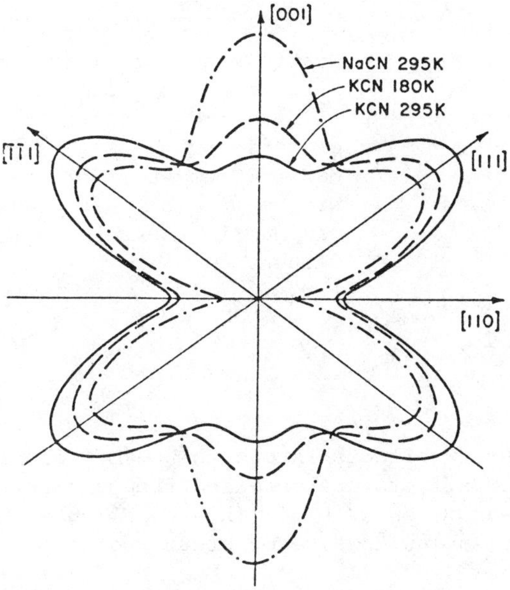


Fig. 4. Angular distribution of CN in the [110] plane of the crystal. Neutron diffraction results from Ref. 12.

On the other hand, F. Lüty[11] has pointed out the empirical fact that near the respective order-disorder transitions in different alkalicyanides NaCN, KCN, RbCN, the critical reorientation rates have the same numerical value. This fact has also been found by NMR results[34]. Assuming an isotropic single particle potential, we have calculated the reorientation rates and find[25] numerical values close to experiment.

A discussion of the transition to the incommensurate phase in $NaNO_2$ has been given in Ref. 21 on the basis of Eq. (4.11). Microscopic theory confirms the conclusions about the existence of an incommensurate phase which were drawn from a phenomenological free energy[22]. In addition microscopic theory allows us to calculate explicitly the relevant coefficients. In particular, since $\ell$ is odd, one finds via Eq. (3.13b) and (4.16c) that the acoustic phonon mediated interaction $C(\vec{k})$ is quadratic in $\vec{k}$. In order to discuss the divergence of the susceptibility, we have to study the denominator of Eq. (4.15) as a function of temperature and wave vector. For a given $T > T_o$, $\chi(\vec{k},T)$ is maximum at a wave vector[21]

$$k_m(T) \propto \left|\bar{v}_{\lambda i}(\kappa)\right| . \tag{4.18}$$

We have seen that the renormalization (4.1) - (4.4) leads to a temperature dependence of the coupling such that $|\bar{v}(\kappa)|$ decreases with $T \to T_o$. We propose that the observed shift of $k_m$ to lower values[35, 36] with decreasing temperature in the disordered phase of $NaNO_2$ is due to this effect.

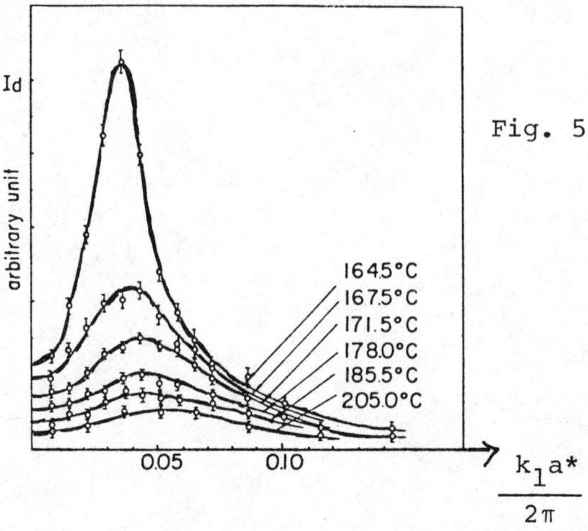


Fig. 5. Static susceptibility X-ray results along $a^*$ direction in para-electric phase of $NaNO_2$ (From Ref. 35). Notice the shift of maximum to lower $\vec{k}$ values.

Notice that these observed precursor effects extend to 30-40°C above $T_o \approx 165$ C. Additional precursor effects are obtained from dielectr. susceptibility measurements. Measurements[37] of $\chi(\vec{k} = o, T)$ show large deviations from a Curie-Weiss law in a temperature region extending to 40°C above $T_o$:

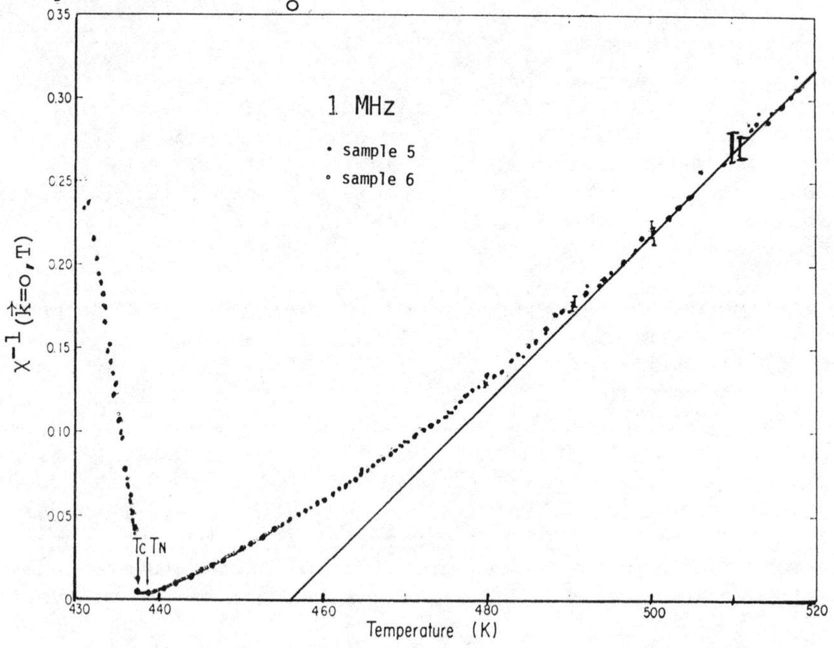


Fig. 6

Examining now the coupling $C(\vec{k}=o)$ mediated by optical phonons, we find by Eqs. (3.14) and (4.16c) that

$$(C^o(\vec{k}=o) - C^s) \propto |\bar{v}(\kappa)|^2, \tag{4.19}$$

(under the assumption that $C^{o}(\vec{k}=o) > C^{s}$). Then a decrease of $|\bar{v}(\kappa)|$ with $T \to T_o$ leads to a behavior of

$$\chi^{-1}(\vec{k}=o) \propto \{T - J(\vec{k}=o) - [C^{o}(\vec{k}=o) - C^{s}]\} , \tag{4.20}$$

in qualitative agreement with experiment. We conclude that the present theory leads to a consistent explanation for both acoustic and optic precursor effects in $NaNO_2$.

DYNAMICS EQUATIONS

Inelastic neutron scattering gives information about the space and time behavior of orientation-orientation, displacement-displacement and displacement-orientation correlation function. Here we want to derive dynamic equation for a model with translation-rotation coupling. The Hamiltonian is taken to be

$$H = T + K + V . \tag{5.1}$$

Here V is the potential described in Sect. 4, Eq. (4.11). The kinetic energy of translations is given by

$$T = \sum_{\vec{k}} \frac{p_\rho(-\vec{k})p_\rho(\vec{k})}{2} , \tag{5.2a}$$

where $p_\rho(\vec{k})$ are the momenta conjugate to the displacements $u_\rho(\vec{k})$. The notation is the same as in Eq. (4.8). The kinetic energy of rotation reads

$$K = \sum_{\vec{k}} \frac{\vec{L}(-\vec{k}).\vec{L}(\vec{k})}{2I} . \tag{5.2b}$$

We restrict ourselves here to one type of molecule with moment of inertia I. We choose secular variables

$$\{A(\vec{k})\} = \{u_\rho(\vec{k}), p_\rho(\vec{k}), \bar{S}_\lambda(\vec{k})\} . \tag{5.3}$$

Here $\bar{S}_\lambda$ is the orthogonalized rotation defined by

$$\bar{S}_\lambda(\vec{k}) = S_\lambda(\vec{k}) - u_\rho(\vec{k})\ D_{\rho\sigma}(\vec{k})\ (u_\rho(\vec{k}), S_\lambda(\vec{k})) . \tag{5.4}$$

Static susceptibilities are defined in classical statistical mechanics by

$$(A(\vec{k}), B(\vec{k})) = \langle A(-\vec{k})B(\vec{k})\rangle/k_BT. \tag{5.5a}$$

The orientational susceptibility (4.15) corresponds to

$$\chi_{\lambda\nu}(\vec{k}) = (S_\lambda(\vec{k}), S_\nu(\vec{k})) , \tag{5.5b}$$

and D is the inverse of the static displacement-displacement susceptibility

$$(D^{-1}(k))_{\rho\sigma} \equiv \langle u_\rho(-\vec{k})u_\sigma(\vec{k})\rangle/k_BT , \tag{5.5c}$$

which is directly obtained from Eq. (4.14). $D \to o$ for $T \to T_o$, $\vec{k} \to \vec{k}_o$.

The dynamic correlation functions which enter the scattering laws are related to the imaginary part of Kubo's relaxation function $\Phi$ by[38]

$$S_{AB}(\vec{k},\omega) = -\omega\,[1 - \exp(-\omega/k_BT)]^{-1}\,\Phi''_{AB}(k,\omega) . \tag{5.6}$$

The relaxation function $\Phi$ is calculated by using the More-Zwanzig projection operator method[39]. Starting from the secular variables (5.4), we obtain the coupled matrix equation[40].

$$\begin{pmatrix} z\underline{1} & -i\underline{1} & \underline{o} \\ i\underline{D} & z\underline{1}+\underline{\Pi} & -\beta X \\ o & -\underline{\beta}^{T} & z\underline{1}+\underline{\Lambda}\;\underline{X} \end{pmatrix} \begin{pmatrix} \underline{\Phi}_{uu} & \underline{\Phi}_{up} & \underline{\Phi}_{u\bar{S}} \\ \underline{\Phi}_{pu} & \underline{\Phi}_{pp} & \underline{\Phi}_{p\bar{S}} \\ \underline{\Phi}_{\bar{S}u} & \underline{\Phi}_{\bar{S}p} & \underline{\Phi}_{\bar{S}\bar{S}} \end{pmatrix} = $$

$$= \begin{pmatrix} \underline{D}^{-1} & \underline{o} & \underline{o} \\ \underline{o} & \underline{1} & \underline{o} \\ \underline{o} & \underline{o} & \underline{X}^{-1} \end{pmatrix} . \tag{5.7}$$

Here $z = \omega + i\varepsilon$, $\varepsilon \to o$, all functions $\underline{\Phi}$, $\underline{\Pi}$, $\underline{\Lambda}$ depend on z and $\vec{k}$. $\underline{X}$ is the inverse bare rotational susceptibility

$$(X^{-1})_{\lambda\nu} \equiv (\bar{S}_\lambda(k), \bar{S}_\nu(k)) , \tag{5.8}$$

$\underline{\beta}$ couples orientational and translational variables:

$$\beta_{\rho\lambda}(\vec{k}) \equiv (p_\rho(\vec{k}), \mathcal{L}\bar{S}_\lambda(\vec{k})) = iD_{\rho\sigma}(\vec{k})\,(u_\sigma(\vec{k}), \bar{S}_\lambda(\vec{k})) . \tag{5.9}$$

Here $\mathcal{L}$ denotes the Liouville operator. One finds

$$(u_\rho(\vec{k}), \bar{S}_\lambda(\vec{k})) = -M^{-1}_{\rho\sigma}(\vec{k})\, v_{\sigma\lambda}(\vec{k})\, \chi_{\lambda\mu}(\vec{k}) \, , \tag{5.10}$$

where $\chi_{\lambda\mu}(\vec{k})$ is the rotational susceptibility (4.11). We quote the exact relation[40]:

$$\underline{\beta}\, \underline{X}\, \underline{\beta}^\tau = \underline{M} - \underline{D} \, . \tag{5.11}$$

Using in addition Eqs. (4.10) and (4.11), we get

$$\underline{X}(\vec{k}) = [\,(\underline{\chi}^o)^{-1} + \underline{C}^s - \underline{J}(\vec{k})] \, . \tag{5.12}$$

Notice that $\underline{X}^{-1}$ does not diverge at the phase transition where $\underline{\chi}(\vec{k})$ diverges and $\underline{D}$ vanishes.

The quantities $\underline{\Pi}$ and $\Lambda$ in Eq. (5.7) are memory kernels which account for dissipative processes. $\underline{\Pi}$ plays the role of a translational viscosity:

$$\Pi_{\rho\sigma}(\vec{k},z) = -\left(Q\mathcal{L}p_\rho(\vec{k}), \frac{1}{(z-Q\mathcal{L}Q)}\, Q\mathcal{L}p_\sigma(\vec{k})\right) . \tag{5.13}$$

Here we use the conventions and notation of Ref. 41. The operator Q denotes the projector onto the space of non secular variables. When the Hamiltonian (5.1) contains only bilinear terms in the secular variables, $Q\mathcal{L}\rho=0$, and $\Pi = o$. Here we formally keep $\Pi$ in our equations; in Sect. VI we shall give an expression of on the basis of a cubic term $V^{TTR}$ in the Hamiltonian. The orientational memory kernel is given by

$$\Lambda_{\lambda\nu}(\vec{k},z) = -\left(Q\mathcal{L}\bar{S}_\lambda(\vec{k}), \frac{1}{(z-Q\mathcal{L}Q)}\, Q\mathcal{L}\bar{S}_\lambda(\vec{k})\right) . \tag{5.14}$$

From the first column of Eq. (5.7) we get

$$\{\underline{1}z^2 - \underline{D} + z\underline{\Pi}(z) - z\underline{\beta}\,\underline{X}\,[\,\underline{1}z + \underline{X}\,\underline{\Lambda}\,(z)]^{-1}\,\underline{\beta}^\tau\}\; \underline{\Phi}_{uu}(z) =$$

$$= \{z + \Pi(z) - \underline{\beta}\,\underline{X}\,[\,\underline{1}z + \underline{X}\,\underline{\Lambda}(z)]^{-1}\underline{\beta}^\tau\}D^{-1} \, . \tag{5.15}$$

In a similar way one obtains from the last column of Eq. (5.7)

$$\{\underline{1}z + \underline{X}\,\underline{\Lambda}(z) - z\underline{\beta}^\tau[\,\underline{1}z^2 - \underline{D} + z\underline{\Pi}(z)]^{-1}\underline{\beta}\,\underline{X}\}\; \underline{\Phi}_{\bar{S}\bar{S}}(z) = X^{-1} \tag{5.16}$$

Similar equations can also be derived for the mixed relation functions $\Phi_{\bar{S}u}$, $\Phi_{u\bar{S}}$. Equations of type (5.15), (5.16) have been derived previously for pseudospin-phonon coupled systems[42] [43] and for orientational polarizabilities coupled to lattice displacements[44]. There the transport coefficients were taken as frequency independent constant: $\Pi(z) \to i\eta$, $X\Lambda(z) \to i\lambda$. For the choice of special directions of the wavevector $\vec{k}$, Eqs. (5.15) (and equivalently (5.16)) are diagonal. The resonances are obtained as a solution of a third order equation in $\omega$. One distinguishes two extreme cases:

Fast relaxation, $\lambda > \sqrt{M}$ (equivalently $\omega\tau<1$, $\tau=\lambda^{-1}$) and small viscosity $\eta < D$. The secular equation is quadratic in $\omega$ with roots

$$\omega_{\pm} = \pm \sqrt{D} - \frac{i}{2} (\eta + \frac{\beta^2 X}{\lambda}) \quad . \tag{5.17}$$

The inelastic scattering law corresponds to a Brillouin doublet which softens for $T \to T_o$, $\vec{k} \to \vec{k}_o$ where $D \to o$. (see also last section). It has been measured in KCN by inelastic neutron scattering[45] [46]. Notice that one finds no critical slowing down of orientational relaxation. The situation is different in case of:

Slow relaxation, $\lambda < \sqrt{M}$, $(\omega\tau>1)$. The resonances are

$$\omega_{\pm} = \pm\sqrt{M} - \frac{i}{2} (\eta + \frac{\lambda\beta^2 X}{M}) \quad , \tag{5.18a}$$

$$\omega_o = -i\lambda(1-\beta^2 X/M) \quad . \tag{5.18b}$$

The scattering law consists of a Brillouin doublet with maxima at the bare phonon frequencies $\pm \sqrt{M}$ and of a central peak. While the phonon frequencies are unaffected, the central peak shows critical slowing down near the phase transition. This follows from Eqs. (5.18b), (5.11). By inelastic neutron scattering one has found that the slow relaxation case is realized in $NaNO_2$[47] [36] $ND_4Br$[48] and tanane[49].

In KCN the experimental data could be explained[45] on the basis of Eqs. (5.15) by taking $\eta = o$ and $\hbar\lambda = 3$meV. The assumption of a constant $\lambda$ is justified the orientational relaxation is essentially a single particle process and much faster than the typical acoustic phonons (0.6 meV). Note that even with $\lambda$ = const, $\eta = o$, one concludes from Eqs. (5.17) and (5.11) that the phonon damping increases with $D \to o$ by approaching the phase transition. This is in qualitative agreement with Brillouin results[50]. In the following figures we show inelastic neutron data of KCN, taken from Ref. 45. Note the cross over from the fast relaxation to the slow relaxation regime obtained by changing the experimental momentum transfer:

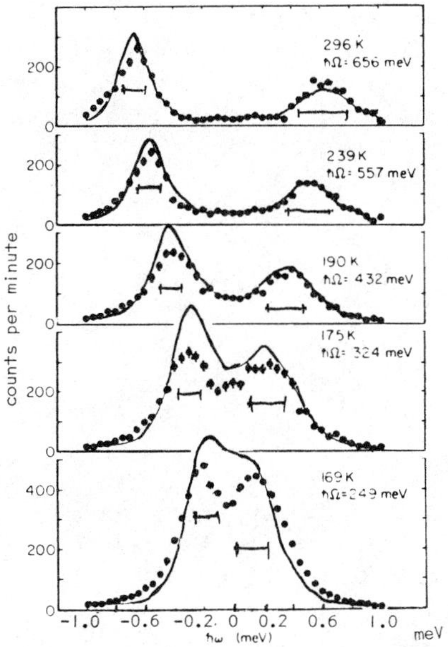


$Q = (1.9, 2.0, 0.0)$

Fig. 7. Neutron data for TA mode of KCN, wave vector $\vec{k}$ equal to $\frac{1}{10}$ of the Brillouin distance.

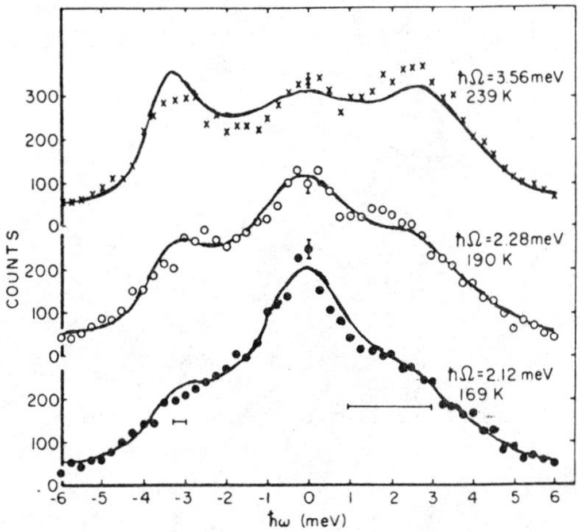


$Q = (-.5, 2.0, 0.0)$

Fig. 8. Neutron data for the same branch with wave vector $\vec{k}$ equal to ½ of the Brillouin zone distance.

In the last part of this section we want to stress the fact that elastic and inelastic scattering laws are most conveniently expressed in terms of symmetry adapted functions. Structural analysis has been applied to the ammoniumhalides[3], solid $CD_4$[4], alkalicyanides[12], adamantane[51], $CBr_4$[52] and other substances. An analysis of ODIC structures including correlations between translations and orientations was developed by Press et al[53]. There the reader finds references to additional work. An application to the study of static critical correlation function was also made for $CD_4$[54]. The use of symmetry adapted functions for the formulation of inelastic neutron scattering laws has been shown in Ref. 55. Recently, a general derivation of elastic and inelastic neutron scattering laws, valid for arbitrary symmetries of the molecule and the site, has been performed on a group theoretical basis[56].

In translation-rotation coupled systems, the inelastic scattering law contains three types of terms: translation-translation, rotation-rotation and in addition mixed translation-rotation dynamic correlation functions. This was first realized in pseudo-spin-phonon coupled systems[48]. A formulation in terms of symmetry adapted functions was given in Ref. 40. As an example, which also shows cleary the advantage of symmetry adapted functions, we quote the coherent inelastic scattering law of a system of dumbbells interacting with acoustic lattice displacements:[57]

$$\frac{d\sigma^2}{d\Omega d\omega} = \frac{q'}{q} \sum_{\vec{k}} \Delta(\vec{Q}+\vec{k}) [ \lambda_o^2 \sum_{\vec{n},\vec{n}'} \exp[ i\vec{Q}.(\vec{X}(\vec{n})-\vec{X}(\vec{n}'))] \delta(\omega) + \tag{5.19}$$

$$+ F^{uu}_{ij}(Q) S^{uu}_{ij}(\vec{k},\omega) + F^{SS}_{\lambda\nu}(\vec{Q}) S^{SS}_{\lambda\nu}(\vec{k},\omega) + 2F^{uS}_{i\lambda} S^{uS}_{i\lambda}(\vec{k},\omega)] \; e^{-2W} \; .$$

Here $\lambda_o$ is the total scattering length per primitive cell for translational motion. $\Delta(\vec{Q}+\vec{k}) = \delta_{\vec{Q}+\vec{k},\vec{K}}$, where $\vec{K}$ is a reciprocal lattice vector. We have written $\exp(-2W)$ for the Debye-Waller factor and introduced the following definitions:

$$F^{uu}_{ij}(\vec{Q}) = \lambda_o^2 Q_i Q_j Nm, \tag{5.20a}$$

$$S^{uu}_{ij}(\vec{k},\omega) = \frac{1}{2\pi} \int dt \; e^{i\omega t} \langle u_i(-\vec{k},t) \; u_j(\vec{k},o)\rangle \; , \tag{5.20b}$$

$$F^{SS}_{\lambda\nu}(\vec{Q}) = 16\pi^2 (\lambda_-)^2 j_2^2(Qd) S_\lambda(\hat{Q}) S_\nu(\hat{Q}) N \; , \tag{5.21a}$$

$$S^{SS}_{\lambda\nu}(\vec{k},\omega) = \frac{1}{2\pi} \int dt e^{i\omega t} \langle S_\lambda(-\vec{k},t) S_\nu(\vec{k},o)\rangle \tag{5.21b}$$

$$F^{uS}_{i\lambda}(\vec{Q}) = i\lambda_- \lambda_o Q_i 4\pi j_2(Qd) S_\lambda(\hat{Q}) N\sqrt{m} \tag{5.22a}$$

$$S^{uS}_{i\lambda}(\vec{k},\omega) = \frac{1}{2\pi} \int dt \; e^{i\omega t} \langle u_i(-\vec{k},t) \; S_\lambda(\vec{k},o)\rangle \tag{5.22b}$$

Here $j_\ell$ denotes the Bessel function, d is the half length of the dumbbell, $\lambda_-/2$ is the scattering length of each extremity of the symmetry dumbbell. $\hat{Q}$ stands for the unit vector in $\vec{Q}$-direction. The important result is that correlation functions of different type such as $S^{uu}$, $S^{uS}$ and $S^{SS}$ have different form factors. One should take advantage of the specific $\vec{Q}$-dependence of these form factors in the planning and interpretation of neutron scattering experiments. The dynamic correlations $S^{uu}$, $S^{uu}$ and $S^{SS}$ are directly obtained from the fluctuation-dissipation theorem (5.6), where the corresponding relaxation functions $\Phi$ are the solutions of the system (5.7). Note that Eqs. (5.7) yield expressions for $\Phi^{\bar{S}\bar{S}}$. Therefrom one obtains easily $\Phi^{SS}$ by means of Eq. (5.4):

$$\underline{\Phi}^{SS}(\vec{k},z) = \underline{\Phi}^{\bar{S}\bar{S}}(\vec{k},z) + \underline{\Phi}^{uu}(\vec{k},z)\,|\underline{D}(\underline{u},\underline{S})|^2 +$$

$$+ 2\,\Phi^{\bar{S}u}(\vec{k},z)\,\underline{D}(\underline{S},\underline{u}) \,. \qquad (5.23)$$

Therefore the dynamic equations (5.7) are the main theoretical basis for the inelastic neutron scattering law (5.19). These equations represent, within the framework of linear response theory, a very general formulation of the dynamics of translation-rotation coupled systems. All the static susceptibilities are determined by the results of Sects. 3 and 4. The remaining problem consists in the evaluation of the memory kernels $\underline{\Delta}(z)$ and $\underline{\Pi}(z)$. That problem will be dealt with in the next section.

## VI. TRANSPORT COEFFICIENTS

We want to calculate explicitly the transport coefficients $\lambda$ and $\eta$, or more correctly $\Lambda(z)$ and $\Pi(z)$ defined by Eqs.(5.14) and (5.13) respectively. As already stated before, $Q\mathcal{L}\bar{S}=\mathcal{L}S$. Therefore we approximate Eq. (5.14) by

$$\Lambda_{\lambda\nu}(\vec{k},z) = -\left(\mathcal{L}S_\lambda(\vec{k}),\ \frac{1}{z-\mathcal{L}}\,\mathcal{L}S_\nu(\vec{k})\right) \qquad (6.1)$$

and calculate the r.h.s. of this expression in single particle approximation, retaining only the terms $K + V^R$ in the Hamiltonian (5.1). The remaining degrees of freedom of the crystal play the role of a heat bath. Within this model, $\Lambda$ becomes independent of $\vec{k}$ and is also diagonal. Applying a continued fraction expansion[58] we obtain the result[59]

$$\Lambda_{\lambda\lambda}(z)X_{\lambda\lambda} = -\langle\omega_\lambda^2\rangle/\big(z + \Sigma_{\lambda\lambda}(z)\big) \,, \qquad (6.2a)$$

with

$$\Sigma_{\lambda\lambda}(z) = -\frac{1}{\langle\omega_\lambda^2\rangle}\,\frac{(\langle\omega_\lambda^4\rangle - \langle\omega_\lambda^2\rangle^2)}{(z+i(\langle\omega_\lambda^4\rangle/\langle\omega_\lambda^2\rangle)^{1/2})} \qquad (6.2b)$$

Here $\langle\omega_\lambda^{2n}\rangle$ represent the 2n-th moment

$$\langle\omega_\lambda^{2n}\rangle = \langle\mathcal{L}^n S_\lambda\ \mathcal{L}^n S_\lambda\rangle/\langle S_\lambda S_\lambda\rangle. \qquad (6.3)$$

For $\lambda$ corresponding to $T_{1n}$, $E_g$ and $T_{2g}$ symmetry of a linear molecule in an octahedral field, the moments have been calculated in Ref. 59. In the high temperature limit, we find for $\ell = 2$, ($T_{2g}$ or $E_g$ representation) in KCN that $\langle\omega^4\rangle = 96x^2$, $\langle\omega^2\rangle = 6\,x$ where $x = \hbar^2 k_B T/I = 7.7\ (\text{meV})^2$ with $T \approx 200\text{K}$. Since z is of the order of a transverse acoustic phonon energy $\hbar\omega \leqslant 1\text{meV}$, we can take $z = o$ in Eqs. (6.2a) and (6.2b), thereby obtaining

$$\hbar\lambda \equiv \hbar\Lambda X/i = \frac{12}{5}\left(\frac{k_B T}{I}\right)^{1/2} = 5.6\ \text{meV}, \tag{6.4}$$

i.e. $\tau = \lambda^{-1} \approx 5 \times 10^{-13}$ sec, a value in fair agreement with Raman[60] and NMR[34] results and also with inelastic neutron scattering. By increasing the momentum transfer $\vec{Q}$ in neutron scattering[45] [46], one leaves the fast relaxation regime $\omega\tau<1$ and reaches the slow relaxation regime as was discussed in the last section. At higher frequencies one should then use the frequency dependent expressions (6.2a) and (6.2b) in the dynamic response equation (5.15) and the corresponding scattering law (5.19).

We have also applied the present theory to describe the dynamics of optical phonons[61]. We have considered equations, similar to (5.7), for the coupling of $\ell = 1$, $T_{1u}$ orientational fluctuations to optical phonons. Since for a Hamiltonian with bilinear coupling $V^{TR}$, $Q\mathcal{L}\vec{p} = o$, we have first taken $\Pi = o$ in the dynamic equations. From the results of Sect. 3, Eq. (3.14) it follows that a non zero coupling to optical phonon exists even at the Brillouin zone center whenever $\ell$ is odd. Therefore we take into account the asymmetry of the CN-molecule in calculating the coupling matrices $v^{(1)}_{\lambda i}(\kappa)$, (3.8b), for $\lambda$ corresponding to $\ell = 1$ and the three components of $T_{1u}$ symmetry. Consequently the coupling matrices $\beta_{\rho\lambda}$ given by Eqs. (5.9) and (5.10) are different from zero. The damping of optical phonons is then, for $\Pi = o$, due to the static coupling of optical displacements to orientational fluctuations. However, even if we take into account the full frequency dependence of $\Lambda(z)$, the calculated spectra show well defined optical phonons. This situation is in contradiction with experiments in KCN where the optical phonons are overdamped[14] [62].

There are two intrinsic mechanisms which contribute to a non zero value of $\Pi$. Firstly the usual third order translational anharmonicities should lead to a phonon damping which is also found in the alkali halides. However their optical phonons are well defined[63]. Secondly, a cubic term due to translation-rotation coupling

$$V^{TTR} = \frac{1}{2}\sum_{\vec{k}}\sum_{\vec{p}} v^{(2)}_{\lambda\rho\sigma}(\vec{k},\vec{p})\ S_\lambda(-\vec{k}-\vec{p})\,u_\rho(\vec{k})\,u_\sigma(\vec{p}) \tag{6.5}$$

leads to non zero currents in expression (5.13) for $\Pi(\vec{k},z)$:

$$Q\mathcal{L}p_\rho(\vec{k}) = i \sum_{\vec{p}} v^{(2)}_{\lambda\rho\sigma}(-\vec{k},\vec{p})\ S_\lambda(\vec{k}-\vec{p})u_\sigma(\vec{p}) \ . \tag{6.6}$$

Here again the summation convention applies for repeated indices $\lambda,\sigma$ etc... Using then mode coupling techniques[64], we arrive at

$$\Pi_{\rho\sigma}(\vec{k},z) = \sum_{\vec{p}} v^{(2)}_{\lambda\rho\rho'}(\vec{k},-\vec{p})\ v^{(2)}_{\lambda\sigma\sigma'}(\vec{k},\vec{p})\ \times$$

$$\times \frac{1}{\pi^2}\int \frac{d\omega_1[1-\exp(-\beta\omega_1)]}{\omega_1(\omega_1 - z)} \int \frac{d\omega' \chi''_{\rho'\sigma'}(\vec{p},\omega')\chi''_{\lambda\lambda}(\omega_1-\omega')}{[1-\exp(-\beta\omega')][1-\exp(-\beta(\omega_1-\omega'))]} \tag{6.7}$$

with $\beta \equiv 1/k_BT$. In obtaining this expression, we did make the factorization of the four point correlation function

$$\langle S_\lambda(-\vec{k}+\vec{p},t)u_{\rho'}(-\vec{p},t)S_\lambda(\vec{k}-\vec{p},o)u_{\sigma'}(\vec{p},o)\rangle \approx$$

$$\langle S_\lambda(-\vec{k}+\vec{p},t)S_\lambda(\vec{k}-\vec{p},o)\rangle\langle u_{\rho'}(-\vec{p},t)u_{\sigma'}(\vec{p},o)\rangle \ . \tag{6.8}$$

The dynamic susceptibilities $\chi''(\omega) = -\omega\Phi''(\omega)$ are then calculated in unperturbed approximation. The orientational susceptibility, taken in single particle approximation, reads[59]

$$\chi''_{\lambda\lambda}(\omega) = \frac{(\mathcal{L}S_\lambda,\mathcal{L}S_\lambda)\ \omega\ \Sigma''(\omega)}{[\omega^2-\langle\omega^2\rangle + \omega\Sigma'_{\lambda\lambda}(\omega)]^2 + [\omega\Sigma''_{\lambda\lambda}(\omega)]^2} \ . \tag{6.9}$$

Here $\Sigma'$ and $\Sigma''$ are the real and imaginary parts respectively of $\Sigma(z)$, Eq. (6.2b) for $z = \omega + i\varepsilon$; $\varepsilon \to o$:

$$\Sigma'_{\lambda\lambda}(\omega) = -\ \omega\left[\frac{y^2_\lambda - \langle\omega^2_\lambda\rangle}{\omega^2 + y^2_\lambda}\right] , \tag{6.10a}$$

$$\Sigma''_{\lambda\lambda}(\omega) = +\ y_\lambda\left[\frac{y^2_\lambda - \langle\omega^2_\lambda\rangle}{\omega^2 + y^2_\lambda}\right] , \tag{6.10b}$$

where $y^2_\lambda = (\langle\omega^4_\lambda\rangle/\langle\omega^2_\lambda\rangle)$. Here $\lambda$ refers to the $\ell = 1$ orientational modes with $T_{1u}$ symmetry. In case of cubic symmetry all three $T_{1u}$ components are equal and we have in the high-temperature limit[59]:

$$(\mathcal{L}S,\mathcal{L}S)_{T_{1u}} = 2x/(3k_BT)\ ;\quad \langle\omega^2\rangle_{T_{1u}} = 2x\ ;\ \langle\omega^4\rangle_{T_{1u}} = 8x^2, \tag{6.11}$$

where $x = \hbar^2 k_B T/I$. Consequently the orientational susceptibility is completely determined. Since $x \approx 8(\text{meV})^2$, $k_B T \approx 20$ meV, the resonances of the orientational susceptibility (6.9) and equivalently of the corresponding orientational relaxation denominator in Eq. (5.15) are rather close to $\omega = o$. We therefore understand why the static coupling to orientational dynamics in Eq. (5.15) does not affect in any noticeable way the optical phonon frequencies near $\hbar\omega \approx 20$ meV. The situation should be different for expression (6.7) for $\underline{\Pi}(k,z)$ since there we have a dynamic coupling in the form of a convolution. In Eq. (6.7) the dynamic displacement-displacement susceptibility is taken in harmonic phonon approximation:

$$\chi''_{\rho\sigma}(\vec{p},\omega) = \frac{\pi}{2} \sum_{\alpha} e^{*}_{\rho}(\alpha|\vec{p})\, e_{\sigma}(\alpha|\vec{p}) \frac{1}{\omega_{\alpha}(\vec{p})} [\delta(\omega-\omega_{\alpha}(\vec{p})) - \delta(\omega+\omega_{\alpha}(\vec{p}))] \tag{6.12}$$

Here $e_{\rho}(\alpha|p)$ and $\omega_{\alpha}(\vec{p})$ are eigenvectors and eigenvalues of the dynamical matrix:

$$M_{\rho\sigma}(\vec{k}) = \sum_{\alpha} e^{*}_{\rho}(\alpha|\vec{k})\,\omega^2_{\alpha}(\vec{k})\, e_{\sigma}(\alpha|\vec{k}) \; . \tag{6.13}$$

Inserting Eqs. (6.9) and (6.10) into Eq. (6.7), and carrying out the integral over $\omega'$, we obtain

$$\Pi_{\rho\sigma}(\vec{k},z) = \sum_{\vec{p}} v^{(2)}_{\lambda\rho\rho'}(\vec{k},-\vec{p})\, v^{(2)}_{\lambda\sigma\sigma'}(-\vec{k},\vec{p}) \frac{1}{2\pi} \int \frac{(1-e^{-\beta\omega_1})}{\omega_1(\omega_1-z)} \times$$

$$\times \frac{e^{*}_{\rho'}(\alpha|\vec{p})\, e_{\sigma'}(\alpha|\vec{p})}{\omega_{\alpha}(\vec{p})} \left[ \frac{\chi''_{\lambda\lambda}(\omega_1-\omega_{\alpha}(\vec{p}))}{(1-e^{-\beta\omega_{\alpha}(\vec{p})})(1-e^{-\beta(\omega_1-\omega_{\alpha}(\vec{p}))})} - \frac{\chi''_{\lambda\lambda}(\omega_1+\omega_{\alpha}(\vec{p}))}{(1-e^{+\beta\omega_{\alpha}(\vec{p})})(1-e^{-\beta(\omega_1+\omega_{\alpha})(\vec{p})})} \right] d\omega_1 \tag{6.14}$$

The resonances of $\chi''_{\lambda\lambda}$ are taken at $\omega_1 = \pm\, \omega_{\alpha}(\vec{p})$. Therefore $\Pi(\vec{k},z)$ should affect the optical phonon frequencies $\omega_{\alpha}(\vec{p})$ much more drastically than does the memory kernel $\Lambda(z)$ in Eq. (5.15). Expression (6.14) becomes more transparent if we separate real and imaginary parts. Since $z = \omega + i\varepsilon$, $\varepsilon \to o$, we write

$$\underline{\Pi}(\vec{k},z) = \underline{\Pi}'(\vec{k},\omega) + i\Pi''(\vec{k},\omega) \quad , \tag{6.15a}$$

with

$$\Pi''_{\rho\sigma}(\vec{k},\omega) = \sum_{\vec{p}} v^{(2)}_{\lambda\rho\rho'}(\vec{k},-\vec{p})\, v^{(2)}_{\lambda\sigma\sigma'}(-\vec{k},\vec{p})\, \frac{e^{*}_{\rho'}(\alpha|\vec{p})\, e_{\sigma'}(\alpha|p)}{\omega_{\alpha}(\vec{p})} \quad \times$$

$$\times \frac{1}{2}\frac{(1-e^{-\beta\omega})}{\omega} \left\{ \frac{\chi''_{\lambda\lambda}(\omega-\omega_{\alpha}(\vec{p}))}{1-e^{-\beta(\omega-\omega_{\alpha}(\vec{p}))}} \left[1 + n(\omega_{\alpha}(\vec{p}))\right] + \right.$$

$$\left. \frac{\chi''_{\lambda\lambda}(\omega+\omega_{\alpha}(\vec{p}))}{1-e^{-\beta(\omega+\omega_{\alpha}(\vec{p}))}}\; n\;(\omega_{\alpha}(\vec{p}))\right\} \quad , \tag{6.15b}$$

where $n(\omega) = [\exp(\beta\omega)-1]^{-1}$. $\underline{\Pi}'$ is obtained from the Kramers-Kronig relation

$$\Pi'_{\rho\sigma}(\vec{k},\omega) = \frac{1}{\pi} ⨍ \frac{d\omega'\Pi''(\vec{k},\omega')}{\omega'-\omega} \quad , \tag{6.15c}$$

where $⨍$ denotes a principal part integral. The expression within braces on the r.h.s. of Eq. (6.15b) has the following meaning: in the first term, an energy transfer $\omega$ between neutron and crystal leads to emission of a phonon of energy $\hbar\omega_{\alpha}(\vec{p})$ and to a transfer of the remaining energy $\hbar(\omega-\omega_{\alpha}(\vec{p}))$ to the rotating molecule, in the second term a phonon of energy $\hbar\omega_{\alpha}(\vec{p})$ is absorbed and energy $\hbar(\omega+\omega_{\alpha}(\vec{p}))$ is transferred to the rotator.

Expressions (6.1a) - (6.2b) and (6.15a) - (6.15c) with (6.9) completely determine the transport coefficients which enter the dynamic equations (5.7). We are still working on a numerical evaluation of the present theory with the aim to explain spectra of optical phonons in the alkali-cyanides[61]. A satisfactory theory should explain quantitatively the overdamping of optical phonons in KCN and NaCN and also the presence of well defined optical phonons in RbCN[65].

In closing this section we should make some comments about the underlying approximations. The basic dynamic equations (5.7) are exact for the given choice of secular variables and within the framework of linear response theory. The main approximation for the calculation of static susceptibilities consists in the use of molecular field theory, Sects. 3 and 4. At the present stage of our knowledge of molecular crystals, a large part of the theoretical efforts has still to be devoted to an understanding of the basic interactions and to the construction of adequate microscopic models. Seen the complexity of these problems, the use of molecular field theory still seems very legitimate. More serious are the approximations made in the derivation of analytical expressions of the transport coefficients. In the present theory, the orientational memory kernel $\Lambda(\vec{k},z)$, Eq. (5.14), is approximated by the single particle expression which is then calculated by a continued fraction method. The most important shortcoming of the result (6.2a,b) is the non ergodicity of the corresponding single particle orientational correlation function[66], a difficulty inherent to correlation functions of a single particle moving in a potential with several minima[67]. However, since the ratio of non ergodicity decreases with increasing temperature[66], and since we apply the theory in the high temperature regime, the errors arising from non ergodicity should be not too severe (less than 10 % in the linewidth of the spectra). Other approximations are made in the calculation of $\Pi(\vec{k},z)$ by mode coupling theory. The quality of our results has not been investigated so far by numerical or analytical methods. Applying a formalisme of the time-dependent Hartree approximation, Hüller and Raich have proposed dynamic equations[68] for the susceptibility $\chi(\vec{k},\omega)$ of translation rotation coupled system. Dynamic effects are taken into account in the form of a single particle response function $\chi(\omega)$. In comparing with the present approach[40], we would like to point out several differences: In the time-dependent Hartree approach[68], one considers dynamic effects exclusively in the single particle approximation, on the other hand the memory kernels in a Mori theory approach, Eqs. (5.7), are collective quantities. Secondly, it is not trivial to incorporate self interaction effects $C^S$ in the static coefficients of the time depedent Hartree theory[69]. Thirdly the dynamic orientational susceptibility is taken in the free rotator approximation. We should recognize however that the time-dependent Hartree theory has the appealing feature of simplicity and that it is useful whenever the model Hamiltonian does not allow a detailed calculation according the lines of Eqs. (5.7). Recently a modified version of the time dependent Hartree theory has been proposed[70], where one replaces the single particle susceptibility by the expression

$$\underline{\chi}(z) = \underline{\Lambda}(z) \left[ \underline{1}z + \underline{\Lambda}(z) X \right]^{-1} \tag{6.16}$$

with $\underline{\Lambda}(z)$ obtained from the continued fraction theory[59]. Notice that Eq. (6.16) follows immediately from Eq. (5.16) if there we neglect translation-rotation coupling and recall[38] relation $\Phi(z) = [\chi° - \chi(z)]/z$, with $\chi° = X^{-1}$.

## Concluding Remarks

We have given a subjective review of some important aspects of static and dynamic properties of orientationally disordered crystals. We have restricted our treatment to orientationally disordered phases. In those high temperature phases, the orientational motion of the molecules can be treated within classical statistical mechanics. We have insisted on the importance of symmetry adapted functions for the description of statics and dynamics of orientatinal disorder. Within that formulation one takes into account adequatly the symmetry of the crystal site and of the molecule. Very often sterical hindrance plays an important role. It is then obvious that large amplitude orientational motions are accompanied by large amplitude center of mass lattice displacements. We have studied the interaction between translational and rotational degrees of freedom. That coupling depends on molecular symmetry and we have illustrated this aspect by studying two specific systems, KCN and $NaNO_2$. The first crystal is a model system for the study of thermoelastic anomalies, the second substance is an insulator with an incommensurate phase. To a large amount, both systems can be treated within a same theoretical description. Their different behavior depends on differences in the translation-rotational coupling and that difference is a consequence of different molecular symmetry. A common aspect to both KCN and $NaNO_2$ is the importance of large lattice anharmonicities. While previous descriptions have been limited to first or second powers in the translational displacements, we have proposed a new treatment which is based on the theory of strongly anharmonic crystals. Especially near structural phase transitions, large amplitude orientational fluctuations lead to large amplitude displacement fluctuations. As a consequence the intermolecular forces are renormalized and become temperature dependent. In systems with sterical hindrance, the force constants or coupling parameters decrease by approaching an orientational phase transition. As a consequence the orientational probability distribution becomes more isotropic in KCN. The decrease of the coupling coefficients can also be related to precursor effects above the incommensurate phase transition in $NaNO_2$.

We have studied dynamic phenomena related to translation-rotation coupling. While the general structure of the dynamic equations is quite transparent, the main problem consists in the evaluation of transport coefficient. We have treated the orientational memory kernel by means of a continued fraction approximation. Finally we have studied the translational memory function within a mode-mode coupling approach. Such a theory should be relevant for the explanation of anomalous phonon damping in orientationally disordered crystals.

REFERENCES


1. A.F. Devonshire, Proc. R. Soc. London, Ser. A 153, 601 (1936).
2. H.M. James and T.A. Keenan, J. Chem. Phys. 31, 12 (1959); T. Yamamoto, Y. Kataoka, and K. Okada, J. Chem. Phys. 66, 2701 (1977).
3. R.S. Seymour and A.W. Pryor, Acta Cryst. Sect. B 26, 1487 (1970).
4. W. Press and A. Hüller, Acta Cryst. Sect. A 29, 252 (1973); idem, Phys. Rev. Lett. 30, 1207 (1973).
5. K.H. Michel and H. De Raedt, J. Chem. Phys. 65, 977 (1976).
6. F.C. von der Lage and H.A. Bethe, Phys. Rev. 71, 612 (1947).
7. C.J. Bradley and A.P. Cracknell, "The Mathematical Theory of Symmetry in Solids" (Clarendon, Oxford, 1972).
8. M. Yvinec and R.M. Pick, J. Phys. (Paris) 41, 1045 (1980).
9. K.H. Michel and K. Parlinski, to be published.
10. W. Press, J. Chem. Phys. 56, 2597 (1972); idem Acta Cryst. A29, 257 (1973).
11. For a review, see F. Lüty, "Defects in Insulating Crystals", ed. by V.M. Turkevich, K.K. Shvarts, Springer-Verlag 1981, p. 69.
12. J.M. Rowe, D.G. Hinks, D.L. Price, S.Susman and J.J. Rush, J. Chem. Phys. 58, 2039 (1973).
13. S. Haussühl, Solid State Commun. 13, 147 (1973); W. Rehwald, J.R. Sandercock and M. Rossinelli, Phys. Stat. Sol. (a) 42, 699 (1977).
14. J.M. Rowe, J.J. Rush, N. Vagelators, D.L. Price, D.G. Hinks and S. Susman, J. Chem. Phys. 62, 4551 (1975).
15. K.H. Michel and J. Naudts, Phys. Rev. Lett. 39, 212 (1977); idem J. Chem. Phys. 67, 547 (1977).
16. K.-D. Ehrhardt and K.H. Michel, Phys. Rev. Lett. 46, 291 (1981); Z. Phys. B-Cond. Mat. 41, 329 (1981).
17. S. Singh and K. Singh, J. Phys. Soc. Japan 36, 1588 (1974); C.W. v.d. Lieth and H. Eysel, J. Raman Spect. 13, 120 (1982).
18. M.L. Klein, I.R. McDonald and Y. Ozaki, Phys. Rev. Lett. 48, 1197 (1982).
19. S. Tanisaki, J. Phys. Soc. Japan 16, 579 (1961). S. Sawada, S. Nomura, S. Fuji and Y. Yoshida, Phys. Rev. Lett. 1, 320 (1958).
20. D. Sahu and S.D. Mahanti, Phys. Rev. B 26, 2981 (1982); ibidem B 29, 340 (1984).
21. K.H. Michel, Phys. Rev. B 24, 3998 (1981).
22. V. Heine and J.D.C. Mc Connell, Phys. Rev. Lett. 46, 1092 (1981), idem J. Phys. C, in press (1984).
23. R.M. Lynden-Bell, I.R. McDonald and M.L. Klein, Mol. Phys. 48, 1093 (1983).
24. D.R. Prince, J.M. Rowe, J.J. Rush, E. Prince, D.G. Hinks, S. Susman, J. Chem. Phys. 56, 3697 (1972).
25. K.H. Michel, Z. Phys. B. Cond. Matter 54, 129 (1984).
26. H. Horner in "Dynamical Properties of Solids", G.K. Horton and A.A. Maradudin (eds.), Vol. I, Chap. 8, North-Holland 1974.

27. L.H. Nosanov, Phys. Rev. 146, 120 (1966).
28. J. Kanamori, J. Appl. Phys. 31, 145 (1960); R. Brout and H. Thomas, Physics (NY), 3, 317 (1967); B.J. Mokross and R. Pirc, J. Chem. Phys. 68, 4823 (1978).
29. B. De Raedt, K. Binder and K.H. Michel, J. Chem. Phys. 75, 2977 (1981).
30. K.A. Müller and T. v. Waldkirch, in "Local Properties at Phase Transitions", K.A. Müller and A. Rigamonti eds., North-Holland (1976).
31. R. Folk, H. Iro and F. Schwabl, Z. Phys. B - Cond. Mat. 25, 69 (1976).
32. H. Jex, M. Müllner, R. Knoth and A. Loidl, Solid State Commun. 36, 713 (1980).
33. K.-D. Ehrhardt, W. Press and G. Heger, Acta Cryst. B 39, 171 (1983).
34. S. Elschner and J. Petersson, Z. Phys. B - Cond. Matt. 52, 37 (1983).
35. Y. Yamada and T. Yamada, J. Phys. Soc. Japan 21, 2167 (1966).
36. D. Durand, F. Denoyer, M. Lambert, L. Bernard and R. Currat, J. Physique 43, 149 (1982).
37. I. Hatta, J. Phys. Soc. Japan, 38, 1430 (1975).
38. R. Kubo, J. Phys. Soc. Japan 12, 570 (1957); L.P. Kadanoff and P.C. Martin, Ann. Phys. (N.Y.) 24, 419 (1963); S.W. Lovesey, "Dynamic Correlations", Condensed Matter Physics (Benjamin, N.Y., 1980).
39. H. Mori, Progr. Theoret. Phys. 33, 423 (1965); R.W. Zwanzig, J. Chem. Phys. 33, 1388 (1960).
40. K.H. Michel and J. Naudts, J. Chem. Phys. 68, 216 (1978).
41. W. Götze and K.H. Michel, in "Dynamical Properties of Solids", G.K. Horton and A.A. Maradudin eds., Vol. I, North-Holland 1974.
42. Y. Yamada, H. Takatera and D. Huber, J. Phys. Soc. Jpn. 36, 641 (1974).
43. J. Feder and E. Pytte, Phys. Rev. B 8, 3978 (1973).
44. E. Courtens, J. Physique Lettres, 37, L 21 (1976).
45. J.M. Rowe, J.J. Rush, N.J. Chesser, K.H. Michel and J.Naudts, Phys. Rev. Lett. 40, 455 (1978).
46. A. Loidl, K. Knorr, J. Daubert, W. Dultz and W.J. Fitzgerald, Z. Physik B - Cond. Mat. 38, 153 (1980).
47. J. Sakurai, R.A. Cowley and G. Dolling, J. Phys. Soc. Japan 28, 1426 (1970).
48. Y. Yamada, Y. Noda, J.D. Axe and G. Shirane, Phys. Rev. B 9, 4429 (1974).
49. J. Laizerowicz, J.F. Legrand and C. Joffrin, J. Physique 41, 1375 (1980).
50. M. Boissier, R. Vacher, D. Fontaine and R.M. Pick, J. Physique 39, 205 (1978).
51. R. Fouret, Transact. Am. Cryst. Assoc. 17, 43 (1981).
52. G. Dolling, B.M. Powell, V.F. Sears, Mol. Phys. 37, 1859 (1979).
53. W. Press, H. Grimm and A. Hüller, Acta Cryst. A 35, 881 (1979).

54. A. Hüller and W. Press, Phys. Rev. Lett. 29, 266 (1972).
55. D.M. Kroll and K.H. Michel, Phys. Rev. B 15, 1136 (1977).
56. R.M. Pick and M. Ivinec, J. Physique 41, 1053 (1980).
57. K.H. Michel and J.M. Rowe, Phys. Rev. B 22, 1417 (1980).
58. H. De Raedt and B. De Raedt, Phys. Rev. B 15, 5379 (1977).
59. B. De Raedt and K.H. Michel, Phys. Rev. B 19, 767 (1979).
60. W. Dultz, Solid State Commun. 15, 595 (1974); D. Fontaine and R.M. Pick, J. Physique 40, 1105 (1979).
61. J.M. Rowe and K.H. Michel, unpublished.
62. H. Happ and W. Sowa, Solid State Commun. 48, 1003 (1983).
63. A.D.B. Woods, B.N. Brockhouse, R.A. Cowley and W. Cochran, Phys. Rev. 131, 1025 (1963); R.A. Cowley, W. Cochran, B.N. Brockhouse and A.D.B. Woods, Phys. Rev. 131, 1030 (1963).
64. K. Kawasaki, Progr. Theoret. Phys. 39, 285 (1968).
65. K.-D. Ehrhardt, W. Press and J. Lefebvre, J. Chem. Phys. 78, 1476 (1983).
66. B. De Raedt and J. Fivez, Z. Physik B 50, 59 (1983); J. Fivez and B. De Raedt, J. Chem. Phys. 79, 3434 (1983); R. W. Gerling and B. De Raedt, J. Chem. Phys. 77, 6263 (1982).
67. M. Suzuki, Physica 51, 277 (1971); Y. Onodera, Progr. Theor. Phys. 44, 1447 (1970).
68. A. Hüller and J.C. Raich, J. Chem. Phys. 77, 2038 (1982).
69. J. Naudts, Z. Physik B 52, 231 (1983).
70. J.C. Raich, H. Yasuda and E.R. Bernstein, J. Chem. Phys. 78, 6209 (1983); H. Yasuda and J.C. Raich, Mol. Phys. 47, 647 (1982).

# CONVECTIVE INSTABILITIES IN LIQUID CRYSTALS OBSERVED BY NEUTRON SCATTERING

T. Riste and K. Otnes

Institute for Energy Technology

2007 Kjeller, Norway

## 1.  INTRODUCTION

We like to consider neutron scattering as an ideal tool for the investigation of the dynamics of condensed matter systems.  The advantages of a probe of atomic dimension and mass, and a de Broglie wavelength of interatomic dimension, are obvious.  The method has its limitations, however, either because of resolution/ intensity problems, or because of the very nature of the phenomenon we are investigating.  The former problem can be remedied by neutron sources of novel designs and higher intensity.  The latter situation is met when the characteristic times that we want to measure are approaching macroscopic values, as for a critical system at $T_c$, or in an externally stressed, nonequilibrium system.  Eventually conventional neutron spectroscopy has to fail when the neutron passage time over a correlation range becomes shorter than the characteristic time that we are trying to measure.  We shall discuss below the merits of a complementary method, a real-time method, which in principle works when the conventional method fails.  It seems worthwhile to seriously consider this method as much of the scientific interest moves to mesoscopic and to nonequilibrium systems.  We have used this method for some years in the study of convection instabilities in nematic liquid crystals.  Probably we have discovered more problems than we have solved, but perhaps a presentation of them is justified at a workshop where we are discussing the problems of tomorrow.

## 2. REAL-TIME NEUTRON SPECTROSCOPY

### 2.1 The Method

The collective properties of a condensed matter system may be described by the Van Hove correlation function G(r,t) which is characterized by a relaxation time $T_o$ (or frequency $\omega_o$) and correlation range $R_o$ (or wavevector $q_o$). In conventional neutron spectroscopy one measures $S(q,\omega)$, the Fourier transform of G(r,t). This method works when a neutron spends sufficient time within $R_o$ to observe a change of the physical state of the scatterer. In principle $R_o^{max} \sim d$, a linear dimension of the sample. With cold neutrons this puts the limit of detectable $T_o$ at $\sim 10^{-5}$ sec, i.e. $\omega_o^{min} \sim 10^5$ Hz for an oscillatory state. Going to this limit may present a problem, however, since the small momentum transfer $h/R_o$ may prevent a separation of the scattered beam from the direct or Bragg-reflected beam.

The real-time neutron method is complementary to the one just described. In this method the temporal changes in the physical state of the system are observed, not by one and the same neutron, but by different neutrons passing successively through the sample. This is possible when the temporal changes are negligible during the passage time of each neutron. For slow neutrons ($v \sim 10^5$ cm $s^{-1}$) and large samples ($d \sim 1$ cm) this gives $\omega_o^{max} \sim 10^5$ Hz. Going to faster neutrons and smaller samples one could push this limit to overlap with spin-echo or neutron gravity spectroscopy.

The application of real-time spectroscopy is, however, limited to systems in which the dynamics gives measurable temporal changes in the scattered intensity at a given angle. Aligned, elongated molecules, as in a liquid crystal, are obvious candidates. The orientational fluctuations imply a correlated motion of the anisotropic scattering power of the elongated molecules. The spectrum of orientational fluctuations extends to macroscopic wavelengths and periods. In the absence of external constraints and fields liquid crystals are in fact in a critical state. The wavelength spectrum extends from macroscopic dimensions to intermolecular distances. Only the long-wavelength end gives intensity changes that are not averaged out.

When applicable, the technique consists of measuring I(t), the coherently scattered neutron intensity as a function of time. The data processing may involve calculation of the autocorreltion function $c(\tau) = \langle I(t) I(t+\tau)\rangle$, its Fourier transform and the power spectrum $P(\omega)$, construction of the phase space portrait $[I(t), \dot{I}(t)]$ and derivation of the fractal dimension of the attractor.

## 2.2 Resolution

We shall in the following denote the Fourier transforms of time signals f(t), g(t) etc. respectively by $F(\omega)$, $G(\omega)$, etc. The symbol x denotes the folding (convolution) operation[1,2]

$$f(t) \times g(t) = \int_{\infty}^{\infty} f(t')g(t-t')dt' \qquad (1)$$

In an actual experiment the signal f(t) is not measured as a continuous function over an infinitely long time. Instead the signal is either a) - sampled repetitively N times at time intervals $\Delta t$ or b) - integrated N times over counting times $\Delta t$. The observed signal $f_o(t)$ and the real signal f(t) are for the respective two cases related by

$$f_o(t) = \sum_{n=-\infty}^{+\infty} f(t)\, \delta(t-n\Delta t) r(t) \qquad (2a)$$

$$f_o(t) = \sum_{n=-\infty}^{+\infty} f(t)\, \delta(t-n\Delta t) \times k(t) r(t) \qquad (2b)$$

k(t) and r(t) are equal to one inside their respective time intervals $\Delta t$ and $N\Delta t$, and are zero otherwise. Hence the Fourier transforms are

$$K(\omega) = \frac{2\sin(\frac{1}{2}\omega\Delta t)}{\omega} \exp\,[-i(\tfrac{1}{2}\omega\Delta t)] \qquad (3)$$

and similarly for $R(\omega)$ when $\Delta t$ is replaced by $N\Delta t$.

Leaving out r(t) in (2a) means that $f_o(t)$ consists of an infinitely long sequence of equidistant pulses. Its Foriers transform $F_\infty(\omega)$ is given by

$$F_\infty(\omega) = \sum_{n=-\infty}^{\infty} F(\omega - 2\pi n/\Delta t) \tag{4}$$

Hence repetitive sampling at intervals $\Delta t$ makes the Fourier transform repeat at frequency intervals $2\pi/\Delta t$ and frame overlap may introduce false frequencies (aliasing). To avoid this complication $\Delta t$ must be chosen such that the maximum frequency of the power spectrum $\omega_{max} < \pi/\Delta t$. Assuming this condition fulfilled, and further $N >> 1$, let us now consider the expression (2b). The convolution theorem gives [1,2]

$$F_o(\omega) = K(\omega)\ [F(\omega) \times R(\omega)] \tag{5}$$

and hence for the power spectrum

$$P_o(\omega) = |F_o(\omega)|^2 = K^2(\omega)\ [F(\omega) \times R(\omega)]^2 \tag{6}$$

Hence we can conclude:

(a) Sampling the signal $f(t)$ at a finite number of interval $N$ gives a broadening of $F(\omega)$ expressed by the resolution function $R(\omega)$.

(b) Integration of $f(t)$ over a counting time $\Delta t$ introduces a form factor $K^2(\omega)$ which makes $P(\omega)$ tend towards zero for increasing $\omega$.

The effect of the resolution and the form factor on the power spectrum of $f(t)$ is demonstrated for $f(t)=\sin(2\ t/T)$ in Fig. 1 with $T = 20$, $\Delta t = 5$, and $N = 20$. In addition to the main peak there are side lobes due to the finite length $N\Delta t$ of the time record. The resolution, defined as the width (FWHM) of the main peak is $\simeq 1/N\Delta t$. Fig. 1a shows sampling without integration over $\Delta t$ (no form factor). With integration over $\Delta t$ (Fig. 1b) the form factor $K^2(\omega)$ reduces the height of the peaks in $P(\omega)$. Aliasing is demonstrated in Fig. 2, in which $\Delta t$ is increased to 14 while other parameters in Fig. 1 are fixed. Curves 2a and 2b are with and without integration of $\Delta t$, respectively. The power spectrum is seen to be symmetric about the socalled Nyquist frequency $\nu_N = 1/2\Delta t \simeq 0.036$ and the aliased frequency is given by $\nu = 2\nu_N - \nu_o \simeq 0.021$.

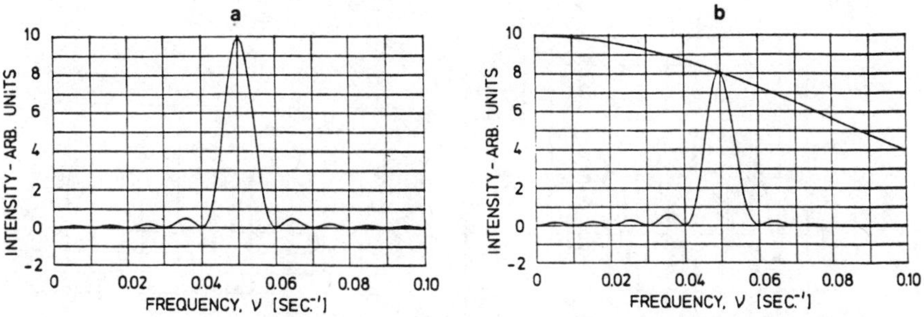


Fig. 1. Power spectrum of $f(t) = \sin(2\pi t/T)$ for $T = 20$, $\Delta t = 5$, and $N = 20$. a and b are without and with form factor, respectively.

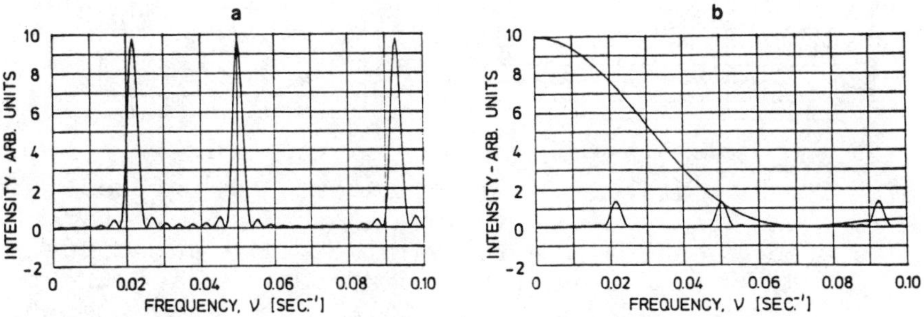


Fig. 2. Power spectrum of $f(t) = \sin(2\ t/T)$ for $T = 20$, $\Delta t = 14$, and $N = 20$. Since $2\Delta t > T$, the spectrum contains an aliased frequency $\nu \simeq 0.021$. a and b are without and with form factor, respectively.

## 2.3 Application

As indicated in the introduction, the slow dynamics of nematic liquid crystals is a candidate for the application of the real-time technique. The first paper on this appeared in 1975[3], in which convective instabilities were studied. For the same system we show in Fig. 3 some more recent data, giving a time record I(t) of the scattered neutron intensity in a regime of periodic convection. The processed autocorrelation function (Fig. 3b) and the power spectrum Fig. 3c) both give rather precisely a frequency $\omega = 0.0017$ for the main mode. Comparing 3a and 3b, we see that the asymmetric shape of the crest in 3 a has been lost in the processing. This is closely related to the phase problem in

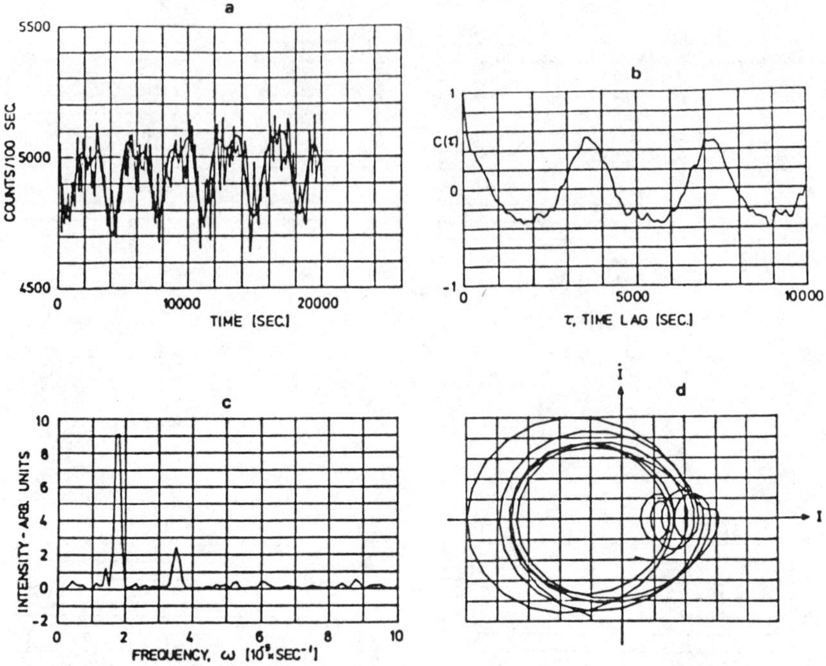


Fig. 3. a - portion of a time record I(t) of scattered neutron intensity for an oscillatory state. b, c and d, are respectively the autocorrelation function, the power spectrum and the phase space trajectory of I(t). The heavy smooth line in a is calculated.

crystallography. In our case we can retain the information about the phases of the different Fourier components if we, instead of $C(\tau)$ and $P(\omega)$, derive the time derivative $\dot{I}(t)$ from I(t) and plot the phase space trajectory of $(I(t), \dot{I}(t))$. In deriving these coordinates some smoothing of I(t) is necessary. To arrive at Fig. 3d from Fig. 3a the Fourier transform $F_I(\omega)$ - leaving out the high frequency part (see Fig. 3c) - was used and the coordinates were determined using

$$\frac{d^n f(t)}{dt^n} \leftrightarrow (i\omega)^n F(\omega) \tag{7}$$

The heavy smooth line in Fig. 3a shows $\langle I(t)\rangle$ calculated from (7).

An interesting observation which has analogies both to phonon linewidths and to modulated spatial structures was observed in a convecting nematic, horizontally aligned[4], at Rayleigh number $R = 3R_c$. The processed data are shown in Fig. 4. The power spectrum looks rather much like a broadened phonon at $\omega_o$ and a relaxational, central mode. The autocorrelation function $C(\tau)$ is apparently modulated, which suggests that the multiplet in $P(\omega)$ around $\omega_o$ can be obtained by a modulation of the phase by some low frequency in the spectrum, of the type $\cos [\omega_o t + a \sin \omega t]$. By this procedure

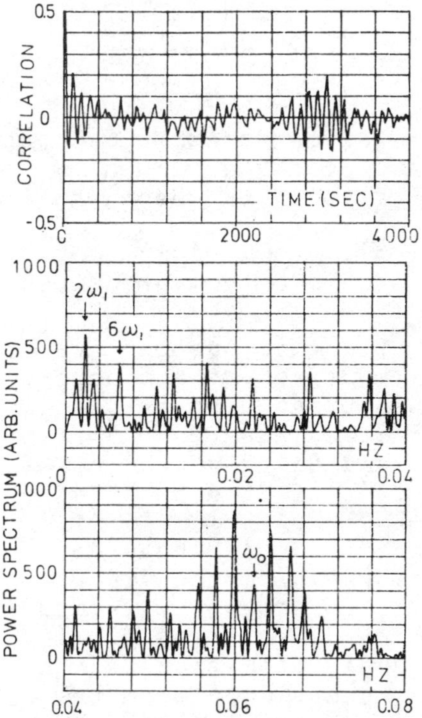


Fig, 4. Processed data from observations at $R = 3R_c$
Top panel: portion of autocorrelation function.
Lower panels: power spectrum

the amplitude of the n'th line in the multiplet is given by $J_n^2(a)$, the square of a Bessel function. Taking $2\omega$, and $6\omega$, in Fig. 4 as phase-modulating frequencies we reproduce the main features of the multiplet. If to this modulated wave we add the slow, modulating waves, we also reproduce the main features of $C(\tau)$. Modeled portions of $C(\tau)$ and $P(\omega)$ are shown

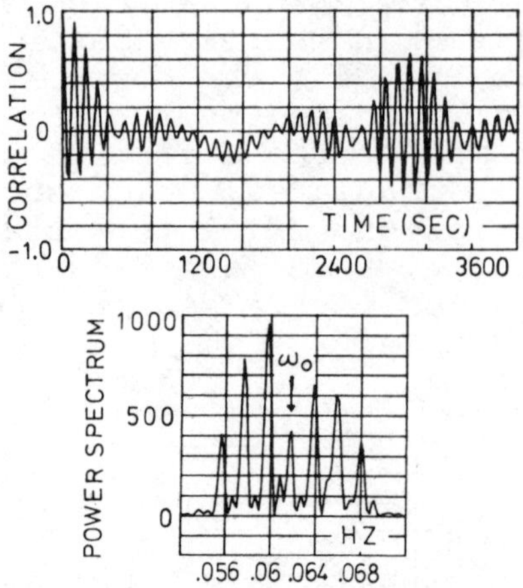


Fig. 5. Autocorrelation function and portion of power spectrum at $\varepsilon = 2$ obtained by modulating phase of $\omega_o$ with frequencies $2\omega_1$ and $6\omega_1$ of Fig. 4.

in Fig. 5, which should be compared with Fig. 4.

Recently much attention has been paid to the dimensions and fractal measures of attractors or phase space trajectories. When transients are dead, the dimension of an attractor becomes less than that of the phase space it sits in, and gives valuable information about the number of degrees of freedom of a dynamical system.

Grassberger and Procaccia[5] have devised a method for deriving a measure $\nu$ of the fractal dimension directly from the time series. The method involves calculations of $C(\ell)$, the number of pairs in the series that are separated by a distance smaller than $\ell$. The dimension $\nu$ is then derived simply from the relation

$$C(\ell) \propto \ell^{\nu} \tag{8}$$

for small $\ell$'s. In another paper they show how noise affects the derivation[6].

We demonstrate this for the time series $\sin 2\pi t/100$, to which we add a random noise of amplitude 0.125. In Fig. 6 we have plotted log C against log $\ell$. The initial slope of 3 just reflects the dimension of the random noise spectrum. The second linear slope of 1.1

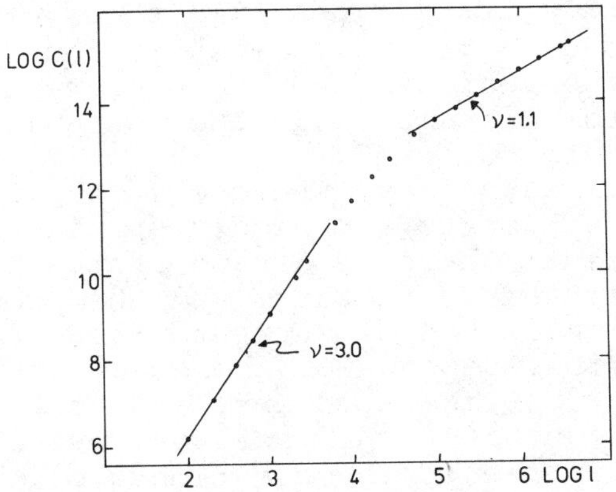


Fig. 6. Derivation of the dimension of a noisy attractor according to the recipe of ref. 6. The attractor is in this case a limit cycle, i.e. a harmonic time series, to which has been added 12.5 per cent of random noise.

is close to the dimension 1.0 expected for a limit cycle, i.e. for our simple time series.

## 3. NEUTRON SCATTERING FROM A THERMOCONVECTIVE NEMATIC

### 3.1 Introduction

Nematic liquid crystals are anisotropic, molecular liquids whose molecules are easily aligned by an external magnetic field[7]. The intensity of the liquid diffraction peaks depends strongly on the direction of alignment. Thus for para-azoxy-anisole (PAA), the substance used in our experiments, the first diffraction peak is strong for a vertical orientation, i.e. normal to the scattering vector, but essentially zero for an alignment parallel to the scattering vector[8]. For a vertically aligned sample the onset of flow is marked by a decreasing intensity of this diffraction peak. This is due to the coupling between orientation and flow[7], which is the basic mechanism that enables one to use neutron scattering in the study of convective instabilities.

In general the development from laminar, time-independent flow occurs through one or several regimes of time-periodic flow. The different regimes are separated by instabilities (bifurcations) which are nonequilibrium phase transitions. In the same way that temperature is a control parameter in equilibrium phase transitions, the vertical temperature gradient ($\Delta T$) is a common control parameter in the nonequilibrium case.

The first instability, the one at the lowest $\Delta T$, marks the transition to a periodic spatial structure of convection rolls. There is a band of possible, allowed wavevectors for this structure, and the process of wavenumber selection is one of current interest[9-10]. This selection possibly depends on the state of the convection, whether steady or time-dependent, and of the aspect ratio of the liquid container. In additon to the spatial structure there is also the question about the temporal structure at the higher instabilities. The formation of spatio-temporal structures is a problem of concern not only for dissipative phenomena in physics, but also in a wider synergetic context[11].

We owe much of our knowledge about instabilties to advances in quantum optics. Much attention has been paid to optical bistability[12], a phenomenon that has both practical and scientific implications. It has become almost the prototype of a first-order, non-equilibrium phase transition. In Fig. 7a is a schematic picture of an absorptive optical bistability in a ring cavity[13]. This figure assumes perfect tuning between the different frequencies: that of the incident laser light, the transition frequency of the atomic absorber, and the round-trip frequency of the major cavity mode. Lugiato et al. have in several papers treated the influence of detuning on the stability of the high-transmission branch[14-16]. The branch may be piecewise stable, periodic or unstable, and the unstable, chaotic regime may develop through periodic doubling. In the periodic regime the amplitude of the oscillation, as a function of the transmitted field, may display the interesting hysteretic behaviour of Fig. 7b. It includes both a first- and a second order phase transition. This phase diagram is deduced under the assumption that only one pair of cavity modes, in addition to the fundamental one, contributes. Relaxing this restriction on the cavity modes, Ikeda[17] found that the ordinary bistable behaviour was drastically modified to give the transmitted field as a multiple-valued function of the incident field. He also found that in some range of this control parameter all the stationary values of the transmitted field became unstable, leading to a chaotic

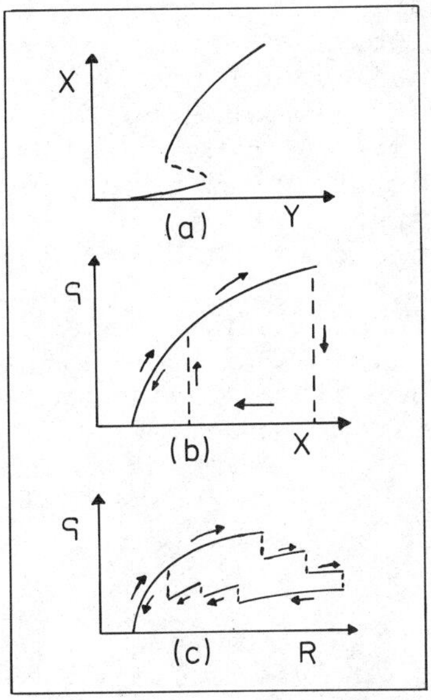


Fig. 7 a and b. Characteristic curves for optical bistability for two different cases, adapted from Lugiato et al., and explained in the text. X and Y are scaled amplitudes of transmitted and incident field, respectively, ρ is the half-amplitude of the oscillations, c is a conjectured characteristic for a multimode system, R denotes the control parameter.

behaviour. Combining now the results of Lugiato et al. and of Ikeda, we arrive at Fig. 7c as a"brave" heuristic (see next section) extrapolation.

Convective instabilties have not been treated to the same degree of sophistication. The justification for carrying the optical work over to convection originates in Haken's theories of non-equilibrium phase transitions[18] and of the analogy between higher instabilities in fluids and lasers[19]. In the latter work he showed that both phenomena could be described in the Lorenz model. The Lorenz equations[20] model the condition in a convective system at control parameters sufficiently high for reaching a higher instability. In this situation the flow pattern will become unstable because of the different rates of fluid rotation and heat dissipation, i.e. because of detuning between flow and heat.

In a laser with a cavity absorber, and in a convecting liquid crystal, the detuning may actually be sufficiently strong that the instability to a time-dependent regime comes at a lower control parameter than the instability to a stationary regime[21]. For the nematic case this was first predicted by Lekkerkerker[22]. He gave a linear theory for a vertically aligned nematic, heated from below. The detuning is here caused by the different diffusion constants of orientation and heat. The former is field dependent, and the detuning and the time-dependent regime disappear at a threshold field for which the diffusion constants are equal. Later theoretical work[23] has raised the question about the lower instability, whether it is continuous or discontinuous. It seems prudent for an experimentalist to approach the question about the nature of the transitions with an open mind.

## 3.2 Experiments

The sample was fully deuterated PAA contained in a parallelepipedic cell of horizontal dimensions 50 and 4 mm and height 25 mm[24,25]. With this geometry one expects two convection rolls with axes parallel to the shorter edge. The nematic range is $118 < T < 135^{\circ}C$, with supercooling to $\sim 100^{\circ}C$.

The top, bottom and side walls are made of Al, Cu and stainless steel, respectively. A vertical temperature gradient is obtained by setting the difference of the power fed to electrical heating elements at the top and bottom of the vessel. The temperature of the bottom plate is controlled and kept constant throughout the experiment. The vertical temperature difference is monitored by thermistors and thermoelements, and stays constant within $\pm 0.01^{\circ}$ for hours. The strength of the magnetic field can be chosen in the range 0 - 1.5 kG.

The scattered neutron intensity was always recorded at $Q \sim 1.8 Å^{-1}$, at the first liquid diffraction peak. It can be shown that the intensity, i.e. its deviation from that of the fully aligned, flow-free state, is proportional to the order parameter of the convective state, the velocity amplitude. This is true as long as the roll configuration is unchanged. With a vertical magnetic field the intensity deviation is negative, hence we mostly use an inverted intensity scale in the figures that follow.

For H = 1.2 kG we observe only the broken curve indicated in Fig. 8, i.e. only one flow regime. For

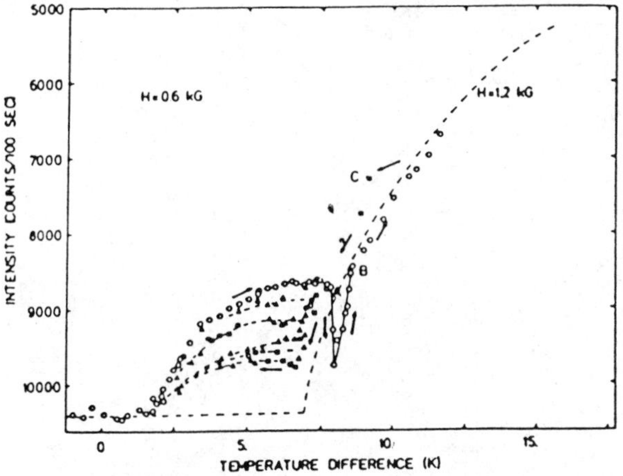


Fig. 8. Neutron intensity versus temperature difference ($\Delta T$) for PAA in a vertical field H = 0.6 kG. Circular and square points are for a complete cyle of increasing and decreasing $\Delta T$, respectively. Triangular points are measured after additional variations of H. Notice intensity dip between A and B. A curve (but no points) obtained at 1.2 kG is also indicated.

H = 0.6 kG we in addition see a lower flow regime with multistability, and separated from the higher regime by a hysteretic transition. Similar runs at other fields gave the phase diagram of Fig. 9, which has a multicritical point at H = 1.1 $\pm$ 0.1 kG. We have lately spent much time to study the behaviour in the multistable regime at H = 0.12 kG. Fig. 10 is a collection of points accumulated in several runs. It should be noticed that the intensity rides on top of a high isotropic background, even so most of the level structure in the multistable regime is well assessed. The reproducibility was excellent, except at the lowest values of $\Delta T$ where anchoring[7] on the sidewalls of the vessel determines the intensity level and in the upper flow regime where slippage on the walls becomes a problem. Hence we have plotted only one set of data for the latter regime.

The most generic route for a monotonic increase and decrease of $\Delta T$ is indicated in Fig. 10. Points on the levels outside of this route were obtained when revers-

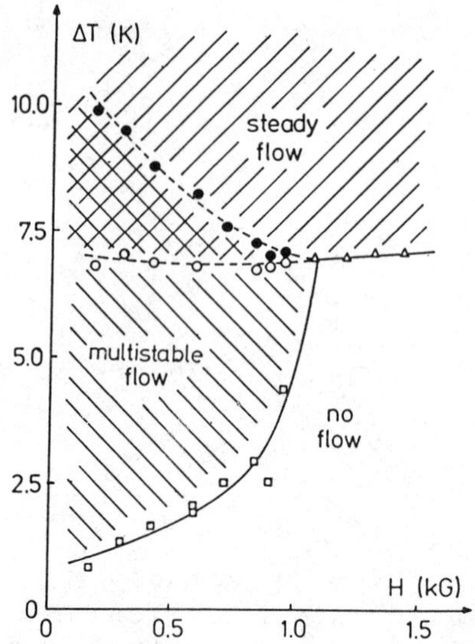


Fig. 9. Phase diagram showing confluence of first-order (broken) and second-order (full-drawn) lines at a multicritical point. First-order lines are upper and lower stabiltiy limits of multistable and steady flow regimes, respectively.

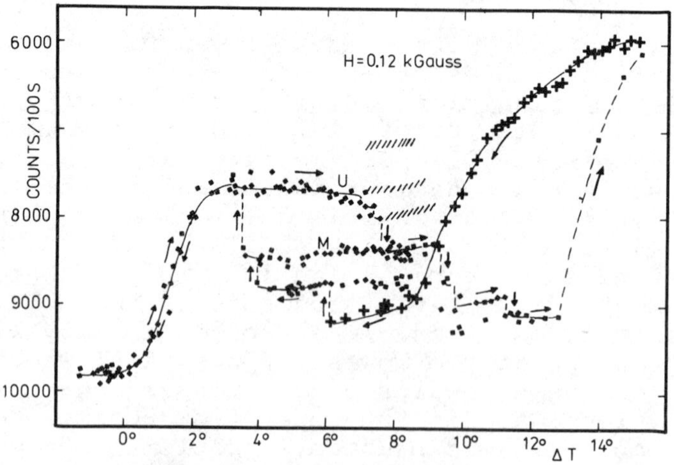


Fig. 10. Neutron intensity versus temperature difference ($\Delta T$) for H = 0.12 kG. Arrows point out the generic route followed at monotonically increasing and decreasing $\Delta T$. For $\Delta T \sim 8^{o}$, and kept constant for long times, also shaded levels are visited.

ing the direction of variation of $\Delta T$. Deviations from this route were observed, but mostly small. An important exception was the observation of a continuous transition from the level marked M to the upper flow regime. The upper levels of the multistable regime loose their stability as T is increased, but they evidently exist as unstable states at higher $\Delta T$'s. This is indicated by the shaded levels that mark the intensities observed in extended runs at constant $\Delta T$. An example of such a run is given in Fig. 11. In the course of about 40 hours the system explores most levels. The levels have different stabilities, as judged from the fluctuations. The transition between the levels is mostly rather abrupt on the time-scale of observation, there are at most a few oscillations between the levels. The sequence in which levels are visited has some similarity to deterministic diffusion[26], although we cannot claim that this is really the case. In a few cases, like in Fig. 12, do we see a diffusive-type increase of the oscillations on the transient before sedimentation at a more stable point.

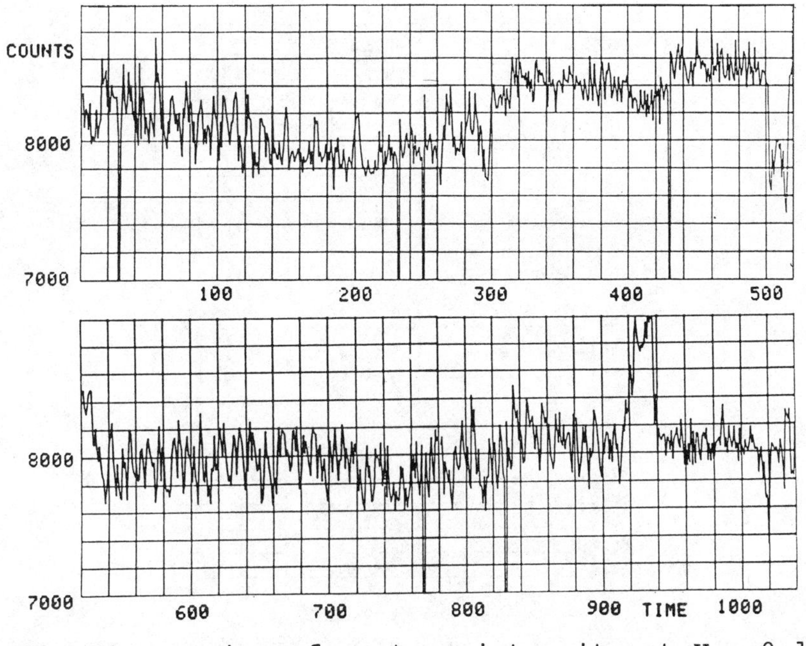


Fig. 11. Time series of neutron intensity at H = 0.12 kG and $\Delta T$=8.1°recorded over 1040 channels, each 100 sec wide. Dips to the base line incidate reactor stops of about 2 hrs. duration.

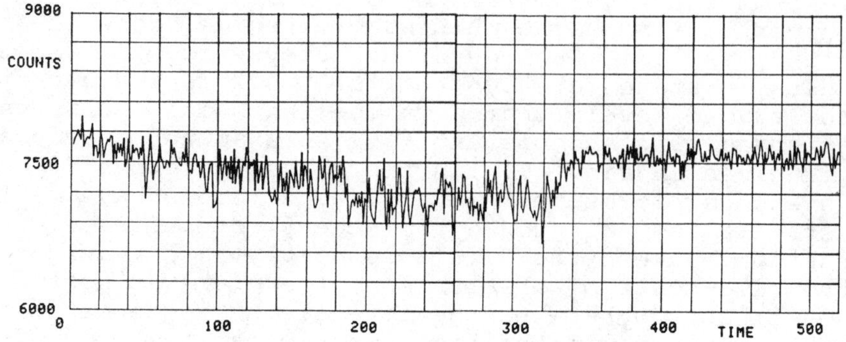


Fig. 12. Neutron intensity versus time (channel width 100 sec.) showing transient fluctuations and sedimentation in a more stable state.

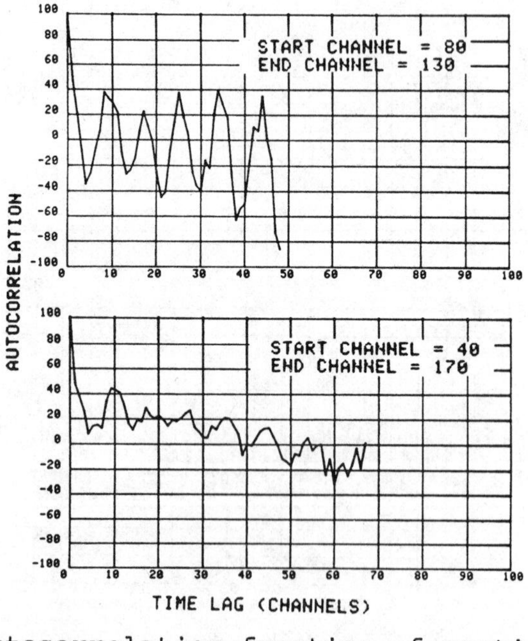


Fig. 13. Autocorrelation function of portions of the time series of Fig. 11. In the upper panel the portion is 5000 sec long, in the lower panel it is 13000 sec.

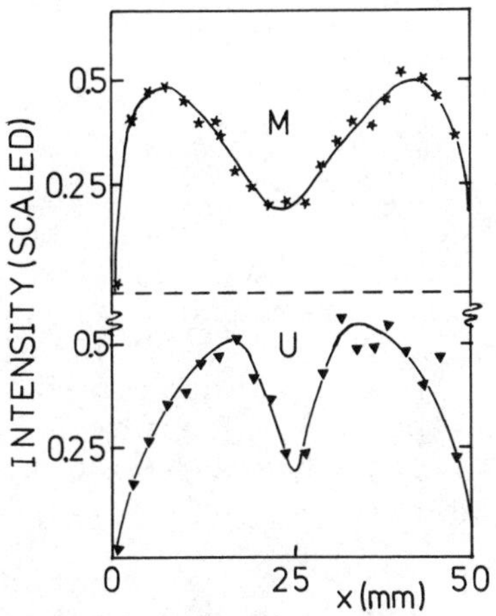


Fig. 14. Horizontal intensity profile along the 50 mm wide sample cell observed at intensity levels marked M and U in Fig. 10. The intensity is scaled such that 0 and 1 denote zero and maximum flow, respectively.

The time series in Fig. 11 is piecewise oscillatory This can be seen from the upper panel of Fig. 13, giving the autocorrelations over some 50 channels a 100 sec. Extending this to 130 channels the correlations are decaying, as seen from the lower panel. This points to an intermittent behaviour. In no case have we with the present setup observed persistent linear oscillations. Earlier experiments with a cell that had space for only one fundamental roll gave more persistent oscillations, as was shown in Fig. 3 above.

To test the meaning of the levels, we mapped out the roll structure by scanning with a narrow neutron beam over the largest face (50 x 25 mm) of the sample cell. Fig. 14 gives the horizontal intensity profile at midheight (i.e. 12.5 mm) for two different levels, marked M and U in Fig. 10. Similar scans were taken at other heights and for other intensity levels. The symmetry of the pattern that we obtained indicates that the main feature perhaps can be explained by a superposition of the two fundamental rolls and its first harmonics. In any case, different intensity levels correspond to different geometry and sense of rotation of the rolls. The profile also showed that there is slippage on the walls in the upper, steady phase.

Referring to Fig. 14, slippage involves non-zero-scaled intensity at x = 0 and 50.

### 3.3 Discussion

Our experiments show two convecting regimes, in agreement with general predictions for double-diffusive systems. Coming from low $\Delta T$'s from the conductive regime, we first encounter a multistable regime whose width decreases with increasing H and vanishes at a multicritical point ($\Delta T_c^m$, $H^m$). Following the nomenclature of equilibrium phase transitions this could be called a bicritical point. The phase diagram of Fig. 9 is, however, incomplete, it neglects the presence of several order parameters in the multistable regime.

The multistable regime is separated from the conductive regime by a continuous transition and from the higher regime by a hysteretic transition. The hysteresis disappear as $H \to H_m$, and for $H > H_m$ there is a continuous transition from the conductive to the upper, convective regime. Perhaps we should make a reservation about this statement: the long build-up time of the flow near the critical gradient is not easily distinguished from a hysteretic behaviour.

The profile measrements referred to above (see Fig. 14) show that for our cell of aspect ratio two, the profile is symmetric around x = 25. This points to basically two rolls, in agreement with predictions on wavenumber selection. There are, however, finer details on the profile that distinguish the different intensity levels of the multistable system. We conclude that the levels correspond to different geometries and senses of rotation of portions of the basic roll. This casts doubt on calculations that rely on truncations that limit a dissipative system to a single or to very few modes. The observation of a novel type of hysteresis in the multistable regime (Fig. 10) lends additional support to this view. The observed hysteretic behaviour is basically of the type predicted for a main resonance plus a pair of sideband modes (Fig. 7b) in optical bistability[15], and agrees with a conjectured extrapolation for a multimode system (Fig. 7c).

According to the general predictions for a double-diffusive system[21] the lower and higher convective regimes should have time-dependentand steady flow, respectively. We have difficulties in checking this prediction. The time dependence that we observe is never one of linear oscillations but rather chaotic. The chaotic attractor is not easily distinguished from a fixed point, due to the ever-existing orientational fluctuations in the nematic state. Excursions from the mean intensity, as measured by the variance, are larger than counting statistics, even in the conductive regime.

Only in the upper end of the multistable regime do we see a distinct time dependence. This is also where the system visits all existing levels in extended runs (Fig. 11). The levels correspond to attractors of different degrees of strangeness: unstable fixed points, quasiperiodic and chaotic attractors. As we have pointed out earlier[27] the long-time observations at a fixed control parameter may be described as a slow, cyclic passage through a number of coexisting attractors, i.e. by several attractors coupled by a much slower attractor. This is exactly a type of crisis predicted very recently by Gu et al.[28] for coupled logistic maps. The hysteresis loop that we observe at lower control parameters is also predicted by the same authors, and also chaotic transients as in Fig. 12. Within some attractors we find oscillatory behaviour of some periods' length, see Fig. 13. The way in which the periodicity is lost may be either through intermittency or some kind of phase modulation, as shown in Figs. 4 and 5. In some earlier experiments on a cell of aspect ratio one, i.e. with room for just one basic roll, we found more persistent oscillations, see Fig. 3. That the stability of the oscillations seems to depend on the aspect ratio is actually predicted by Knobloch et al.[29].

It would be nice to have a complete picture of the evolution of the attractors through the multistable regime. The fractal dimension $\nu$ is a useful measure of the evolution. As shown above in Fig. 6, the derivation of $\nu$ is made difficult by the presence of even a small amount of noise. Our attractor of Fig. 3 suffers from this deficiency. An embedding in five dimensions still shows only five-dimensional noise. A true value of $\nu$ could be compared with calculations by Velarde and Antoranz[30]. They use a generalized five-mode Lorenz model for two-component convection, and their calculations, as revised by Farmer[31], give a dimension of 4.05. Such a high value indicates a weak coupling between the modes, and the dimension is what one would expect from the product of two independent attractors.

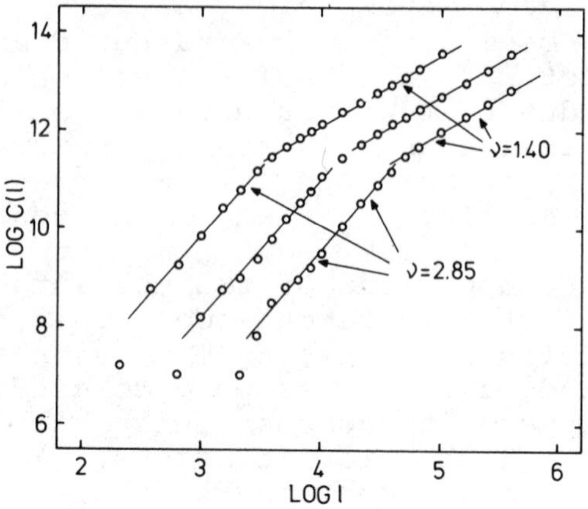


Fig. 15. Derivation of the dimension ν of the smoothed curve of Fig. 3 by an embedding in six dimensions.

If we now, instead of the noisy signal of Fig. 3, use the smoothed curve of the same plot, we get rid of the noise. Following the procedure of Fig. 6 we arrive at Fig. 15, which represents an embedding in six dimensions (6D) for three portions of the curve. The plot gives the values of ν, 1.40 and 2.85 for small and large distances ℓ, respectively. The smaller value remains the same for embedding in 5D and 7D, whereas the higher value increases slightly with D, probably due to noise. We tend to believe that both values are real, reflecting the fact that our system consists of two coupled attractors.

## 3.4 Concluding remarks

Our experimental studies of convection phenomena have taken a very long time, but there are still a number of questions to which we do not have a complete answer: are there critical fluctuations connected with any of the instabilities, is the second instability one that involves breaking of time-symmetry, are there soft nonlinear modes etc. Our experiments were made with a 2 MW reactor. Going to a stronger neutron source would enable one to use smaller samples. This would shorten the time constants that are proportional to the square of the sample dimension. The waiting time for the flow to stabilize might remain the same, however, and surface effects might become more of a problem.

## REFERENCES


1. D.C. Champeney, Fourier Transforms and Their Physical Applications, Academic Press, London and New York (1973).
2. R.K. Otnes and L. Enochsen, Applied Time Series Analysis, John Wiley and Sons, New York (1978).
3. H.B. Møller and T. Riste, Phys. Rev. Lett. 34, 996 (1975).
4. K. Otnes and T. Riste, Phys. Rev. Lett. 44, 1490 (1980)
5. P. Grassberger and I. Procaccia, Phys. Rev. Lett. 50, 346 (1983)
6. A. Ben.Mizrakhi, I. Procaccia and P. Grassberger, Phys. Rev. 29A, (1984)
7. P.G. de Gennes, The Physics of Liquid Crystals, Clarendon Press, Oxford (1974).
8. R. Pynn, K. Otnes and T. Riste, Sol. St. Commun. 11, 1365 (1972).
9. P. Manneville, Phys. Lett. 95A, 463 (1983).
10. L. Kramer, E. Ben-Jacob, H. Brand and M.C. Cross, Phys. Rev. Lett. 49, 1891 (1982)
11. H. Haken, Synergetics - An Introduction, Springer-Verlag, Berlin (1977).
12. L.A. Lugiato, P. Mandel, S.T. Dembinski and A. Kossakowski, Phys. Rev. A18, 238 (1978).
13. L.A. Lugiato and L.M. Narducci in Nonequilibrium Cooperative Phenomena in Physics and Related Fields (eds. M.G. Velarde and M. Scully) Plenum (to be published).
14. V. Benza and L.A. Lugiato, Z. Phys. B35, 383 (1979).
15. V. Benza and L.A. Lugiato, Z. Phys. B47, 79 (1982).
16. L.A. Lugiato, V. Benza, L.M. Narducci and J.D. Farina, Z. Phys. B49, 351 (1983).
17. K. Ikeda, Opt. Commun. 30, 257 (1979).
18. H. Haken, Z. Phys. B21, 105 (1975).
19. H. Haken, Phys. Lett. 53A, 77 (1975).
20. E.N. Lorenz, J. Atmos. Sci. 20, 130 (1963).
21. S. Chandrasekhar, Hydrodynamics and Hydrodynamic Stability, Clarendon Press, Oxford (1961).
22. H.N.W. Lekkerkerker, J. Phys. Lett. 38, L-277 (1977).
23. E. Dubois-Violette and M. Gabay, J. Physique 43, 1305 (1982).
24. K. Otnes and T. Riste, Helv. Phys. Acta 56, 837 (1983).
25. T. Riste and K. Otnes, Multicritical Phenomena (eds. R. Pynn and A.T. Skjeltorp) Plenum, New York and London (1984).
26. T. Geisel and J. Nierwetberg, Phys. Rev. Lett. 48, 7 (1982).

27. T. Riste, Nonequilibrium Cooperative Phenomena in Physics and Related Fields (eds. M.G. Velarde and M. Scully), Plenum (to be published).
28. Y. Gu, M. Tung, J.M. Yuan, D.H. Feng and L. Narducci, Phys. Rev. Lett. 52, 701 (1984)
29. E. Knobloch, N.O. Weiss and L.N. da Costa, J. Fluid Mech. 113, 153 (1983) and references therein.
30. M.G. Velarde, Nonlinear Phenomenon at Phase Transitions and Instabilities, (ed. T. Riste) Plenum, New York and London (1982) p. 205.
31. J.D. Farmer, Evolution of Order and Chaos (ed. H. Haken), Springer-Verlag, Berlin (1982) p. 228.

# NEUTRON SCATTERING AS A PROBE OF MAGNETIC SUPERCONDUCTORS AND INTERMEDIATE VALENCE COMPOUNDS

S. K. Sinha

Exxon Research and Engineering Company
Annandale
New Jersey 08801

## 1. INTRODUCTION

Owing to the fortunate circumstances of their possessing a magnetic moment, neutrons have been for over 25 years a central tool in our microscopic investigations of magnetic phenomena in condensed matter. This is likely to be the case for the foreseeable future as well. In this article, I shall attempt to review what neutron scattering has revealed in recent years about two rapidly expanding areas of magnetism. The first is the field of the so-called "magnetic superconductors" which got its impetus from the discovery of certain families of ternary compounds [1,2] in the late seventies. The second is the field of intermediate valence and so-called "Kondo-type" magnetic materials, which have also been vigorously studied since the mid-seventies. The physics in both of these areas is not completely understood at present, but neutrons have turned out to be a very useful probe in these investigations.

## 2. THE PROBLEM OF THE CO-EXISTENCE OF SUPERCONDUCTING AND MAGNETIC LONG-RANGE ORDER

After nearly 20 years of attempts to understand the mutual effects of superconductivity and magnetism on each other, the existence of long-range order (LRO) of both kinds in the same material was finally established recently with the advent of the families of ternary compounds, which have come to be known as "magnetic superconductors" [1,2]. The general use of this term does not necessarily imply a simultaneous co-existence of long-range magnetic order and superconductivity. Nevertheless, in spite of the pair-breaking effects introduced by the presence of magnetic ions in a superconducting crystal, the very existence of superconductivity in crystals containing dense lattices of magnetic ions is itself initially very surprising. For these materials, this is believed to be due to the relative weakness of the exchange spin-flip interaction between the f-shell electrons on the magnetic ions and the conduction electrons. The best studied families of these compounds are the ternary borides with the chemical formula $RRh_4B_4$ (where R is a rare earth ion) and the Chevrel phase compounds of the type $RMo_6S_8$ or $RMo_6Se_8$. There exist by now several review articles on these extremely interesting classes of materials, which the reader may consult for more detailed experimental information [3,4].

While neutrons unfortunately cannot couple directly to the superconducting order parameter, they are an ideal probe for studying the magnetic order parameter at a microscopic level. Thus, they can be used to study the kind of magnetic order parameter which exists, its temperature dependence, critical fluctuations and magnetic excitations in such systems. We first consider neutron diffraction studies carried out on the antiferromagnetic superconductors [3-10]. For all of these materials (see Table 1) below $T_N$, superconductivity and antiferromagnetic LRO coexist all the way down to the lowest temperatures. While there are undoubtedly some

Table I

Magnetic structures of antiferromagnetic superconductors as determined by neutron diffraction

| Compound | $T_c$ | $T_N$ | $\|\vec{q}\|$ | Direction of $\vec{q}$ | Moment Direction | Max. Ordered Moment | Reference |
|---|---|---|---|---|---|---|---|
| $GdMo_6Se_8$ | 5.6K | 0.75K | (1/2,0,0) $2\pi/a$ | [100] | -- | -- | 7 |
| $DyMo_6S_8$ | 2.05K | 0.4K | (1/2,0,0) $2\pi/a$ | [100] | [111] | 8.77$\mu$B | 5,6 |
| $TbMo_6S_8$ | 2.05K | 1.0K | (1/2,0,0) $2\pi/a$ | [100] | [111] | 8.28$\mu$B | 6 |
| $NdRh_4B_4$ | 5.4K | 1.5K | 0.135Å$^{-1}$ | [100] | [001] | | |
| | | 1.0K | 0.139Å$^{-1}$ | [110] | [001] | 3.4$\mu$B | 8 |
| $TmRh_4B_4$ | 9.8K | 0.7K | 0.238Å$^{-1}$ | [101] | [010] | 6.6$\mu$B | 9 |

dramatic effects of the magnetic LRO on properties such as the critical field $Hc_2$ [11,12], the superconducting and magnetic order parameters do not interfere too strongly with each other, presumably because their associated characteristic wavevectors lie in different regions of reciprocal space. Hamaker et al [10] have studied, by powder diffraction, the compound $Ho(Rh_{0.3}Ir_{0.7})_4B_4$, where a continuous transition occurs to an antiferromagnetic state at 2.7 K from a normal paramagnetic state. Superconductivity then sets in at a lower temperature of $T_c$ = 1.34 K and coexists with the magnetic LRO for lower temperatures. The onset of superconductivity has no discernible effect on the magnetic ordering, which is estimated to be long range. While the antiferromagnetic Chevrel-phase compounds such as $DyMo_6S_8$, $TbMo_6S_8$ and $GdMo_6S_8$ have conventional up-down antiferromagnetic structures, the compounds such as $NdRh_4B_4$ and $TmRh_4B_4$ have rather long wavelength modulated antiferromagnetic structures. It is possible that the latter may be incipient ferromagnets forced into antiferromagnetic structures by the presence of the superconductivity, as discussed below. (The fact that $TmRh_4B_4$, for instance, goes ferromagnetic [9] when the superconductivity is quenched by a magnetic field and exists metastably in the ferromagnetic state even when the field is lowered considerably adds some support to this view).

The problem of the coexistence of ferromagnetic LRO and superconductivity is a more subtle one, since the competing interactions both occur at small wavevectors. In the case where the exchange interaction J(q) between the spins is mediated by the conduction electrons, it is proportional to the generalized susceptibility function $\chi(q)$ of the conduction electrons. In the normal state $\chi(q)$ will have a maximum at $q = 0$. However, as discussed by Anderson and Suhl [13] in the superconducting state $\chi(q)$ must vanish at $q = 0$, reflecting the singlet nature of the superconducting electron ground state. (This is only strictly true in the case where spin-orbit coupling is neglected for the conduction electrons, but stays at least approximately true for real systems). This dip in $\chi(q)$ at small wavevectors occurs on a scale of the inverse coherence length of the superconductor. The net result is a broad maximum at a finite wavevector (typically of order 0.2 $Å^{-1}$) which would cause the material in the superconducting state to go into some kind of modulated magnetic structure. As discussed above, some of the observed antiferromagnetic ternary borides may be realizations of the Anderson-Suhl state.

There is, however, an even more important effect, namely, the electromagnetic competition between ferromagnetism and superconductivity, as has been discussed by several authors [14-18]. The Meissner effect causes the crystal to expel all magnetic flux inside it, which is difficult to do if the spins are all lined up ferromagnetically, producing a macroscopic $\underline{B}$-field. From the electromagnetic point of view, there are two ways in which the tendency to ferromagnetism may reach a compromise with superconductivity. One is for the spins to produce a long-wavelength spinal or other modulated structure so that the B-field averages to zero over a length-scale comparable to the London penetration depth, while still allowing neighbouring spins to stay approximately parallel to each other. The other is for the superconductor to allow flux from aligned spins to penetrate in the form of a vortex lattice.

Calculations have been carried out [14,16,18] based on a Ginzburg-Landau expression for the Free Energy, expressed in terms of both the superconducting and the magnetic order parameters and all the electromagnetic interactions between them. These predict an intermediate phase which could occur between the superconducting (paramagnetic) and the ferromagnetic normal phases. This was predicted to be either a spiral or a vortex lattice, depending on the choice of parameters used in the free energy expression. In both cases, the wavelength of the periodic structure was of the order of $\lambda_s = (\lambda_L \lambda_M)^{1/2}$ where $\lambda_L$ is the London penetration depth of the superconductor, and $\lambda_M$ a characteristic length for the normal magnetic interaction; $\lambda_s$ comes out to be $\sim 100$ Å. An alternative method of obtaining this phase was that used by Tachiki, Umezawa and their collaborators [15,17] where the J(q) arising from the dipole-dipole magnetic interaction between spins is screened by supercurrents, producing an effective J(q) with a <u>minimum</u> at $q = 0$ and a maximum at $q \sim 1/\lambda_s$. Later, single-ion anisotropy effects were included by Greenside, Blount and Varma [19] and a linearly polarized sinusoidal structure was also predicted. (Note that only the spiral magnetic phase can coexist with a <u>uniform</u> superconducting order parameter).

In the next section, we will review to what extent neutron scattering experiments have confirmed such predictions and the new puzzles which such experiments have revealed.

## 3. NEUTRON DIFFRACTION STUDIES OF FERROMAGNETIC SUPERCONDUCTORS

The best studied ferromagnetic superconductors are the ternary boride $ErRh_4B_4$ and the Chevrel phase compound $HoMo_6S_8$. These are also called "reentrant superconductors" because as the temperature is lowered, the material ultimately goes into a ferromagnetic normal state and superconductivity is destroyed. (As stated in the previous section, there are "almost ferromagnetic" superconducting compounds such as $DyMo_6S_8$, $NdRh_4B_4$ and $TmRh_4B_4$, which exhibit a metastability of the ferromagnetic normal phase or choose a long-wavelength antiferromagnetic structure coexisting with superconductivity. $HoMo_6Se_8$ is yet another recently discovered compound of this type [20]).

The earliest measurements on $ErRh_4B_4$ were all carried out on polycrystalline samples. The superconducting transition temperature $Tc_1$ is at 8.67 K. Heat capacity and resistivity measurements gave a reentrant transition at $Tc_2 = 0.93$ K, with hysteretic behaviour characteristic of a first-order transition. Early neutron scattering measurements [21] revealed that the low temperature phase had an ordered basal-plane ferromagnetic moment of 5.6 $\mu_B$/Er, about 30% smaller than the values determined by Mossbauer [22] measurements (8.3 $\mu_B$/Er). Later neutron scattering results [23] (also on polycrystalline samples) showed a peak in the small angle scattering with $q_s = 0.06$ Å$^{-1}$ occurring near $Tc_2$ between temperatures of 0.65 K and 1.1 K. This peak was originally taken as evidence of the intermediate spirally ordered magnetic phase predicted theoretically and discussed in the last section. However, the spontaneous vortex lattice state could also not be ruled out completely. The small angle peak showed strong temperature hysteresis and so did the ferromagnetic peak, but in the opposite sense, ie. the small angle peak was larger on cooling, while the ferromagnetic intensity stayed large on warming.

Since the neutron diffraction intensity from a magnetic structure is proportional to the square of the magnetization component perpendicular to the neutron scattering wave vector Q, observation of a small angle satellite necessarily implies that the modulated structure in the intermediate phase must have a non-zero transverse component. However, the lack of further detailed information about the magnetic structure of this phase, the fact that the polycrystalline measurements showed some sample dependence, and the possible smearing out of the transitions due to sample inhomogeneities pointed to the need for single crystal measurements, which will be described next. The latter measurements have shown that the behaviour in the "intermediate phase" is in fact more subtle and complex than the above pictures of a simple pure spiral or vortex type intermediate phase separating the superconducting paramagnetic and normal ferromagnetic phases. More recently, single crystal samples of $ErRh_4B_4$ were grown by D G Hinks at Argonne National Laboratory and pieces were used for resistivity and magnetization measurements by Crabtree and collaborators [24] and for neutron diffraction [25,26].

The resistive behaviour near $Tc_2$ is shown in detail in Fig. 1. A hysteresis in temperature of about 50 mK is observed, indicating a first order transition at $Tc_2$. The temperatures at which the transitions occurred were not perfectly reproducible, varying by as much as 20 mK from run to run. However, the magnitude of the resistivity change at the transition and the hysteresis between warming and cooling were reproducible, even though the absolute temperatures were not. Such a non-reproducibility is consistent with a firs-order transition, since the exact temperature at which nucleation occurs may depend on sample history.

There have been many detailed magnetization measurements carried out on the single crystal by Crabtree and his collaborators [24], which will not be discussed here in detail.

The neutron measurements were carried out at the HFIR Reactor at Oak Ridge National Laboratory. Diffraction measurements were carried out in the a*-c* plane of the reciprocal lattice, with the (010) a-axis perpendicular to this scattering plane. For the field-dependent measurements described below, the field was applied along the a-axis. From the point of view of a quantitative analysis of the results, the principal difficulties in the single crystal measurements are due to the neutron absorption in the Rh nuclei, (the boron used in the sample being isotopically depleted in the

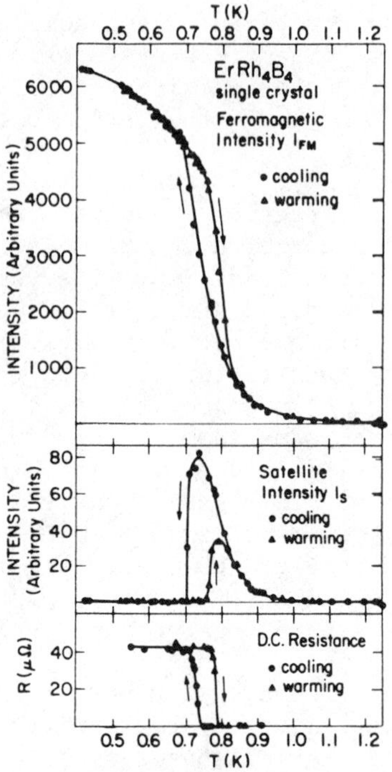


Fig. 1: Behaviour of the intensities of the ferromagnetic and the satellite (MM) reflections and of the D.C. resistance as a function of temperature for the $ErRh_4B_4$ single crystal

heavily-absorbing $B^{10}$ nuclei) and due to secondary extinction effects in the very strong Bragg peaks (both magnetic and nuclear) from the sample. The latter preclude a highly accurate determination of the ordered ferromagnetic moments at low temperatures.

Magnetic satellites were found in the a*–c* plane around the Bragg reflections at the positions $\underline{Q} \equiv \underline{\tau} + q_s$, where $q_s = \pm 0.042\ \underline{a}^* \pm 0.55\ c^*$. Because of the ratio of c/a for $ErRh_4B_4$, these correspond to a nearly perfect square of points in reciprocal space. Two much weaker satellites were also seen at ± 0.055 c* but these are attributed to the equivalent other set of four satellites ± 0.042 b* ± 0.055 c* (b* being the (vertical) reciprocal lattice vector (010) equivalent to the a* vector in the scattering plane) picked up by the relaxed vertical resolution of the diffractometer. The magnitude of the modulation wavevector corresponds to a modulation wavelength of 91.8 ± 2.7 Å, consistent with the peak seen in the small-angle scattering from the polycrystalline samples [23]. Ferromagnetic intensity was also observed growing at the nuclear Bragg positions below T $\lesssim$ 1.2 K. The temperature dependence of the ferromagnetic and satellite intensities, together with the resistivity is shown for both cooling and warming curves in Fig. 1. The equilibration time (with changes in temperature) for the magnetic intensities was found to be of the order of a few minutes or less, and the results are completely reproducible, provided the temperature is changed in only one direction during the measurements.

From Fig. 1, certain conclusions may be drawn. There is a complete correlation of the modulated magnetic moment with bulk superconductivity (with hysteresis) at $Tc_2$. The rapid rise in the ferromagnetic component and the normal behaviour of the order parameter below $Tc_2$, accompanied by the disappearance of the satellite and of bulk superconductivity in the sample, confirms a "re-entrance" to a normal ferromagnetic state in a first-order transition. Nevertheless, there is clearly a temperature region where ferromagnetic order co-exists with the magnetically modulated (MM) phase. At 0.73 K, where the satellite is nearly at its maximum intensity, high-resolution diffraction scans (with a diffractometer resolution of 0.005 $Å^{-1}$ (FWHM)) showed no broadening of either the main ferromagnetic reflection or the satellite peak, indicating coherence of both structures on length scales of the order of $\gtrsim$ 1000 Å in real space. At higher temperatures, ($T \gtrsim 0.88$ K), both the ferromagnetic intensity and the satellite intensity show a temperature dependence more characteristic of critical scattering or short-range order with an upward curvature. The ferromagnetic intensity is essentially saturated below T = 0.4 K.

A search was also made for harmonics of the basic satellite peak, ie. at $2\underline{q}_s$, $3\underline{q}_s$ and also at other points in the square grid suggested by the satellite positions, at $(3\underline{\zeta} + \underline{\eta})$, $(\underline{\zeta} + 3\underline{\eta})$, etc. (where $\underline{\zeta} = 0.042\ \underline{a}^*$, $\underline{\eta} = 0.055\ \underline{c}^*$). None were found within a statistical uncertainty of 2% of the principal satellite peak.

From the observed intensities, magnetic structure factors were obtained for the satellite peaks and the ferromagnetic intensities (although, as discussed above, extinction effects made the latter subject to large uncertainties at low temperatures). The ferromagnetic intensity per unit cell from the $Er^{+3}$ ions in $ErRh_4B_4$ for neutron scattering wate vector $\underline{Q}$ (equal to a reciprocal wave vector) is proportional to

$$I = (\gamma e^2/2m_e c^2)^2\ [f(Q)]^2\ |F(Q)|^2 \qquad (1)$$

where f(Q) is the magnetic form factor for $Er^{+3}$ (determined from other neutron measurements [27]) and $F(\underline{Q})$ is the magnetic structure factor, given by

$$|F(\underline{Q})|^2 = 4\mu^2 \, [1-(\hat{Q}\cdot\hat{\mu})^2] \tag{2}$$

where $\mu$ is the ferromagnetic moment, and $\hat{\underline{\mu}}$ and $\hat{\underline{Q}}$ are unit vectors along $\underline{\mu}$ and $\underline{Q}$. In Eq. (1), $(\gamma e^2/2m_e c^2)$ is the magnetic scattering length per Bohr magneton and is numerically equal to $0.2695 \times 10^{-12}$cm. The first measurements on the single crystal were made with a neutron wavelength of 2.35 Å, and indicated that practically all the ferromagnetic moment was oriented along the vertical a axis (010) (normal to the scattering plane), as indicated by the equality of the magnetic structure factor at the (100), (002), and (101) Bragg reflections, but the low-temperature moment yielded only 4.8 $\mu_B$ per $Er^{+3}$ ion, considerably lower than the 5.6 $\mu_B$ obtained from the powder diffraction measurements of Moncton et al. [21]. Later measurements [28] at wavelengths of 0.8 Å (and thus subject to much less severe extinction effects) indicate that there is in fact a distribution of ferromagnetic domains along both easy (a) axes, in the scattering plane as well as normal to it (but with the majority of the domains still aligned along the normal to the plane), with a saturated ferromagnetic moment of $\sim$ 5.3 $\mu_B$, indicating still some extinction at these wavelengths. (The polycrystalline diffraction data are not subject to extinction effects owing to the random orientation of the grains). The extinction effects indicate that the lower temperature part of the ferromagnetic intensity shown in Fig. 1 is probably smaller (by about 30%) than the true curve determined by the square of the real ferromagnetic order parameter.

For the MM structure, the weakness of the satellite peaks renders extinction effects inoperative as a source of error. The direction of propagation of the modulation (which is at roughly 45° to the c-axis)makes a spiral structure unlikely, since the magnetic anisotropy locking the moments in the basal (001) plane is very large [24]. A non-right angle spiral with the moments in the basal plane is also ruled out since the magnetic structure factor of satellites around (200), (002), and (101) are found to be equal within experimental error; whereas a basal-plane spiral would require the (002) satellite structure factor (where $\underline{Q}$ is everywhere normal to the moments) to be twice as large as the (200) satellite structure factor (where the $[1-(\hat{Q}\cdot\hat{\mu})^2]$ factor gives an average of one half (see Eq. (2)). The alternative possibility of a plane polarized sinusoidal MM structure with the ordered moment everywhere along the easy a* axis which is normal to the direction of propagation of the modulation, is consistent with the data. The plane polarized form of the MM structure draws attention to the role of single-ion magnetic anisotropy in determining the structure for magnetic LRO in the superconductivity phase. The role of magnetic anisotropy and the possibility of a linear sinusoidal wave structure was first discussed by Greenside, Blount, and Varma [19] and has been incorporated now into most of the theoretical treatments of the problem [29-32]. The reason for the rather strange direction of $q_s$ is not yet understood and may depend on Fermi-surface effects which determine (together with the superconducting interactions) the effective $J(\underline{q})$.

The existence of magnetic LRO of both $\underline{q} = 0$ (ferromagnetic) and $\underline{q} = \underline{q}_s$ (the MM structure) in the intermediate temperature region $0.7 \text{ K} < T < 0.88 \text{ K}$ (which also implies a coexistence between a ferromagnetic component and superconductivity) is an important but puzzling result. There are two possibilities: either both the ferromagnetic and sinusoidal components arise from the same microscopic regions of the sample, or the sample divides spontaneously into normal ferromagnetic domains and superconducting domains containing the MM structure. One example of the first possibility is the regular vortex lattice [16], where by symmetry both components of $\underline{q}_s$ perpendicular to the moment directions (eg. an (010) axis) would coexist with a ferromagnetic component throughout a particular domain. We tend not to favour such an interpretation for the following reasons: (a) there seems to be no dependence of $q_s$ on the spontaneous magnetization or an applied field as would be expected by the flux-quantization condition for a vortex lattice, (b) critical field measurements by Behroozi et al. [24] indicate that below 1 K $ErRh_4B_4$ behaves like a type-I rather than a type-II superconductor, (c) the various satellite intensities around a given Bragg point in the a*-c* plane are not always equal, depending on the history of warming and cooling the sample, which would imply that there are at least domains corresponding to the different (equivalent) directions of propagation of the modulation, and (d) the strong hysteresis making the ratio of modulated to ferromagnetic intensities history-dependent also argues against at least a conventional vortex lattice phase. If we assume that the sample is heterogeneously divided between volumes of ferromagnetic and sinusoidal order, and further assume that the latter is equally distributed among domains with different but equivalent $q_s$ vectors (an assumption not always justified as discussed above), we arrive at the following ratio of the squared structure factors

$$|F_{MM}|^2/|F_{FM}|^2 = \frac{1}{16}\left(\frac{\mu_{MM}}{\mu_{FM}}\right)^2 \frac{V_{MM}}{V_{FM}} \tag{3}$$

where $\mu_{MM}$, $V_{MM}$, and $\mu_{FM}$, $V_{FM}$ are the amplitude of the ordered moment and the total crystal volume associated with the MM and ferromagnetic structures, respectively. Thus, in order to determine the two ratios individually, we must assume that an externally variable parameter (such as a magnetic field) converts some of the crystal volume from $V_{MM}$ to $V_{FM}$ without affecting the ordered moments in the respective regions. Field-dependent measurements of the intensity ratio yielded, at 0.72K, values of $(V_{MM}/V_{FM}) \simeq 0.2$ and $\mu_{MM}/\mu_{FM} \simeq 1.4$.

We now briefly discuss the field-dependent measurements which are still preliminary. Fig. 2 shows the effect at T = 0.78 K of a magnetic field applied along the (010) axis on the intensities of a magnetic satellite from the (002) Bragg point and of the (101) Bragg peak. One notices that the satellite intensity decreases very rapidly with increasing field, and is essentially gone by $H_{app}$ = 0.045 Tesla. (It should be remembered that owing to the irregular shape of the crystal used, the demagnetization field cannot be calculated very accurately. Hence, the applied field values quoted here are not to be taken as the actual internal field). The position of the satellite peak does not shift, except to slightly larger wave vectors just before it vanishes. The ferromagnetic intensity, on the other hand, is initially very little affected by the field, and then slowly increases. Fig. 3 shows the behaviour of the (101) intensity at 1 K as a function of increasing and decreasing field. It may be seen that the ferromagnetic intensity is not affected up to a critical field (corresponding to $Hc_1$) but on applying and then decreasing the field to zero, a finite ferromagnetic intensity remains. This may be due to small ferromagnetic domains residing in regions of trapped flux, or the "forced ferromagnetism" referred to earlier. Fig. 4 shows the corresponding curve at T = 2 K. One may see effects at both $Hc_1$ and $Hc_2$. At $Hc_2$ one sees a jump in the ferromagnetic intensity with increasing field consistent with the findings of Behroozi et al. [24] of a discontinuity in the magnetization at $Hc_2$ in the lower temperature part of the H-T phase diagram.

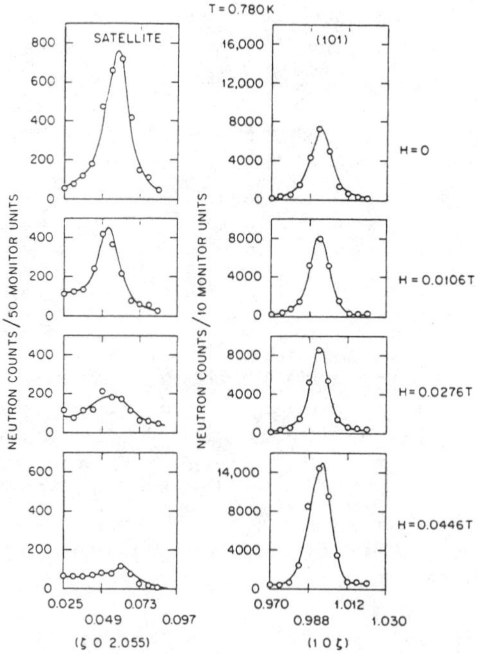


Fig. 2:
Scans through the satellite and ferromagnetic (101) reflections at T=0.78K for various values of applied field H (in Tesla)

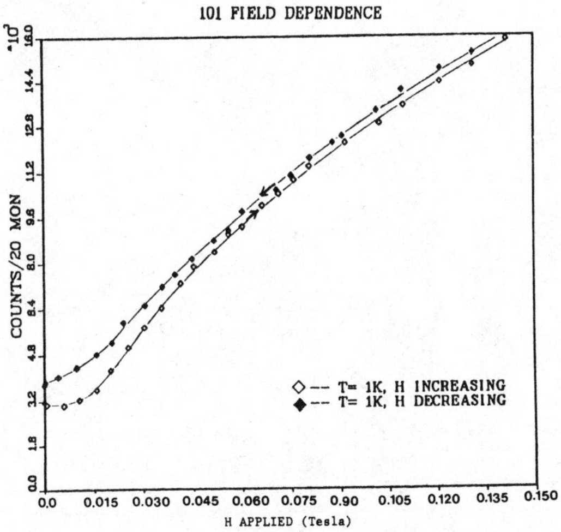


Fig. 3: Ferromagnetic intensity at T=1 K as a function of applied field

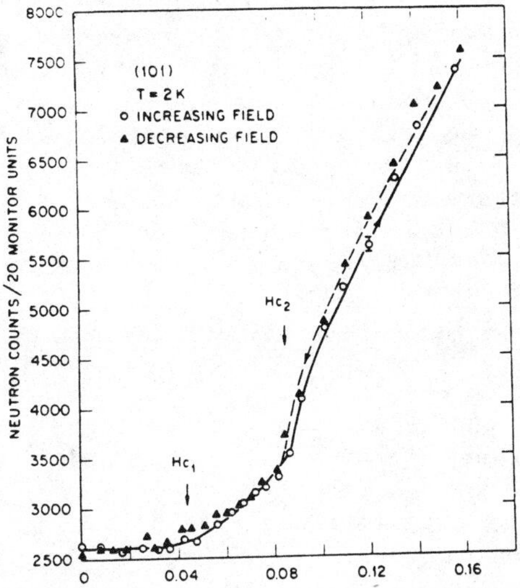


Fig. 4: Ferromagnetic intensity at T=2 K as a function of applied field

This jump in the ferromagnetic intensity decreases with increasing temperature, also consistent with the results of Behroozi et al. and indicating the presence of a tricritical point, where the transition at $Hc_2$ changes from first to second-order. At applied fields greater than 0.07 Tesla, the temperature hysteresis of the ferromagnetic intensity has disappeared, again indicating the finite range of the first-order phase boundary in H-T space (see Fig. 5).

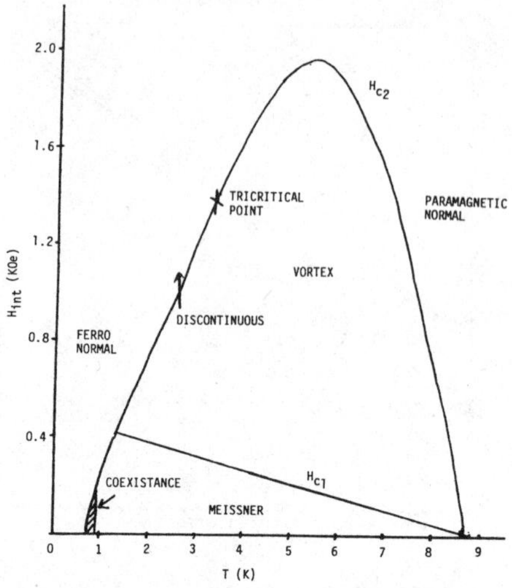


Fig. 5: Rough sketch of phase diagram for $ErRh_4B_4$ as a function of internal field and temperature

Finally, we discuss, briefly, recent more detailed studies of the higher temperature region of the "co-existence" phase (between 0.88 K and 1.1 K) at Risø National Laboratory, Denmark [38]. These measurements, which are still preliminary, show that in this temperature regime both the ferromagnetic intensity and the satellite peak show broadening in wave vector space (see Fig. 6) corresponding to short-range fluctuations. As the temperature is lowered, these correlations develop longer and longer ranges, until true LRO sets in. Thus, short-range order seems to be the reason for the upward curvature of the intensities shown in Fig. 1 in this temperature regime. In this regime, there is no temperature hysteresis of the intensities, and the ferromagnetic and satellite intensities seem to grow proportionally to each other. This is consistent with critical fluctuations in the superconducting paramagnetic phase, but the occurrence of ferromagnetic as well as sinusoidal fluctuations seems strange and unanticipated by theory. In this regime, the peak of the scattering at the satellite position also seems shifted to slightly smaller wave vector.

## 4. COMPARISON WITH THEORY

For the ferromagnetic superconductor $ErRh_4B_4$, it thus appears that there are (in zero applied field) four regimes as the temperature is lowered below $Tc_1$. First, there exists a superconducting paramagnetic phase. At lower temperatures, short-range correlations develop, corresponding to both ferromagnetic and sinusoidal order. (Long wavelength ferromagnetic fluctuations will not in principle be prevented by superconductivity provided their wavelength is not

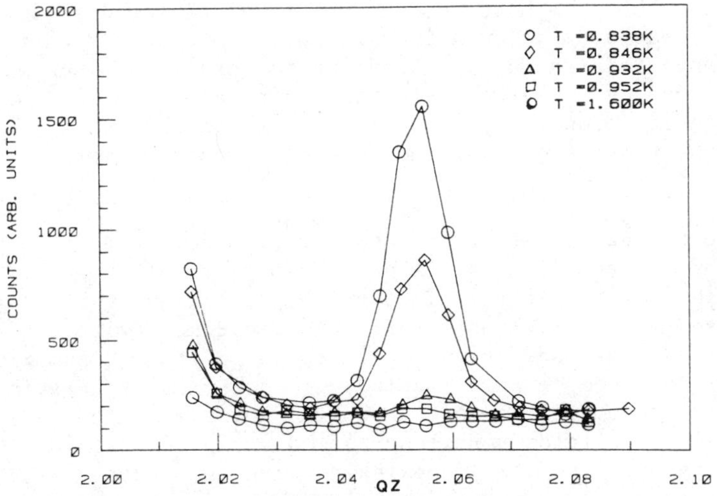


Fig. 6: Scans through satellite positions and in wings of ferromagnetic peak as a function of temperature showing broadening effects at high temperature

greater than $\lambda_L$ which is presumably very large on the scale of the resolution used in these experiments). At about 0.88 K, LRO of both the ferromagnetic and modulated components sets in, although the exact coincidence of the transition for both these components is still not established. Finally, there is a lower first-order transition, with temperature hysteresis, to a normal ferromagnetic phase.

Although the existence and wave vector of the MM structure is consistent with the theoretical ideas based on the electromagnetic coupling with the superconducting order parameter, the actual coexistence of both the ferromagnetic and MM components (and bulk superconductivity) is as yet not very well understood. The transverse plane polarized nature of the MM structure may be understood in terms of the electromagnetic coupling between the moments. For the reasons given in the last section we do not tend to favour the picture of the coexistence region as that of a conventional vortex lattice phase. Hu [33] has proposed that this phase is in fact due to an antiferromagnetic vortex lattice, with a very long-wavelength periodicity (and therefore the q=0 component presumably corresponds to intensity at a finite but very small wave vector unresolved from the Bragg point). In this model the MM structure is simply due to oscillatory magnetism between the vortex cores. There is as yet no direct proof of the correctness of this model, and some other predictions of this model (relating to the existence of harmonics and satellites at $(3\,\underline{\zeta} + \underline{\eta})$ and $(3\,\underline{\eta} + \underline{\zeta})$ are contradicted by experiment. The lack of observable harmonics also seems to rule out the antiphase domain structure predicted by Bulaevskii et al [34] based on calculations including both the electromagnetic and exchange interactions. According to this model, one would expect higher order harmonics of the main satellite (m = odd integer) with intensity proportional to $(1/m^2)$. Bulaevskii et al [34] explain this discrepancy as due to the presence of irregular domain walls.

More recently some other theoretical models have been proposed to account for the coexistence phase. Several of these are due to Tachiki and collaborators. In one [29], a laminar vortex structure is proposed (which implies a ferromagnetic domain with a single linear sinusoidal modulation superimposed on it) and numerical calculations based on this model do in fact predict behaviour qualitatively similar to that shown in Fig. 2. A more recent model by Tachiki [30] yields very long-period antiphase domains (with periodicity $\sim 5\lambda_L$ to $10\lambda_L$) coexisting with the shorter wavelength MM structure in the superconductivity phase just above $Tc_1$. Since $\lambda_L \sim 1000$ Å or greater, the former would account for the "ferromagnetic" LRO observed as well as the ferromagnetic fluctuations in the paramagnetic phase. Yet another recent proposal by Tachiki et al [31] suggests inhomogeneities (eg. due to strains) in the crystal as the reason for local variations of $Tc_2$ and hence for the coexistence of a true normal ferromagnetic and superconducting MM phases. This does not seem consistent with the growth of the "ferromagnetic" and sinusoidal fluctuations proportionately in the paramagnetic phase, and the smooth increase but sudden cut-off of the satellite intensity at 0.71 K.

Machida and collaborators [32] have proposed a quite different picture for the coexistence phase. According to these authors this phase is a realization of the so-called Fulde-Ferrell state where superconductivity exists in a finite molecular exchange field but with a finite $q_o$ pairing to minimize the free energy. The superconducting order parameter thus has a modulation wave vector $q_o$ which, by second-order exchange coupling between the conduction electrons and f-shells, induces a sinusoidal MM structure for the local spins $q_s = 2q_o$. While their original theory predicted a first-order transition from a paramagnetic superconducting state to one with coexisting ferromagnetic and sinusoidal components, and a second-order transition below that to a normal ferromagnetic phase (in contradiction to experiment), more refined versions of this theory which include higher-order couplings between the two components in the Ginzburg-Landau free energy expression can approximately reproduce the observed behaviour. Their most recent calculations predict several harmonics of the satellites just at the temperature ($\sim$ 1.2 K) where the magnetic correlations begin to set in. These are probably so weak at this temperature that they would in any case be difficult to observe. However, one prediction (a tendency to a smaller $q_s$ when the satellite appears) seems to be borne out experimentally (in the regime where the satellite peak is also broadened, however). This theory completely ignores the electromagnetic interactions and thus the Meissner effect poses a problem here, as yet unresolved.

For the ferromagnetic superconductor $HoMo_6S_8$ [35,36], the situation is probably very similar to that in $ErRh_4B_4$. Again, both ferromagnetic and MM fluctuations occur just above $Tc_2$ in the superconducting paramagnetic phase, below which a MM phase exhibiting LRO appears and also ferromagnetic order. The actual coexistence region is not quite so well defined as yet, but clearly within it hysteretic effects are observed [36]. A detailed determination of the structure and direction of propagation of the MM phase awaits the availability of single crystals of this material.

A word should be said about domain wall effects in the ferromagnetic regimes. In both the single crystal and polycrystalline samples, diffuse scattering is seen around $\underline{q} = 0$ in the ferromagnetic phase. This is not due to critical scattering because for single crystal $ErRh_4B_4$ this component could not be fit with the usual Lorentzian scattering function in reciprocal space. It has been interpreted as due to domain walls in the ferromagnetic phase. Its shape can be fit with either a Gaussian centered at $\underline{q} = 0$ [26] or with a $(1/q^4)$ law characteristic of small-angle scattering from sharply-defined but smoothly curved surfaces (Porod scattering). Finally, the question of why the spontaneous ordered ferromagnetic moment at low temperatures is so much less than the $Er^{3+}$ free-ion moment, or that determined by Mossbauer measurements [22] is still a mystery, and apparently cannot be explained by crystal-field effects [46].

## 5. NEUTRON SCATTERING FROM INTERMEDIATE VALENCE MATERIALS

The intermediate valence (IV) materials consist mainly of metals, alloys or compounds containing the rare earth elements Ce, Yb, Sm, Eu and Tm. In addition, certain actinide compounds also exhibit IV-like behaviour. The physics of the IV compounds and the related class of compounds known as "Dense Kondo Systems" has been one of the major areas of activity in the field of magnetism over the last few years.

Intermediate valence compounds are characterised by a quantum fluctuation of the f-shell on a rare-earth ion between $f^n$ and $f^{n-1}$ configurations. This is due to the fact that the energy difference between these configurations

$$\Delta E = E(f^n) - E(f^{n-1}) - \mu$$

(where $\mu$ is the chemical potential or Fermi energy in the case of a metal) is comparable to the hybridization V between electrons in the f-orbitals and the conduction band. As a result ground-state measurements, or measurements which probe over long time-scales (eg. Mossbauer, NMR, etc.) are consistent with a non-integral number of f-electrons on the f-shell. Nevertheless, these f-electrons cannot be treated simply as itinerant or band electrons, because there are important effects arising from the very large Coulomb correlations between the f-electrons, as well as the strong spin-orbit coupling on the f-shells. The result is a very difficult many-body problem and significant theoretical progress has only been made in treating the case of a <u>single</u> rare-earth impurity in a crystal, rather than the full lattice of such ions.

The IV materials are characterised by having the Fermi level trapped in a region of very high density of electronic states, resulting in large values of the bulk susceptibility ($\sim$ 1-8 x $10^{-3}$ emu/mole) and of the electronic heat capacity ($\sim$ 200-2000 mjoules/K/mole). These materials also show considerable lattice softening effects [38] in their elastic moduli. One of the real mysteries regarding these materials is why they do not in general order magnetically. (The only exception is TmSe). The associated so-called "Dense Kondo Systems" are very similar to the above materials except that $\Delta E$ is larger and more negative, ie. the "f-level" lies further below the Fermi energy. For these materials, the measured valence comes out to be very close to being integral. Nevertheless, there is in general still an enormously large density of states at the Fermi energy, $E_f$, and a weakening of the tendency to magnetic order.

An important point to note is that the fluctuation time between f-configurations in these materials is given by

$$\tau = 2\hbar/(V^2\rho)$$

where $\rho$ is the density of conduction electron states at $E_F$. This result is simply obtained by perturbation theory assuming V is the perturbation causing a fluctuation of the f-electron from a given f-shell. For most IV compounds, $\tau$ comes out to be of the order of $\sim 10^{-12}$-$10^{-13}$ secs, which shows that inelastic neutron scattering is in fact a useful probe for such fluctuations. The neutrons actually see the spin fluctuations of the f-electrons, but in the IV regime, a hopping of the f-electron also implies a spin fluctuation, so that spin and charge fluctuations occur together. In the Kondo-like regime, charge fluctuations are considerably decreased, but spin fluctuations still occur, and so neutrons can probe both kinds of systems. In addition, the charge fluctuations can couple both adiabatically and dynamically to the phonons of an IV crystal and this leads to further areas in which inelastic neutron scattering (INS) can be applied to the study of the physics of these materials. It is not our purpose here to review the physics of IV materials. The interested reader may find several comprehensive works on the subject [39-42]. In this article, we shall simply discuss certain recent INS studies of such systems and attempt to relate them to the (still incompletely understood) theory of elementary excitations in these materials.

At this stage, and for the purposes of this workshop, it seems worth pointing out the strengths and weaknesses of INS as a probe of electronic structure in paramagnetic metals. Photoemission is currently one of the most commonly used techniques to study electronic densities of states. The expression for valence band photoemission is

$$\rho_v(\omega) = \sum_n | < \phi_n^{N-1} | C_f | \phi_o^N > |^2 \delta(\omega - E_o^N + E_n^{N-1}) \qquad (4)$$

where $\phi_n^{N-1}$ denotes the $n^{th}$ excited state of the system with one less electron, $C_f$ is the f-electron annihilation operator, and $E_n$ is the corresponding energy. Eq. (4) is easily seen to be proportional to the spectral distribution of the one-electron Green's function for f-electron, ie. Im $G_f(\omega)$. In principle an exact many-body calculation of this function should yield agreement with experiment, but should not be taken as rigorously proportional to the "one-electron density of states (DOS)", owing to the relaxation effects when the hole is left behind after photoemission (shake-up or shake-down processes). Nor should we necessarily expect good agreement with calculations of excited DOS from band calculations, as the Density Functional Theorem says that such calculations are only in principle exact for the ground state of the system. An inelastic scattering experiment, on the other hand, where the electron does not leave the system but makes a transition from one state (below $E_f$) to another (slightly above $E_f$) is a gentler spectroscopy, where one may hope to find some simpler connection with ground-state properties, since the system is not violently changed during the process. As is well-known, the scattering function $S(\underline{Q},\omega)$ for momentum transfer $\underline{Q}$ and energy loss $\omega$ of a general probe is given in the Born approximation by

$$S(\underline{Q},\omega) \sim [V(\underline{Q})]^2 [1-e^{-\beta\omega}]^{-1} \text{Im}\chi(\underline{Q},\omega) \qquad (5)$$

where $[V(\underline{Q})]^2$ represents a $\underline{Q}$-dependent interaction of the probe with the particles of the system, and $\chi(Q,\omega)$ is a generalized susceptibility. Eq. (5) applies for neutrons, high-energy electrons and X-rays. In the first case, $\chi(Q,\omega)$ represents a spin susceptibility, whereas in the latter two cases it represents a charge susceptibility. In a purely non-interacting band picture, both are related to a joint density of electronic states from occupied to unoccupied states across momentum transfer Q and energy $\omega$.

Let us consider in more detail the cross-section for unpolarized neutron scattering from a paramagnetic system of electrons, neglecting electron-electron interaction effects. The cross-section may be written [43]

$$\frac{d^2\sigma}{d\Omega dE} = \frac{k}{k_o}[1-e^{-\beta\omega}]^{-1}\left(\frac{\gamma e^2}{m_e c^2}\right)^2/(g\mu_B)^2 f^2(\underline{Q})\,\mathrm{Im}\chi^{+-}(\underline{Q},\omega) \tag{6}$$

where $k_o$ and k are the incident and scattered neutron wave vectors, $f(\underline{Q})$ is a factor which averages over matrix element effects and can be written [44] as a sum of a spin and an orbital contribution

$$f(\underline{Q}) = f_{spin}(\underline{Q}) + f_{orb}(\underline{Q}) \tag{7}$$

and

$$\mathrm{Im}\chi^{+-}(\underline{Q},\omega) = 2(g\mu_B)^2\Sigma_k n_k(1-n_{k+Q})\delta(\omega+E_k - E_{k+Q}) \tag{8}$$

where $n_k$ is the occupation number for electrons in state k (including an implicit band index) with a given spin and with energy $E_k$, etc.

For scattering from a powder sample, $\underline{Q}$ is averaged over all directions and thus if the volume over which Q is averaged is large enough compared to a Brillouin zone (we assume that the form factor f(Q) is slowly varying so that a finite range of |Q| is also averaged over in the experiment), we may write

$$\mathrm{Im}\chi^{+-}(\underline{Q},\omega) = \frac{2}{N}(g\mu_B)^2{}_{k,k'}\Sigma\; n_k(1-n_{k'})\delta(\omega+E_k - E_{k'}) \tag{9}$$

where N is the total number of unit cells in the sample.

It may then easily be shown that

$$\mathrm{Im}\chi^{+-}(\underline{Q},\omega) = \tfrac{1}{4}(g\mu_B)^2 \int n(\varepsilon)[1-(\varepsilon+\omega)]g(\varepsilon)g(\varepsilon+\omega)d\varepsilon \tag{10}$$

where $g(\varepsilon)$ is the density of electronic states at energy $\varepsilon$, and $n(\varepsilon)$ is the Fermi function. Thus, in this approximation, a paramagnetic scattering experiment from a powder sample probes the joint DOS of the electrons, a quantity which may readily be computed from the predictions of band theory. The above applies to the case of weakly interacting electrons where strong exchange and correlation effects are absent. Unfortunately, insertion of typical numbers for normal metals into Eq. (10) and (6) yields miniscule scattering cross-sections, eg. the partial differential cross-section comes out to be of the order of $10^{-4}$ barns/steridian/eV. For the case of transition metals, the number is larger but still only $\sim 10^{-3}$ barns/steridian/eV. On the other hand, we note from Equation (10) that such scattering cross-sections will (for scattering mainly across the Fermi Level) be roughly proportional to $|g(E_f)|^2$ so that with the large increase in $g(E_f)$ for the IV class of materials, enormous gains in scattering cross-section occur. In addition, the free-electron g-factor in Eq. (10) has to be replaced for strongly spin-orbit coupled rare earth ions by the Landé g-factor of the rare earth ion, leading to an even grea er increase in cross-section. For these reasons, the paramagnetic scattering from these systems does in fact lend itself to observation more easily than for ordinary metals, and the cross-sections can become almost of the order of phonon cross-sections in such cases.

Nethertheless, there are severe experimental problems arising from two main sources: (a) The kinematic restrictions on the INS process (ie. the need to conserve energy and momentum for particles of the neutron mass in scattering off light particles like the electrons) force one to use rather large incident neutron energies to stay at reasonably small Q while still transferring large amounts of energy. (The need to stay at small Q is dictated by the fact that the form-factor $f(\underline{Q})$ drops rapidly with $\underline{Q}$, owing to the spatial distribution of the electron wave functions). Thus, for example to transfer even 200 meV of energy at a $\underline{Q}$ of 2.5 $Å^{-1}$, it is necessary to come in with an incident neutron-energy of 1 eV. (b) The contamination due to phonon, multiphonon and multiple (ie. Bragg + phonon) scattering. The first problem can be alleviated by going to spallation sources where large incident fluxes of high energy neutrons are available. The second problem is usually treated by carrying out a second experiment on a "blank" sample, where the phonon frequency spectrum is believed to be the same, but the electronic scattering from the f-electrons is absent. Thus, La or Y compounds may be substituted for the corresponding Ce compounds, etc. This procedure is fraught with danger, however, as the nuclear scattering properties of the "blank" or "pure phonon" sample may not be the same as that of the IV sample, and thus the multiple scattering corrections will in general be different. To some extent this may be circumvented by choosing the "blank" sample so that the product $\mu t$ is the same in the two cases, where $\mu$ is the inverse attenuation length due to all scattering processes, and t is the thickness of the sample in the direction of the beam. Significantly different absorption coefficients, however, can destroy the validity of this result.

## 6. EXPERIMENTAL INS DATA ON IV SYSTEMS

By now a fairly large number of INS studies have been carried out. Some of the many studies of paramagnetic inelastic scattering include studies of Ce-Th alloys [45], Ce compounds ($CeSn_3$ [46], $CePd_3$ [46], $CeAl_2$ [47], $CeCu_2Si_2$ [48]) Yb compounds ($YbAl_3$ [49], $YbCu_2Si_2$ [46], YbCuAl [50]) and from the compound TmSe [51]. In addition, phonon dispersion relations have been studied in α-Ce [52], $CeSn_3$ [53], $Y_xSm_{1-x}S$ [54], and TmSe [55], among others.

The earlier measurements showed that for most of these compounds, the paramagnetic scattering from polycrystalline samples consisted of a broad quasielastic line, which did not show any change in shape with Q but only a decrease in overall intensity consistent with a 4f form-factor. Fig. 7 shows the INS spectrum at 240 K for $CePd_3$ [46]. The phonons are removed by subtracting the corresponding spectrum of $YPd_3$ suitably weighted by a factor taking into account variations of cross-section for the two materials. The width of this quasielastic line is in general one to two orders of magnitude larger than the quasielastic line width in the corresponding integral valent compound, (eg. $CePd_3$ and $TbPd_3$, as shown in Fig. 8), and is generally very weakly temperature dependent as opposed to the linear T-dependence of the (Korringa-like) line width in stable moment systems owing to interactions between the local moments and the conduction electrons. Another difference compared to paramagnetic scattering from "normal" rare earth systems is the lack of any crystal-field transitions.

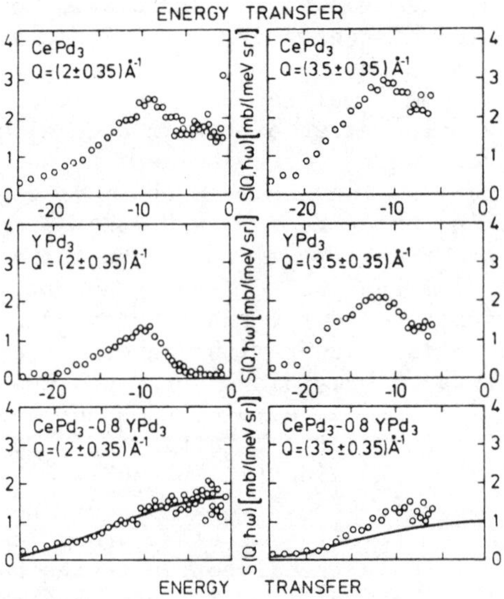


Fig. 7: Energy spectra for $CePd_3$, $YPd_3$, and for their difference at fixed momentum transfer Q

Inelastic (as opposed) to quasielastic peaks have been seen in single-crystal TmSe (at about 10 meV energy transfer) by Grier and Shapiro [51], and in polycrystalline samples by Holland-Moritz, et al. The single crystal results show (Fig. 9) that the intensity of this peak is strongly q-dependent (where q is now the reduced wave-vector in the zone) being maximum at the zone boundary point $2\pi/a$ (1,0,0). The intensity decreases rapidly with increasing temperature. Inelastic peaks have also been inferred by Holland-Moritz et al. [46]

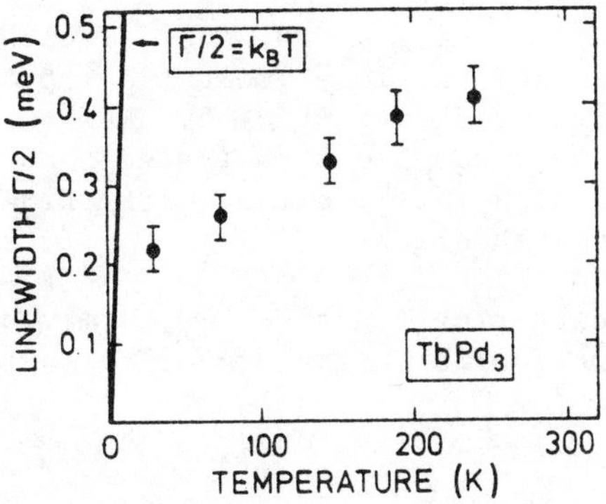


Fig. 8a: Temperature dependence of the quasielastic linewidth for $TbPd_3$

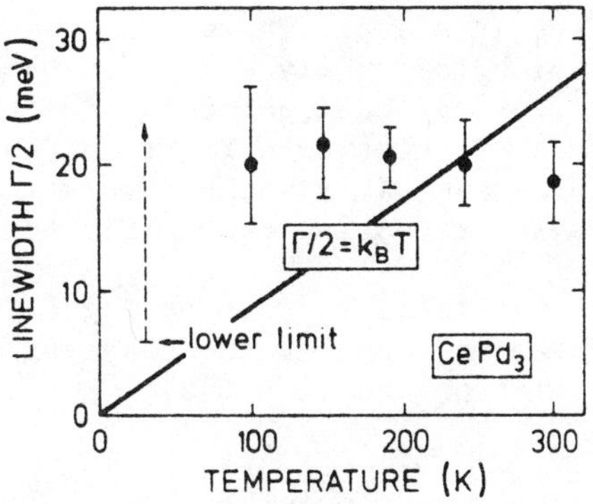


Fig. 8b: Temperature dependence of the quasielastic linewidth for $CePd_3$ (single Lorentzian fits)

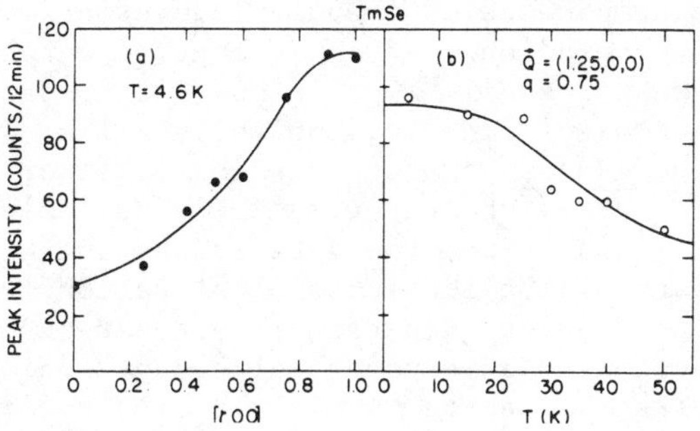


Fig. 9: The (a) Q dependence and (b) temperature dependence of the inelastic peak intensity for TmSe. The lines are drawn as guides to the eye

in polycrystalline samples of $YbCu_2Si_2$ from measurements at different incident energies. They interpret these as practically unresolved and broadened crystal-field transitions. Recently, Murani [50] has found evidence for a broadened 40 meV transition at 5 K in $CeSn_3$ which broadens and disappears into a quasielastic peak at higher temperatures, and for inelastic peaks in YbCuAl and $YbCl_3$. Some of these peaks sit very close to peaks in the phonon density of states but, if the subtraction procedure is conceded as accurate, must nevertheless be electronic in origin.

## 7. DISCUSSION

We have seen that IV compounds show rather broad quasielastic paramagnet scattering, with evidence of inelastic peaks in some cases. We discuss briefly what this tells us about the f-electrons in the IV state. There are two ways of looking at this data (neither of which is totally correct); namely, from the local moment point of view, or the band point of view as discussed in Section 5. The strong correlation effects on the f-shell coupled with f-electron hopping make IV systems a very interesting intermediate correlation case and a testing ground for many-body theories.

For local spins, the magnetic scattering function $S(\underline{Q},\omega)$ may be written in terms of the space and time Fourier transform of the spin correlation function [43],

$$\frac{1}{N}\sum_{i,j} < S_i^+(0)S_j^-(t) > e^{-i\underline{Q}\cdot\underline{R}_{ij}}.$$

If, as some of the polycrystalline measurements suggest, there is no $\underline{Q}$-dependence to $S(Q,\omega)$ (apart from the form-factor), one may conclude that the spins are fluctuating in an uncorrelated manner. While this may be true in some cases, and would lend experimental support to the theoretical speculation that an IV lattice can be approximated by a set of independent f-shell "impurities", it is dangerous to base this conclusion on evidence from polycrystalline data, since it is variation with reduced wavevector which is important, and this is clearly averaged over in the experiments. The inelastic spectrum is then proportional (apart from the detailed balance factor) to the time Fourier transform of the single-spin relaxation function $< S_i^+(0)S_i^-(t) >$. This Fourier transform is often taken to be a Lorentzian corresponding to an exponential decay. Values of $\Gamma$ derived from quasielastic spectra thus yield values for the "fluctuation time" on an f-shell and cause it to lose all memory of its previous spin orientation. As stated above, these fluctuation times come out to be typically of the order of $10^{-12}$ - $10^{-13}$ seconds, and fairly independent of temperature, owing to the quantum nature of the fluctuations on the 4f shell. An integral over energy then also yields the value of $<S^2>$ and hence for the 4f occupation on the f-shell and values for the valence of several IV compounds have been determined with this method [46]. (It should be noted that the above result stays true even if the local moment picture is not valid, provided an average over the reduced wavevector is made). Values for $<n_f>$, the average number of f-electrons obtained by this method for a number of IV compounds [46], are in approximate agreement with the values obtained by other methods, such as X-ray L-edge absorption spectroscopy, lattice constant measurements and photoemission. It should also be noted that a check on the quantitative accuracy of neutron $S(Q,\omega)$ determinations is the rigorous sum rule for the bulk magnetic susceptibility, which from Eq. (6), can be shown to be

$$\chi = \left(\frac{g\mu_B m_e c^2}{\gamma e^2}\right)^2 \lim_{Q\to 0} \int d\omega \, \frac{[1-e^{-\beta\omega}]}{\omega} \, S(Q,\omega) \qquad (11)$$

Values from neutron scattering measurements, where such quantitative determinations have been made [50], are in general in good agreement with measured bulk susceptibilities.

Our present theoretical understanding of the fluctuations from the local point of view is based on a picture of the IV crystal lattice as a collection of "dense impurities" (ie. non-interacting ions) embedded in a sea of conduction electrons with which they interact via an on-site hybridization mechanism, ie. the dynamics is governed by the Anderson Hamiltonian. Treatments of the dymanical relaxation of the local moments (assuming a very large or effectively infinite on-site Hubbard interaction parameter U for the f-electrons) by Kuramoto [56], Muller-Hartmann [57] and others yield semi-quantitative agreement for the behaviour of the paramagnetic scattering and magnetic susceptibility at high temperatures. On the other hand, as discussed above, at low temperatures often inelastic features (rather than a single broad peak centered on $\omega = 0$) are observed. It is not clear whether these can be explained within the single impurity picture. In any case, at low temperatures, one expects coherence to develop between the f-orbitals on the lattice of rare-earth ions. A solution of the dynamics would require a rigorous solution of the Anderson Lattice problem, which is not available. Nevertheless, on theoretical and experimental grounds one expects for T less than some characteristic temperature (often referred to as the Spin Fluctuation Temperature) that the electrons develop so-called "Fermi Liquid" behaviour in spite of the large Coulomb correlations between the f-electrons. Mean-field treatments of the Anderson Lattice using a Green's function approach, by Fedro and Sinha [58-60], Huber [61], Brandow [62] and others have yielded a picture of the low-lying one-electron excitations in such a system as being band-like states resulting from a flat multiplet f-band hybridized with conduction bands. In this picture, one obtains an $S(Q,\omega)$ similar to that given by Eqs. [6]-[8] above, but with the correlation effects resulting in an enhancement of the susceptibility [58], similar to exchange enhancement in transition metals. In addition, one obtains hybridization "gaps" in the band-structure, as shown schematically in Fig. 10. The theory predicts the $E_F$ is trapped in a high density of states region close to the gap. The central peak then arises from intraband transitions across $E_F$ while inelastic structure arises from transitions across the gap (or gaps, in the general case). Phenomenological Fermi liquid theories have also been used to explain the central peak, but obviously not the inelastic structure, since $S(Q,\omega)$ from a simple Fermi liquid has no such structure. In the "coherent band" models for the scattering, the scattering function clearly depends on the reduced wavevector q, although this can only show up in single crystal experiments. Thus, the 10 meV excitation observed in TmSe [51] can be explained in terms of an excitation across the hybridization gap [58]. This model successfully explains both the $\underline{q}$ dependence and temperature dependence of the observed peak (Figs. 9a and 9b). The reason why the peak occurs predominantly at the zone boundary can be seen from Fig. 10, where one may see that this wavevector is the one which connects the high density of states regions in the upper and lower "sub-bands". Similarly, one may speculate that the broadened inelastic peaks observed in other IV compounds can also be due to splitting up of the densities of states associated with the f-bands by relatively small hybridization splittings or combinations of hybridization and pure crystal field splittings. There are, however, some recent measurements [63] which indicate that in TmSe there is an inelastic peak even when Tm is considerably diluted (to the 5% level) by Yttrium. This is hard to reconcile with a coherent gap model for this excitation. Some single crystal measurements, would be extremely desirable. For the "Kondo-like" materials, the situation is slightly different. Here the unperturbed f-level lies much further below $E_F$, but calculations of the one-electron Green's function both for the lattice [59,60] and the single impurity [64,65] beyond the mean field-type theories, have shown that a sharp resonance arises at temperatures below a characteristic temperature $T_k$ in the f-like DOS at $E_F$. This would then account for both the high susceptibility and electronic heat capacity of these materials and the presence once again of a central peak in the scattering. The theories predict, however, that the resonance in $E_F$ sharpens as T is lowered and so one would expect a Γ which decreases with T. This is indeed what is observed for some of these materials.

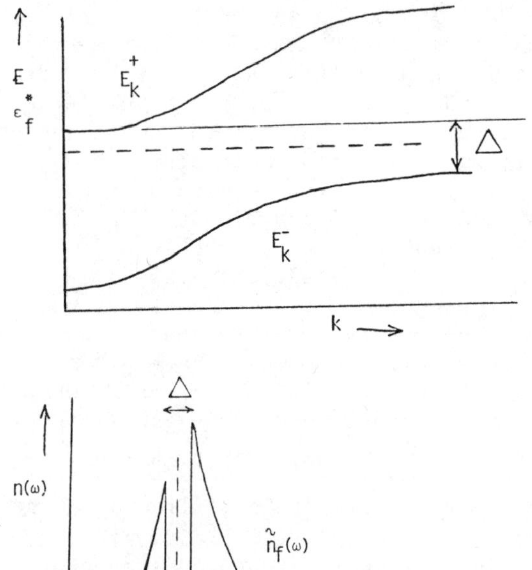


Fig. 10a: Schematic of k-dependence of positions of poles of the one-electron Green's function from the mean-field theory

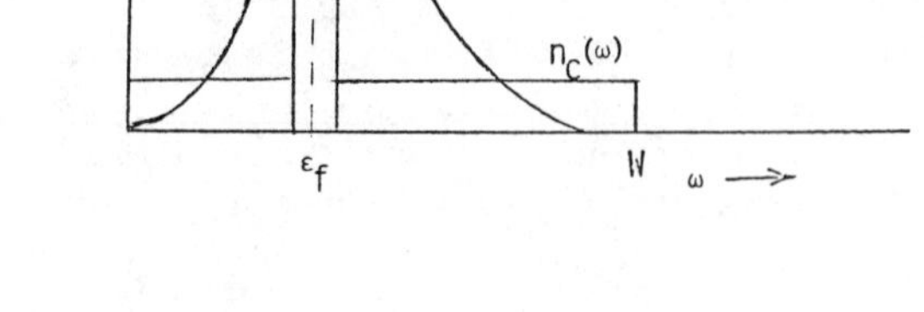


Fig. 10b: Effective densities of "f-like" and conduction-like states for our simple model obtained from the mean-field theory

In summary, while some broad features of the dynamical response of f-electrons in IV systems have been obtained by inelastic neutron scattering, detailed studies of Q-dependence of this scattering need to be carried out on single crystals. This will help to settle the issue of IV as a purely "local" property of the rare-earth ion or as depending more on coherently propagating states in the crystal, as indicated by, for instance, de Haas van Alphen measurements of the Fermi Surface in $CeSn_3$ [66]. This issue clearly needs also to be resolved theoretically, and the theory of the excitations of the Anderson Lattice both at high and low temperatures and in both the IV and Kondo regimes needs to be put on firmer footing.

## 8. ACKNOWLEDGEMENTS

I am indebted to my collaborators D. G. Hinks, H. A. Mook, G. W. Crabtree, O. A. Pringle, J. Kjems and A. J. Fedro for much of the work presented here and to them and colleagues too numerous to mention here for many helpful discussions on these systems.

## 9. REFERENCES

1. B. T. Matthias, E. Corenzwit, J. M. Vandenburg and H. E. Barz, Proc. Nat. Acad. Sci. U.S.A. 74, 1334 (1977).
2. M. Ishikawa and Ø. Fischer, J. de Physique C67, 1379 (1978).
3. G. K. Shenoy, B. D. Dunlap and F. Y. Fradin (Eds.), Proc. Int. Conf. on Ternary Superconductors (North-Holland, Amsterdam, 1981).
4. M. B. Maple and Ø. Fischer (Eds.), Superconductivity in Ternary Compounds, Vols. I and II, Topics in Current Physics (Springer, Berlin, Heidelberg and New York, 1982).
5. D. E. Moncton, G. Shirane, W. Thomlinson, M. Ishikawa and Ø. Fischer, Phys. Rev. Lett. 41, 1133 (1978).
6. W. Thomlinson, G. Shirane, D. E. Moncton, M. Ishikawa and Ø. Fischer, J. Appl. Phys. 50, 1981 (1979).
7. M. B. Maple, L. D. Woolf, C. F. Majkrzak, G. Shirane, W. Thomlinson and D. E. Moncton, Phys. Lett. 77A, 487 (1980).
8. C. F. Majkrzak, D. E. Cox, G. Shirane, H. A. Mook, H. C. Hamaker, H. B. Mackay, Z. Fisk and M. B. Maple, Phys. Rev. B26, 245 (1982).
9. C. F. Majkrzak, S. K. Satija, G. Shirane, H. C. Hamaker, Z. Fisk and M. B. Maple, Phys. Rev. B27, 2889 (1983).
10. H. C. Hamaker, H. C. Ku, M. B. Maple and H. A. Mook, Sol. St. Comm. 43, 455 (1983).
11. M. Ishikawa and Ø. Fischer, Solid State Comm. 24, 747 (1977).
12. M. B. Maple, H. C. Hamaker and L. D. Woolf, Chapter 4 or Ref. 4.
13. P. W. Anderson and H. Suhl, Phys. Rev. 116, 898 (1959).
14. E. I. Blount and C. M. Varma, Phys. Rev. Lett. 42, 1079, (1979).
15. H. Matsumoto, H. Umezawa and M. Tachiki, Solid State Comm. 31, 157 (1979).
16. C. G. Kuper, M. Revson and A. Ron, Phys. Rev. Lett. 44, 1545 (1980).

17. M. Tachiki, H. Matsumoto, T. Koyama and H. Umezawa, Solid State Comm. 34, 19 (1980).
18. R. A. Ferrel, J. K. Bhattacharjee and A. Bagchi, Phys. Rev. Lett. 43, 154 (1979).
19. H. S. Greenside, E. I. Blount and C. M. Varma, Phys. Rev. Lett. 46, 49 (1981).
20. J. W. Lynn, J. A. Gotaas, R. W. Erwin, R. A. Ferrell, J. K. Bhattacharjee, R. N. Shelton and P. Klavins, Phys. Rev. Lett. 52, 133 (1984).
21. D. E. Moncton, D. B. McWhan, J. Eckert, G. Shirane and W. Thomlinson, Phys. Rev. Lett. 39, 1164 (1977).
22. G. K. Shenoy, B. D. Dunlap, F. Y. Fradin, S. K. Sinha, C. W. Kimball, W. Potzel, F. Probst and G. M. Kalvius, Phys. Rev. B21, 3886 (1980).
23. D. E. Moncton, D. B. McWhan, P. H. Schmidt, G. Shirane, W. Thomlinson, M. B. Maple, H. B. MacKay, L. D. Woolf, Z. Fisk and D. C. Johnston, Phys. Rev. Lett. 45, 2060 (1980).
24. G. W. Crabtree, F. Behroozi, S. A. Campbell and D. G. Hinks, Phys. Rev. Lett. 49, 1342 (1982); F. Bahroozi, G. W. Crabtree, S. A. Campbell and D. G. Hinks, Phys. Rev. B27, 6849 (1983).
25. S. K. Sinha, G. W. Crabtree, D. G. Hinks and H. A. Mook, Phys. Rev. Lett. 48, 950 (1982).
26. S. K. Sinha, G. W. Crabtree, D. G. Hinks and H. A. Mook, Physica 109 + 110B, 1693 (1982); S. K. Sinha, G. W. Crabtree, D. G. Hinks and H. A. Mook, in Novel Materials and Techniques in Condensed Matter, (G. W. Crabtree and P. Vashishta, Eds.), North-Holland, New York, Amsterdam, Oxford, P. 133 (1982).
27. C. Stassis, H. W. Deckman, B. N. Harmon, J. P. Desclaux and A. J. Freeman, Phys. Rev. B15, 369 (1977).
28. S. K. Sinha, J. Kjems, H. A. Mook, O. A. Pringle and D. G. Hinks, to be published.
29. M. Tachiki, Physica 109 + 110B, 1699 (1982).
30. M. Tachiki, J. Magnetism Mag. Materials, 31-34, 479 (1983).
31. M. Tachiki, B. D. Dunlap, G. W. Crabtree, Phys. Rev. B28, 5342 (1983).
32. K. Machida, J. Phys. Doc. Jap. 51, 3462 (1982); ibid. 52, 2181 (1983); K. Machida and T. Matsubara, J. Magnetism Mag. Materials 31-34, 507 (1983); K. Machida, Proceedings this Symposium.
33. C.-R. Hu and T. E. Ham, Physica 108B, 1041 (1981) and private communication.
34. L. N. Bulaevskii, A. I. Buzdin, D. I. Khomskii and S. V. Panjukov, Solid State Comm. 46, 133 (1983).
35. J. W. Lynn, G. Shirane, W. Thomlinson, R. N. Shelton and D. E. Moncton, Phys. Rev. B24, 3817 (1981).
36. J. W. Lynn, J. L. Raggazoni, R. Pynn and J. Joffrin, J. Phys. (Paris) Lett. 42, L45 (1981).
37. B. D. Dunlap, private communication.
38. T. Penney and R. L. Melcher, J. Physique Colloq. 37, C4-273 (1976).

39. C. M. Varma, Rev. Mod. Phys. 48, 219 (1976).
40. J. M. Lawrence, P. S. Riseborough and R. D. Parks, Rep. Prog. Phys., 44, 1 (1981).
41. L. M. Falicov, W. Hanke and M. B. Maple, eds., "Valence Fluctuations in Solids" (North-Holland, 1981).
42. P. Wachter and H. Boppart, eds., "Valence Instabilities" (North-Holland, 1982).
43. W. Marshall and S. W. Lovesey, "Theory of Thermal Neutron Scattering" (Clarendon Press, Oxford), 1971.
44. J. F. Cooke and J. A. Blackman, Phys. Rev. B26, 4410 (1982).
45. S. M. Shapiro, J. D. Axe, R. J. Birgeneau, J. M. Lawrence and R. D. Parks, Phys. Rev. B16, 2225 (1979).
46. E. Holland-Moritz, D. Wohlleben and M. Loewenhaupt, Phys. Rev. B25, 7482 (1982).
47. M. Loewenhaupt, B. R. Rainford and F. Steglich, Phys. Rev. Lett. 42, 1709 (1979).
48. S. Horn, M. Loewenhaupt, E. Holland-Moritz, F. Steglich, H. Scheuer, A. Benoit and J. Flouqet, Phys. Rev. B23, 3771 (1981).
49. A. P. Murani, private communication.
50. A. P. Murani, Phys. Rev. B28, 2308 (1983); J. Phys. C33, 6539 (1983); A. P. Murani, W. C. M. Mattens, F. R. de Boer and G. H. Lander, Phys. Rev. B, to be published (1984).
51. S. M. Shapiro and B. H. Grier, Phys. Rev. B25, 1457 (1982).
52. C. Stassis, private communication.
53. C. Stassis and C. N. Loong, to be published.
54. H. A. Mook, R. M. Nicklow, T. Penney, F. Holtzberg and M. W. Schaefer, Phys. Rev. B18, 2925 (1978).
55. H. A. Mook and F. Holtzberg, in Ref. 41, p. 113.
56. Y. Kuramoto, J. Magn. and Mag. Mat. 31-34, 463 (1983); Z. Physik 37, 299 (1980).
57. E. Muller-Hartmann, in "Electron Correlations and Magnetism in Narrow-Band Systems", T. Moriya, Ed., Springer Berlin Heidelberg New-York (1981).
58. A. J. Fedro and S. K. Sinha, in Ref. 41, p. 321 (1981).
59. A. J. Fedro and S. K. Sinha, in Ref. 42, (1982).
60. A. J. Fedro and S. K. Sinha, in "Moment Formation in Solids" (Proceedings of NATO Institute, Vancouver, 1983, W. J. Buyers, editor), to be published.
61. D. L. Huber, Phys. Rev. B28, 860 (1983).
62. B. H. Brandow, in Ref. 42 (1982).
63. E. Holland-Moritz and M. Praeger, J. Mag. Megn. Mat. 31-34, 395 (1983).
64. C. Lacroix, J. Appl. Phys. 53 (3), 2131 (1979).
65. Y. Kuramoto, to be published.
66. W. R. Johanson, C. W. Crabtree, D. D. Koehling, A. S. Edelstein and O. D. McMasters, Phys. Rev. Lett. 46, 504 (1981).

# USES OF RESONANCE SCATTERED NEUTRONS

G. T. Trammell

Physics Department
Rice University
Houston, TX 77251 USA

## INTRODUCTION

With the advent of neutron spallation sources giving high intensities in the several eV range it is worthwhile to consider the potential uses of neutron resonance scattering for condensed matter investigations since there are several dozens of resonances in this energy range among the heavy nuclei. In the table we list some of these strong scattering resonances (taken from BNL 325). There are two noteworthy features of resonant neutron scattering: the scattering amplitudes near resonance are 1-2 orders of magnitude larger than non-resonant (potential) amplitudes, and the phase of the amplitude increases from 0 to $\pi$, as one increases the energy of the incident neutron from well below the resonance to well above it.

The nuclear resonances effective in the thermal neutron region $^{157}$Gd (0.03 eV), $^{149}$Sm (0.1 eV) and $^{113}$Cd (0.18 eV) are relatively weak scatterers ($\sigma_{sc}/\sigma_{abs}=\Gamma_n/\Gamma_\gamma$). But because $\Gamma_n$ has an overall increase with energy $\propto\sqrt{E_0}$ the resonances in $\gtrsim$ 1 eV region are moderately strong scatterers. Thus $\sigma_{sc}/\sigma_{abs} \approx$ 7% for the $^{240}$Pu (1 eV), and $\approx$ 13% for the $^{185}$Re (2 eV), resonances. For the Gd, Sm and Cd resonances, on the other hand, these scattering probabilities are only 0.6, 0.8, and 0.5%, respectively. Thus for neutrons incident on samples of $^{240}$Pu or $^{185}$Re near resonance up to 7% or 13% of them will reemerge from the compound nucleus as scattered neutrons, and may be analyzed for information on the samples, while less than 1% would be available for that purpose for the thermal neutron resonances.

Since the resonance amplitudes are 1-2 orders of magnitude

Table 1. Some Low Energy Resonance Parameters

| Isotope | I | E(eV) | $\Gamma_\gamma$(meV) | $\Gamma_n$(meV) | $f_o$($10^{-12}$cm) |
|---|---|---|---|---|---|
| $^{103}$Rh | 1/2 | 1.26 | 156 | 0.8 | 2.5 |
| $^{109}$Ag | 1/2 | 5.2 | 140 | 12 | 17 |
| $^{113}$Cd | 1/2 | 0.18 | 113 | 0.65 | 6 |
| $^{123}$Sb | 1/2 | 21.6 | 64 | 32 | 34 |
| $^{123}$Te | 1/2 | 2.3 | 104 | 10 | 29 |
| $^{124}$Xe | | 5.2 | 100 | 8 | 17 |
| $^{149}$Sm | | 0.1 | 63 | 0.5 | 12 |
| $^{145}$Nd | 7/2 | 4.4 | 48 | 13 | 60 |
| $^{157}$Gd | 7/2 | 0.03 | 100 | 0.65 | 19 |
| $^{159}$Tb | 3/2 | 11 | 100 | 9 | 13 |
| $^{165}$Ho | 7/2 | 3.9 | 66 | 2.5 | 9 |
| $^{167}$Er | 7/2 | 6 | 70 | 24 | 52 |
| $^{167}$Tm | 1/2 | 3.9 | 70 | 12 | 36 |
| $^{177}$Hf | 9/2 | 1.1 | 67 | 2.1 | 14 |
| | | 2.4 | 60 | 9 | 42 |
| $^{181}$Ta | 7/2 | 4 | 56 | 3.6 | 15 |
| $^{182}$W | | 4 | 46 | 1.5 | 8 |
| $^{185}$Re | 5/2 | 2 | 52 | 6.5 | 38 |
| $^{191}$Ir | 3/2 | 5.4 | 84 | 6.6 | 16 |
| $^{197}$Au | 3/2 | 4.9 | 124 | 15 | 24 |
| $^{234}$U | | 5.8 | 26 | 4.4 | 32 |
| $^{240}$Pu | | 1 | 34 | 2.5 | 31 |
| N | | | | | 1 |
| $^2$H,O,C | | | | | 0.6 |
| H | | | | | 1.7(inc),0.4(co) |

$$f_o = \lambdabar \Gamma_n/(\Gamma_\gamma + \Gamma_n),$$

$$f_{co} = \frac{2j_i+1}{2(2\ I+1)} f_o, \qquad f_{coh} = \frac{f_{co}}{\frac{2(E_R-E)}{\Gamma} - i}$$

larger than the non-resonance ones, a small concentration of resonance atoms will give a large interference term with the waves scattered from the other atoms in the scattering cross-section. Furthermore, since the phase of the resonance amplitude may be varied by changing $E_o$, this affords a means of extracting this interference term, and, as we discuss this can give valuable information concerning the structures and <u>motions</u> of multiatomic condensed matter systems.

The use of a resonant scatterer (or for x-rays, an anomalous scatterer) as a "phasor" to obtain the phase of the "structure factor" of e.g. a molecular crystal, leading to the molecular structure determination, is a well-known procedrue.[1] However, by analyzing the <u>inelastically</u> scattered neutrons from, e.g., a molecule containing resonant atoms we can, in some cases, determine the amplitudes and phases of the motions of the various atoms in the molecule when, e.g., it is executing simple harmonic motion (speaking semi-classicially).

Biological molecules are important physical systems. While the determination of their (time-averaged) structures is rapidly advancing, their motions, or their dynamical structures, the knowledge of which is important in order to understand their functions, has not (to my knowledge) been investigated experimentally. For that purpose, I think that resonant neutron "phasors" can play an important role.

There are, of course, no low energy resonances among the light atoms which are the principal constituents of biological molecules, so if resonant phasors should be used to determine some of the motions of these molecules, they would have to be chemically attached.

In the next section, I develop the formulae for the scattering from systems containing resonant atoms and discuss (in an idealized case neglecting Doppler broadening, etc.) how the dynamical structure can be determined by using these atoms as phasors. In the last section, I consider a more realistic case and discuss how we might determine the individual atoms' vibrational amplitudes and phases for various normal modes of molecules.

## PHASORS. IDEALIZED CASE

Let $b_i$ be the (coherent) non-resonant (potential) scattering amplitude of the ith atom. Then according to the well-known van Hove formula[2,3] the differential scattering cross section from a small sample (for which multiple scattering is negligible) is given by (we ignore the incoherent scattering)

$$\frac{d^2\sigma}{d\Omega d\omega} \sim S(\underline{Q},\omega) = \sum_{i,j} \int_{-\infty}^{\infty} e^{+i\omega t}\, b_i b_j \langle e^{-i\underline{Q}\cdot\underline{r}_i(t)} e^{i\underline{Q}\cdot\underline{r}_j} \rangle \, dt, \tag{1}$$

where $\hbar\omega = E_o - E$, $\underline{Q} = \underline{k}_o - \underline{k}$, and the sum is over the atoms of the sample. Writing

$$b(\underline{r},t) = \Sigma b_i\, \delta(\underline{r} - \underline{r}_i(t)), \tag{2}$$

we have

$$S(\underline{Q},\omega) = \int e^{-i\underline{Q}\cdot\underline{r}}\, e^{+i\omega t} \left[ \langle b(\underline{r}+\underline{x},t)\, b(\underline{x},0)\rangle d\underline{x} \right] \cdot d\underline{r}dt. \tag{3}$$

Knowing $S(\underline{Q},\omega)$ (from measurements of $d^2\sigma/d\Omega d\omega$) we obtain by Fourier inversion

$$G_{bb}(\underline{r},t) \equiv \int d\underline{x}\, \langle b(\underline{r}+\underline{x},t)\, b(\underline{x},0)\rangle. \tag{4}$$

The correlation function $G_{bb}$ may be viewed as the "time dependent Patterson map"[4,1] of the scattering density $b(\underline{r},t)$. The Patterson map usually refers to the elastic scattering $t \to \infty$ limit of (4), which gives the time averaged correlation function

$$G_{bb}(\underline{r}) = G_{bb}(\underline{r},\infty) = \int d\underline{x} < b(\underline{r}+\underline{x})><b(\underline{x})>. \tag{5}$$

Hereafter, I shall restrict my considerations to the determination of molecular structures and motions and shall suppose the sample consists of a crystal of these molecules.

Except for very simple molecules $G_{bb}(\underline{r})$ cannot be solved to yield the molecular structure $<b(\underline{r})>$ [the phase of the structure factor problem in crystallography]. $G_{bb}$ has "peaks" at all the $N(N-1)+1$ interatomic vectors of the N-atom molecule. Roughly, it is the "N-ply" exposed image obtained by translating the molecule N times bringing in turn each of the N atoms to a fixed point. [For small molecules, with knowledge of some chemistry, and e.g., if we know $<b(\underline{r})>$ is negative, say, then we can determine $<b(\underline{r})>$ to within an entiomorph[1] from $G_{bb}$, but when N becomes large it is practically impossible to determine an unknown structure from the Patterson map ]. Similarly $G_{bb}(\underline{r},t)$, (4), is incapable of yielding the detailed structure and motion, the dynamical structure, of any but the simplest systems. The strategy resorted to in crystallography in order to determine $<b(\underline{r})>$ is to vary b in a known way $b \to b + F$ so that $G_{bb}$, (4), becomes

$$G_{b+F,b+F} = \int <b(\underline{r}+\underline{x},t)b(\underline{x})>d\underline{x} + \int <F^{*}(\underline{r}+\underline{x},t)F(\underline{x})>d\underline{x} + \int <b(\underline{r}+\underline{x},t)F(\underline{x})>d\underline{x} + \text{c.c.} \tag{6}$$

If then F is varied in a known way leaving b unchanged, the interference term

$$G_{bF}(\underline{r},t) = \int <b(\underline{r}+\underline{x},t)F(\underline{x})>d\underline{x} + \text{c.c.}, \tag{7}$$

can be extracted from an analysis of the scattering data. If, for example, F represents the scattering amplitude from one extra reference scatterer (or one per molecule for the molecular crystal) then (7) allows us to obtain the positions and motions of the other scatterers relative to the one reference scatterer rather than the positions and motions of all the scatterers relative to all the others as given by (4).

If the sample contains resonance scatterers then we obtain from $d^2\sigma/d\Omega d\omega$ a function characterizing the sample, $W(\underline{k},\underline{k}_o)$,[5,6] which contains two extra parameters in addition to $\underline{Q}$ and $\omega$, e.g. $\phi$ and $\Delta$ where $\phi$ is the azimuthal angle of the plane $(\underline{k}_o,\underline{k})$ around $\underline{Q} = \underline{k}_o-\underline{k}$ and $\Delta=E_R-E_o$, where $E_R$ is the resonance energy. In this case (1) is replaced by

$$\frac{d^2\sigma}{d\Omega d\omega} \sim W(\underline{k},\underline{k}_o), \tag{8}$$

with

$$W = \int e^{+i\omega T} \langle [\sum_j b_j \, e^{-i\underline{Q}\cdot\underline{r}_j(T)} + \sum_b f_{co} \int_o^\infty e^{-i\underline{k}_o\cdot\underline{R}_b(T-t')} g^*(t')\, dt' \cdot e^{i\underline{k}\cdot\underline{R}_b(T)}] \cdot [\sum_i b_i \, e^{i\underline{Q}\cdot\underline{r}_i} + \sum_a f_{co} \, e^{-i\underline{k}\cdot\underline{R}_a} \int_o^\infty g(t) e^{i\underline{k}_o\cdot\underline{R}_a(-t)} dt \,] \rangle dT, \tag{9}$$

where $f_{co}$ is the coherent scattering amplitude at resonance,

$$g(t) = i \, \gamma \, e^{-i(\Delta - i\gamma)t} \tag{10}$$

and we have used, $\gamma = \Gamma/2$, $\underline{R}_a(t)$ for the position operator of the a'th resonant scatterer and $\underline{r}_i(t)$ that of the i'th non-resonant scatterer.

The resonance terms come from using the Lamb formula[7,8]

$$F_{fo} = \sum_n \frac{\langle \chi_f | e^{-i\underline{k}\cdot\underline{R}} | \chi_n \rangle \langle \chi_n | e^{i\underline{k}_o\cdot\underline{R}} | \chi_o \rangle}{\Delta - i\gamma + \varepsilon_n - \varepsilon_o} \gamma f_{co}, \tag{11}$$

$$= f_{co} \int_o^\infty g(t) \; \langle \chi_f | e^{-i\underline{k}\cdot\underline{R}} \, e^{i\underline{k}_o\cdot\underline{R}(-t)} | \chi_o \rangle dt,$$

where the $\chi$'s represent the crystalline vibrational states and $\varepsilon$'s are their energies. (11) represents the scattering amplitude for $0 \rightarrow f$. Adding the resonant terms $F_{fo}$ to the nonresonant terms $b_{fo} = b\langle \chi_f | e^{i\underline{Q}\cdot\underline{r}} | \chi_o \rangle$, and proceeding in the Van Hove manner to obtain $d^2\sigma/d\Omega d\omega$, yields (8) and (9).

The form of the resonance terms in (9) and (11) show, speaking semi-classically, that there is a mean time delay, $T \sim h\Gamma/|\Delta - i\gamma|^2$, between the virtual absorption and re-emission of the neutron incident upon the resonant nucleus. In an earlier paper[8] I have discussed the effects of this time delay on scattering from bound resonance scatterers in the various regimes from $T \ll \omega_m^{-1}$ (fast collisions) to $T \gg \omega_m^{-1}$ (slow collisions), where $\omega_m$ is a characteristic vibration frequency of the resonant atom in the molecule. The low energy neutron widths, $\Gamma$, (see table) cover a range of about 30-150 meV. Since the effective $\omega_m$'s for these heavy atoms are of the order of 10 meV, these resonance collisions range from medium fast to very fast.

Now while the $b_i$'s are real (negative for hard sphere potential scattering in the convention used here) the phase of the resonance scattering terms ("a" terms in (9)) increases from 0 to $\pi$ as $E_o$ increases from well below the resonance ($\Delta >> \Gamma$) to well above the resoance ($-\Delta >> \Gamma$). It is this feature which will allow us to extract the resonance-non-resonance interference terms from (9).

Consider a sample consisting of resonant and non-resonant scatterers. Then measurement of $d^2\sigma/d\Omega d\omega$ gives us W, eqn (8),(9). For simplicity I suppose $\gamma$ sufficiently large (fast collision limit) that I may set $R(-t) = R_a(0)$ and $R_b(T-t') = R_b(T)$ in (9) [ In the next section I shall allow for recoil and Doppler broadening effects ]. W, which then becomes a function only of $\underline{Q}$, $\omega$, and $\Delta$, may be written

$$W = \int e^{+i\omega t} e^{-i\underline{Q}\cdot\underline{r}} [G_{bb}(\underline{r},t) + \frac{\gamma^2 f_{co}^2}{\Delta^2+\gamma^2} G_{RR}(\underline{r},t) \tag{12}$$

$$+ \frac{\gamma f_{co}}{\sqrt{\Delta^2+\gamma^2}} (e^{i\phi}G_{bR}(\underline{r},t) + e^{-i\phi}G_{Rb}(\underline{r},t) ] d\underline{r}dt,$$

where

$$G_{bb}(\underline{r},t) = \int \langle b(\underline{r}+\underline{x},t)b(\underline{x})\rangle d\underline{x} , \tag{13i}$$

$$G_{RR}(\underline{r},t) = \int \langle \rho_R(\underline{r}+\underline{x},t)\rho_R(\underline{x})\rangle d\underline{x}, \tag{13ii}$$

$$G_{bR}(\underline{r},t) = \int \langle b(\underline{r}+\underline{x},t)\rho_R(\underline{x})\rangle d\underline{x}, \tag{13iii}$$

$$G_{RB}(\underline{r},t) = \int \langle \rho_R(\underline{r}+\underline{x},t)b(\underline{x})\rangle d\underline{x}, \tag{13iv}$$

and

$$\rho_R(\underline{r},t) = \sum_a \delta\ (\underline{r}-\underline{R}_a(t)), \tag{14}$$

where the sum is over the resonant atoms, and finally

$$\phi = \tan^{-1} \gamma/\Delta. \tag{15}$$

If we denote by G the contents of the square bracket in (12) then using (15)

$$G = G_{bb} + \frac{\gamma^2 f_{co}}{\Delta^2+\gamma^2}[ f_{co}G_{RR}+2i(G_{bR}-G_{Rb}) + \frac{2\Delta}{\gamma}(G_{bR}+G_{Rb})] . \tag{16}$$

Now from (12) and (16) we obtain

$$W-(\underline{Q},\omega,\Delta) = \frac{2\ \Delta\gamma f_{coh}}{\Delta^2+\gamma^2}\int e^{-i\underline{Q}\cdot\underline{r}+i\omega t}(G_{bR}+G_{Rb})d\underline{r}dt, \tag{17}$$

$$W_+(\underline{Q},\omega,\Delta) = \int e^{-i\underline{Q}\cdot\underline{r}+i\omega t}\{G_{bb}+\frac{\gamma^2 f_{co}}{\Delta^2+\gamma^2}[f_{co}G_{RR}+2i(G_{bR}-G_{Rb})]\}d\underline{r}dt, \quad (18)$$

where

$$W_\pm(\underline{Q},\omega,\Delta) \equiv \frac{1}{2}[\,W(\underline{Q},\omega,\Delta) \pm W(\underline{Q},\omega,-\Delta)] \,. \quad (19)$$

By measuring $W(\underline{Q},\omega,\Delta)$ and $W(Q,\omega,-\Delta)$at two different $\Delta$'s and using(17) and (18), and inverting the Fourier integrals, we can obtain $G_{bR}(\underline{r},t)+G_{Rb}(\underline{r},t)$, and $f_{co}G_{RR}(\underline{r},t)+2i[G_{bR}(\underline{r},t)-G_{Rb}(\underline{r},t)]$. If, for example, there is just one resonance scatterer in the molecule then $G_{RR}(\underline{r},t)$, representing the self correlation function, will be concentrated near the origin while the $G_{bR}-G_{Rb}$ terms will rather be significant at interatomic distances. In this way, for this case of very fast resonant collisions, we can determine $G_{bR}$ by eqn. (13iii) and $G_{Rb}$ by eqn. (13iv).

In practice, before determining the atomic motions in the molecule, one would, of course, first determine the time averaged structure by measuring the $\omega = 0$ components appearing in the Bragg peaks. Using the resonance scatterer as phasor as described above would give us

$$G_{bR}(\underline{r}) = \int \langle b(\underline{r}+\underline{x})\rangle\langle\rho_R(\underline{x})\rangle d\underline{x} \quad (20)$$

Taking the center of the $\langle\rho_R(\underline{x})\rangle$ in the molecule as the origin would then give us $G_{bR}(\underline{r}) = \langle b(\overline{\underline{r}})\rangle$, where the bar means that the $\langle b(\underline{r})\rangle$ distribution is convoluted with the zero point and thermal spread of the phasor, which is Gaussian in the linear regime.

For the determination of the time averaged molecular structure by Bragg scattering, it is usually the case that x-ray analysis, with heavy atomic replacements, or using anamolous scatterers as phasors,[1] is easier, cheaper and faster, because of intensity advantages, than using neutrons to do the job. But determining the molecular motions at frequencies in the 1-10 meV range by inelastic scattering is far beyond the resolution obtainable with present day x-ray monochromaters. (See, however, the reports by J. B. Hastings[9] and Bruno Dorner[10] on the work underway to develop x-ray monochromators yielding resolutions $\delta E \approx 1$ meV in the 1 Å x-ray region).

We should also mention that the correlation functions (13) $G_{bR}(\underline{r},t)$ are not real (except at the classical level) but have an imaginary part of quantum origin. [ See Marshall and Lovesey[3] for a good discussion of correlation functions. ] It follows from the hermiticity of the b and $\rho_R$ operators and the thermal averages involved that e.g. $G^*_{bR}(\underline{r},t) = G_{Rb}(-\underline{r},-t)$, and from time reversal invariance that $G_{bR}(\underline{r},t) = G_{Rb}(-\underline{r},t)$.

## Allowances for Doppler Broadening and Recoil

The extraction of $G_{bR}(\underline{r},t)$ from the inelastic scattering data requires that we know the scattering amplitude as a function of $\Delta$ of the reference wave (resonant) scatterer.

In the previous section we supposed $\gamma$ was sufficiently large that all effects of the motion of the resonator were negligible. For the actual resonances, however, Doppler broadening at room temperature is comparable to $\gamma$ and also the recoil term is appreciable. In order to gauge these effects, in this section I consider the problem of determining the amplitudes and phases of the individual atoms' motions corresponding to a normal mode of the molecule. I shall also assume that the temperature is low enough so that linear analysis is appropriate. Since there is little dispersion for optical modes of large molecules in crystals I shall neglect the intermolecular correlations. I suppose the time average structure of the molecule is known and denote by $\underline{r}_i$ and $\underline{U}_i$ the equilibrium position and displacement of the non-resonant scatterers and $\underline{r}_a$ and $\underline{U}_a$ those of the resonant scatterer.

Now from (9)

$$W = W_{bb} + W_{bR} + W^{*}_{bR} + W_{RR}, \tag{21}$$

where

$$W_{bb} = \int e^{+i\omega T} \sum_{i,j} b_j b_i e^{-i\underline{Q}\cdot(\underline{r}_j-\underline{r}_i)} \langle e^{-i\underline{Q}\cdot\underline{U}_i(T)} e^{i\underline{Q}\cdot\underline{U}_j}\rangle dT, \tag{22i}$$

$$W_{bR} = \int e^{+i\omega T} \sum_i b_i f_{co}\, e^{-i\underline{Q}\cdot(\underline{r}_i-\underline{r}_a)} \tag{22ii}$$

$$\cdot \langle e^{-i\underline{Q}\cdot\underline{U}_i(T)} e^{-i\underline{k}\cdot\underline{U}_a} \int_0^\infty dt\, g(t) e^{i\underline{k}_o\cdot\underline{U}_a(-t)}\rangle dt,$$

$$W_{RR} = \int e^{+i\omega T} f^2_{co} \langle \int_0^\infty g^{*}(t') e^{-i\underline{k}_o\cdot\underline{U}_a(T-t')} dt'\, e^{i\underline{k}\cdot\underline{U}_a(T)} \tag{22iii}$$

$$\cdot\ e^{-i\underline{k}\cdot\underline{U}_a} \int_0^\infty g(t) e^{i\underline{k}_o\cdot\underline{U}_a(-t)} dt\rangle\, dT.$$

Now using the theorem[3]

$$\langle e^{A} e^{B} e^{C} \ldots\rangle = \exp \tfrac{1}{2} \langle A^2+B^2+C^2+\ldots 2AB+2AC+2BC\rangle , \tag{23}$$

the bracketed term in $W_{bR}$ is

$$\langle bR\rangle = \int_0^\infty \exp \tfrac{1}{2} \langle -(\underline{Q}\cdot\underline{U}_i)^2-(\underline{Q}\cdot\underline{U}_a)^2-2(\underline{k}\cdot\underline{U}_a)\underline{k}_o\cdot(\underline{U}_a-\underline{U}_a(-t))\rangle \tag{24}$$

$$\cdot\exp \langle\underline{Q}\cdot\underline{U}_i(T)\underline{Q}\cdot\underline{U}_a-\underline{Q}\cdot\underline{U}_i(T)\underline{k}_o\cdot(\underline{U}_a-\underline{U}_a(-t))\rangle\cdot g(t)dt$$

The Doppler broadening and the recoil terms are contained in the $\underline{U}_a-\underline{U}_a(-t)$ term in the first exponential in (24) and are obtained by Taylor series expansion retaining terms $\sim t^2$:

$$\langle\underline{k}_f\cdot\underline{U}_a k_o\cdot(\underline{U}_a-\underline{U}_a(-t))\rangle \stackrel{\sim}{\sim} \langle\underline{k}\cdot\underline{U}_a\underline{k}_o\cdot\dot{\underline{U}}_a\rangle t - \frac{t^2}{2}\langle\underline{k}\cdot\underline{U}_a k_o\cdot\ddot{\underline{U}}_a\rangle ,$$

$$= \frac{i}{\hbar} Rt + \langle\underline{k}\cdot\underline{p}_a\underline{k}_o\cdot\underline{p}_a\rangle \frac{t^2}{2Ma} , \tag{25}$$

$$\doteq i\ Rt/\hbar + R\langle E\rangle t^2\hbar^{-2},$$

where

$$R = (\underline{p}_o\cdot\underline{p})/2Ma, \tag{26}$$

is the so called recoil energy, and

$$\langle E\rangle = \frac{1}{3}\,\underline{p}_a^2/Ma \quad = 2\langle T\rangle/3, \tag{27}$$

where $\langle T_a\rangle$ is the mean kinetic energy of the resonant atom, $\langle E\rangle \doteq k_BT$ if $T \gg \Theta_D$.

Now define

$$\bar{g}(t) = g(t)\ e^{-\langle\underline{k}\cdot\underline{U}_a\underline{k}_o\cdot(\underline{U}_a-\underline{U}_a(-t))\rangle} \tag{28}$$

and using (25) and (10)

$$\bar{g}(t) \doteq i\gamma e^{-i(\Delta+R-i\gamma)t}\ e^{-R\langle E\rangle t^2/\hbar^2} \tag{29}$$

The Gaussian in (29) is the Doppler broadening term stemming from the zero-point + thermal motion of the resonator. Assuming $R \geq 0$ (see discussion on pp. 1050-1051 of [8] for $R < 0$), from (29) we obtain[7,8]

$$\int_0^\infty \bar{g}(t)dt= \int_{-\infty}^\infty \frac{\gamma\ \exp(-x^2/4R\langle E\rangle)}{\sqrt{\pi 4R\langle E\rangle}\,[\Delta+R+x-i\gamma]}\,dx. \tag{30}$$

(30) is proportional to the elastic scattering amplitude for the

resonator, exhibiting the effect of recoil and the Doppler broadening. The increased width of the resonance is of course not of the nature of the natural width but corresponds to a continuous distribution of poles, with half width $\approx 2\sqrt{R\langle E\rangle}$, on the line $\Delta = +i\gamma$. Sometimes, however, this effect is represented by a single pole by replacing (30) by $\gamma[\Delta+R-i\gamma]^{-1}$, $\bar{\gamma}= \gamma+2\sqrt{R\langle E\rangle}$. In any case, $2\sqrt{R\langle E\rangle}$ is called the Doppler width. For $\underline{p}_o\cdot\underline{p} = p_o^2$ we have at resonance $R = 4$ meV, $2\sqrt{RE} = 22$ meV, $\gamma = 18$ meV for $^{240}$Pu and $R = 10$ meV, $2\sqrt{RE} = 34$ meV, $\gamma = 30$ meV for $^{185}$Re.

The argument of the second exponential in (24) can be represented by

$$\langle\ \rangle = \sum_s \underline{Q}\cdot\underline{v}_j^s \frac{\hbar}{2\sqrt{M_j M_a'}\,\omega_s} \underline{v}_a^s \cdot \Big[ (\bar{n}_s+1)(\underline{k}_o e^{-i\omega_s(T+t)} - \underline{k}\, e^{-i\omega_s T}) + \bar{n}_s(\underline{k}_o\, e^{+i\omega_s(T+t)} - \underline{k}\, e^{i\omega_s T}) \Big] , \tag{31}$$

where $\omega_s$ is the frequency of the s'th normal mode of the molecule and $\underline{v}_j^s$ is the unit vector in the 3N dimensional space of the displacements of the N atoms, $\sum_j \underline{v}_j^s\cdot\underline{v}_j^{s'} = \delta_{ss'}$, representing the s'th mode, and the sum is over the 3N normal modes. The n'th term in the expansion of the exponential in powers of (31) represents n phonons created or absorbed by the neutron.

The contribution of the one phonon term to $\langle bR\rangle$ in (24) is proportional to

$$\int_0^\infty \bar{g}(t)\langle\ \rangle\, dt \doteq \sum_s \underline{Q}\cdot\underline{v}_j^s \frac{\hbar}{2\sqrt{M_j M_a'}\,\omega_s} \underline{v}_a^s\cdot \Big[(\bar{n}_s+1)e^{-i\omega_s T} \left( \frac{\underline{k}_o}{\Delta+R+\omega_s-i\bar{\gamma}} - \frac{\underline{k}_f}{\Delta+R-i\gamma} \right) + \bar{n}_s\, e^{i\omega_s T}\left(\frac{\underline{k}_o}{\Delta+R-\omega_s-i\bar{\gamma}} - \frac{\underline{k}}{\Delta+R-i\bar{\gamma}}\right)\Big] , \tag{32}$$

where for convenience I have represented the Doppler broadening by $\gamma \to \bar{\gamma}$ discussed above. For the $\omega_s$'s small compared to $\bar{\gamma}$(or $|\Delta+R-i\bar{\gamma}|$) we may neglect $\omega_s$ in the resonance denominators in (32). We saw above that $\bar{\gamma} \approx 40$ and 65 meV for the $^{240}$Pu and $^{185}$Re resonances respectively, and since for $\omega_s$'s of this order or larger the displacement amplitudes of the heavy resonant atoms $\sim \sqrt{(\hbar/2M_a\omega_s)\bar{n}_s}\, \underline{v}_a^s$ are very small and give only a small contribution to (32), we may then approximate (32) by

$$\int_0^\infty \bar{g}(t)\langle\ \rangle dt \doteq \sum_s \underline{Q}\cdot\underline{v}_j^s \frac{\hbar\, \underline{Q}\cdot\underline{v}_a^s}{2\sqrt{M_j M_a}\ \omega_s}\left[(\bar{n}_s+1)e^{-i\omega_s T}+\bar{n}_s e^{i\omega_s T}\right] \tag{33}$$

$$\cdot(\Delta+R-i\bar{\gamma})^{-1}.$$

$$\doteq \langle \underline{Q}\cdot\underline{U}_j(T)\underline{Q}\cdot\underline{U}_a\rangle/(\Delta+R-i\bar{\gamma})$$

Using similar arguments we may put

$$\int_0^\infty \bar{g}(t)\langle\ \rangle^n dt \doteq \langle(\underline{Q}\cdot\underline{U}_j(T)\underline{Q}\cdot\underline{U}_a)^n\rangle/(\Delta+R-i\bar{\gamma}). \tag{34}$$

With these approximations $W_{bR}$(22ii) becomes

$$W_{bR} \doteq \frac{\gamma f_{co}}{\Delta+R-i\bar{\gamma}}\int e^{i\omega T}\sum_j b_j e^{-i\underline{Q}\cdot(\underline{r}_j-\underline{r}_a)} \tag{35}$$

$$\cdot e^{-\frac{1}{2}\langle+(\underline{Q}\cdot\underline{U}_j)^2+(\underline{Q}\cdot\underline{U}_a)^2-2\underline{Q}\cdot\underline{U}_j(T)\underline{Q}\cdot\underline{U}_a\rangle}\, dT.$$

$$\doteq \frac{\gamma f_{co}}{\Delta+R-i\bar{\gamma}}\int e^{i\omega t-\underline{Q}\cdot\underline{r}}G_{bR}(\underline{r},t)dtd\underline{r}\ ,$$

where $G_{bR}$ is given by (13iii).

The term $G_{RR}$ may be treated in a similar approximate manner; then the procedure suggested in the previous section may be used to deduce $G_{bR}(\underline{r},t)$ from the scattering data by making measurements at several values of $\Delta$. In analyzing the data, eqn (30) for $\int \bar{g}(t)dt$, rather than $\gamma(\Delta+R-i\bar{\gamma})^{-1}$ written in (35), should be used.

$G_{bR}$ cannot be used to get the high frequency motions of the light atoms, since the heavy atoms will not respond appreciably to the high frequencies, and so there is negligible correlation in the light-heavy atoms' motions at these high frequencies. However, the correlation should be appreciable for frequencies $\lesssim$ 10-20 meV, perhaps.

From (26) $R = \underline{p}_o\cdot\underline{p}/2Ma$, R and R<E> can then be determined experimentally by absorption measurements on the sample (which via the optical theorem depends only on $\underline{p} = \underline{p}_o$).

## CONCLUSION

I have presented an approximate theory indicating that we can determine the low frequency motions of the atoms in, e.g., molecules, by using resonant scatters as "phasors." It would be highly desirable to check the theoretical conclusions (based on approximations) experimentally.


## ACKNOWLEDGEMENTS

This theory stems from earily work done in collaboration with R. W. Word, and was supported under the NSF Grant DMR 80-15706.


## REFERENCES


1. T. L. Blundell and L. N. Johnson, "Protein Crystallography" Academic Press (1976).
2. L. van Hove, Phys. Rev.95:249 (1954); 95:1374 (1954).
3. W. Marshall and S. Lovesey, "Theory of Thermal Neutron Scattering," Oxford Univ. Press (1971).
4. A. L. Patterson, Phys. Rev. 46:372 (1934).
5. R. E. Word "Use of Resonant Neutron Scattering for the Study of Condensed Matter," Ph.D. Thesis, Rice University (1979).
6. R. E. Word and G. T. Trammell, Phys. Rev. B 24:2430 (1981).
7. W. E. Lamb, Phys. Rev. 55:190 (1939).
8. G. T. Trammell, Phys. Rev. 126:1045 (1962).
9. J. B. Hastings, Proceedings of the Los Alamos Workshop on High Energy Excitations in Condensed Matter (1984).
10. B. Dorner, ibid.

# INELASTIC NEUTRON SCATTERING FROM LATTICES, MOLECULAR CRYSTALS AND POWDERS

M. Warner and J.M.F. Gunn

Rutherford Appleton Laboratory
Chilton, Didcot, Oxon, OX11 0QX, UK

ABSTRACT

New neutron sources provide epithermal neutrons well suited in energy to investigate the internal motions of molecules. With these higher energies also come high momenta. Do extreme limits of these two factors yield apparent quasi-free behaviour for the scattering centres? Asymptotic analysis for model systems of harmonic lattices, isolated oscillators and oscillators in lattices answer this qualitative question. Exact numerical results reveal how good approximate methods actually are. A lot of structure in oscillator lines can result from kinematic effects of the lattice environment and is not revealed by approximate treatments. This is a possible artifact to be first removed from experimental interpretation before more subtle explanations of splittings and shifts are invoked.

The advantages offered by high momentum transfer (neutron) spectroscopy over usual spectroscopic techniques are explained, as are the problems thereby encountered when then dealing with powders. A systematic technique involving the use of displacement tensors in spherical coordinates yields simple, exact results rather than the usual expansions or numerics.

## 1. INTRODUCTION

There are many materials in which essential chemical and physical properties are revealed in the motion of the constituent hydrogen atoms. Neutron scattering affords a sensitive method for studying hydrogen motion because the scattering cross-section for a bound proton is very large compared with that for most other elements, ie. there is often a high contrast factor for scattering from hydrogeneous materials. In addition, the single scattering of neutrons is a weak process that can be described within the first Born approximation. In consequence, neutron scattering provides information on the motion of hydrogen atoms which is undistorted by the experimental probe. For these reasons there are many studies of hydrogen motion in solids using thermal neutrons from steady-state reactors[1,2].

Advanced, pulsed neutron facilities have the great advantage of providing intense beams of hot neutrons. Thus, it is feasible to study dynamic processes in which the neutron energy change, $\omega$, is as large as a few electron volts. The concomitant change in the wave vector of the neutron, $\underline{Q}$, might also be large compared with values readily obtained in thermal neutron scattering. Since protons in solids usually have a characteristic frequency of oscillation which is large compared with the Debye frequency, we anticipate that the study of hydrogen motion in materials will benefit greatly from the use of hot neutrons from advanced pulsed sources. Questions such as the anharmonicity of the proton potential and intrinsic damping processes might be addressed in greater detail than has been possible hitherto with thermal neutron scattering[3].

A prerequisite to the interpretation of data is a thorough understanding of the behaviour of the scattering from harmonically bound protons as an initial model system. We report some exact theoretical studies of the dynamic structure factor, as a function of $\underline{Q}$ and $\omega$, for the scattering of neutrons in several contexts, all arising from the nature of new neutron sources.

At high Q and $\omega$ one intuitively believes that interaction times and lengths are so short that the binding of the scattering centre will not be seen, that is the particle will appear free. Indeed for a particle in a lattice interacting with many degrees of freedom this is so and we reproduce the analytic pathway to this result. For a single isolated oscillator free particle behaviour is seen asymptotically provided energy resolution is not too fine. Careful analysis[3] of the "short-time" method shows that the energy level structure actually survives as Q and $\omega \to \infty$.

The path to asymptopia has pitfalls both for lattices and oscillators. As Q and $\omega$ increase higher and higher multiphonon and multi-quanta processes are excited. Expansion techniques become inadequate in a regime where multiphonon effects give $S(Q,\omega)$ considerable structure before the Gaussian (free particle) result attains. We compare the exact results[3] (avoiding expansion) with approximate schemes such as that of Sjölander[4]. For oscillators (for example hydrogen in metals or protons in molecules) high Q can mean that the expansion techniques for powder averaging line intensities can be in gross error. This too will be an important problem with new sources and we shall show a systematic and analytic method[5] by which this problem can be solved.

In fact, for many cases the objects to which the scattering centre is bound is itself bound within a crystal, for example protons in the potential field of atoms in a metallic crystal or of molecules in a molecular crystal. Our last concern is to describe the expected line intensity of the oscillator as a result of the lattice environment. We find[6] that the lines can develop considerable structure as Q increases, a cautionary tale for those who would wish to immediately ascribe such features to anisotropy, multi-site potentials or anharmonicity.

Before attacking the various problems mentioned we define the scattering from oscillators and lattices. We shall limit ourselves to incoherent scattering, a good approximation for the strongest scatterers, protons. The correlations measured are therefore those of a given particle with itself in time.

The neutron scattering cross-section is related simply (by kinematical factors and the square of a nuclear length) to what Lovesey[6] calls the scattering law $S(\underline{Q},\omega)$. The scattering law in turn is determined by a time correlation function $F(\underline{Q},t)$ such that

$$S(\underline{Q},\omega) = \frac{1}{2\pi} \int_{-\infty}^{\infty} dt\, e^{-i\omega t} F(\underline{Q},t) \tag{1.1}$$

and

$$F(\underline{Q},t) = \left\langle e^{i\underline{Q}\cdot\hat{\underline{r}}(0)}\, e^{-i\underline{Q}\cdot\hat{\underline{r}}(t)} \right\rangle \tag{1.2a}$$

where 0 denotes a thermal average at temperature T over the matrix elements of the operator 0. In the above $\underline{r}(t)$ is the position operator of the scattering centre. These averages and matrix elements can be taken trivially for a harmonic system[6]. The average can be taken into the exponent yielding

$$\exp\{-\frac{Q^2}{2}\langle \hat{r}^2(0)\rangle + \frac{Q^2}{2}\langle \underline{\hat{r}}(t).\underline{\hat{r}}(0)\rangle\} \tag{1.2b}$$

The first (static) term is the mean square displacement leading to the exponent of the Debye-Waller factor. It determines line intensities and is discussed below. The second, time dependent term governs the eventual energy shifts.

Taking these averages over the motion in the case of a harmonic oscillator of fundamental frequency $\omega_o$ and mass m leads to

$$F_o(Q,t) = \exp\{\frac{Q^2}{2m\omega_o} G_o(\omega_o,t)\} \tag{1.3}$$

where we suppress the vector nature of Q by taking a one-dimensional or isotropic oscillator until our discussion of powders and where

$$G_o(\omega_o,t) = (\cos(\omega_o t)-1)\coth(\omega_o/2T) + i\sin(\omega_o t) \tag{1.4}$$

For a lattice with a normalized density of states $Z(\omega)$ characterised by a frequency $\omega_\ell$ we can express everything in terms of a reduced frequency $u = \omega/\omega_\ell$ with $Z(\omega) \rightarrow 1/\omega_\ell Z(u)$ to get, within the incoherent approximation[6], $F_L$:

$$F_L(Q,t) = \exp\{\frac{Q^2}{2m\omega_\ell} G_L(\omega_\ell,t)\} \tag{1.5}$$

with G now generalized to $G_L$

$$G_L(\omega_\ell,t) = \int_0^\infty du\{(\cos(u\omega_\ell t)-1)\coth(\frac{\omega_\ell u}{2T}) + i\sin(u\omega_\ell t)\}\frac{Z(u)}{u} \tag{1.6}$$

It is convenient to introduce dimensionless quantities for the lattice or oscillator as follows:

$$\frac{Q^2}{2m\omega_o} \text{ or } \frac{Q^2}{2m\omega_\ell} = \kappa^2 \tag{1.7}$$

$\kappa$ is the ratio of the momentum transfer to the mean momentum associated with the harmonic mode or modes as appropriate. Likewise the temperature scale is set by the energy $\omega_o$ or $\omega_\ell$ to give:

$$\omega_o/T \text{ or } \omega_\ell/T = 1/T^* \qquad (1.8)$$

If frequencies $\omega$ in $S(Q,\omega)$ are reduced by $\omega_o$ or $\omega_\ell$ to u (equivalent to taking $t' = \omega_o t$ or $\omega_\ell t$ as the variable in (1.1)) we get the completely reduced expressions:

$$S(Q,u) = \frac{1}{\omega} \int_{-\infty}^{+\infty} dt' \, e^{-iut'} \exp\{\kappa^2 G(t')\} \qquad (1.9)$$

with the prefactor $1/\omega$ denoting $1/\omega_o$ or $1/\omega_\ell$ as appropriate. Henceforth, for S and G as well, o and L will denote oscillator and lattice respectively. For the oscillator we get $S_o$ with G given by $G_o$ derived from (1.4):

$$G_o(t') = (\cos t') \coth(1/2T^*) + i \sin t' - 2W/\kappa^2 \qquad (1.10)$$

defining 2W for the oscillator, see (1.17); and, equivalently for the lattice case, $S_L$ and $G_L$:

$$G_L(t') = \int_{-\infty}^{\infty} \frac{du'}{u'} Z(u')n(u')e^{iu't'} - 2W/\kappa^2 \qquad (1.11)$$

where $G_L(t)$ has been slightly rewritten with the definition that $Z(u)$ is even and that $n(u') = 1/(\exp(u'/T^*)-1)$ is the Bose temperature factor. The exponent of the Debye-Waller factor, 2W has been extracted. It is discussed before (1.3) and later at (1.12). The prime will be dropped on t. Equations (1.9) - (1.11) represent the starting point of both the approximate and numerical analyses that are to follow.

The response function can be calculated exactly for an oscillator. Consider first a free particle in equilibrium, for which[6], on expanding the correlation function G(t) for inertial motion

$$F(Q,t) = \exp\{\frac{Q^2}{2m}(it - Tt^2)\} \qquad (1.12)$$

and

$$S(Q,\omega) = \left(\frac{m}{2\pi TQ^2}\right)^{\frac{1}{2}} \exp\{-(\omega - Q^2/2m)^2/2TQ^2/m\} \qquad (1.13)$$

where m is the mass of the particle. We use units in which $h = k_B = 1$. From (2.1) we see that the spectrum is centred about the of G(t) about a saddle point in the complex plane. What dominates is the short time behaviour about this point, whence the connection with the physical picture sketched above.

One important feature is the position of the maximum $\omega_{max}$ of $S(Q,\omega)$ at fixed Q. For a normalised Gaussian function the maximum occurs at the value of the first moment

$$\overline{\omega} = \int d\omega \; \omega S(Q,\omega) = Q^2/2m \tag{2.1}$$

where the second equality is an exact sum rule for any system, satisfied in particular by the free particle Gaussian (1.13).

It turns out that the Sjölander approximation (2.5) discussed later has departures from Gaussian form and that, although the moment relation is preserved, consequently the maximum is shifted (to lower frequencies). The asymptotically correct form, as we show in the appendix, does indeed have its maximum at $Q^2/2m$, a property evidently related to its Gaussian character.

In the appendix we show that the scattering from a lattice is asymptotically Gaussian;

$$S(Q,u) = (2/\pi\Gamma^2)^{\frac{1}{2}} \exp(-(u-\kappa^2)^2/2\Gamma^2) \tag{2.2}$$

which is characterised by a maximum at the recoil energy $u=\kappa^2$ equivalent to $\omega=Q^2/2m$, and a width $\Gamma$ given by

$$\Gamma^2 = \kappa^2 \int du' \; Z(u') \; n(u')u' \tag{2.3}$$

in contrast to the free particle result, that is, the lattice gives the individual particles an effective temperature by confining them

$$T' = \tfrac{1}{2} \int_0^\theta du' \; Z(u')u' \coth(u'/2T^*) \tag{2.4}$$

equal to the actual temperature as $T \ll \omega_{max}$, the upper extent of the lattice modes.

The main result of a careful asymptotic expansion is that the point of expansion of the exponent G(t) is not t=0 as one might expect (except at the maximum of S(Q,u)), but varies in position along the imaginary t-axis as the reduced frequency u is varied. In the appendix we show how the expansion about the saddle point allows the necessary Fourier transformations (1.11) and (1.9) to be performed and the result (2.1) yielded.

recoil energy of the particle, $Q^2/2m$, and that the Doppler broadening also results in a characteristic width, $\Gamma$, of $(Q^2T/m)^{\frac{1}{2}}$. For the oscillator we can rearrange (1.10) to give :

$$G_o(t) \equiv -\coth(1/2T^*) + \frac{\cosh(it+1/2T^*)}{\sinh\,(1/2T^*)} \qquad (1.14)$$

The Fourier transform of F(Q,t) is readily obtained using the generating function for modified Bessel functions of the first kind, $I_n$.

$$\exp\{a(x+x^{-1})/2\} = \sum_n I_n(a)\; x^n$$

The result is,

$$S_o(Q,u) = \exp\{-2W\} \sum_{-\infty}^{\infty} I_n(z)\; \exp(n/2T^*)\; \delta(u-n) \qquad (1.15)$$

The delta function expresses conservation of energy on exciting ($n > 0$) or annihilating ($n < 0$) many quanta. The argument of the Bessel function is,

$$z = Q^2/\{2m\omega_o \sinh(\omega_o/2T)\} \equiv \kappa^2/\sinh\,(1/2T^*) \qquad (1.16)$$

and the exponent of the Debye-Waller factor is

$$2W = (Q^2/2m\omega_o)\,\coth(\omega_o/2T) \qquad (1.17)$$

The expansion (1.15), and a similar one obtained from (1.11) for lattice quanta, are increasingly dominated by higher order terms as $Q^2$, or more revealingly, $\kappa^2$, is increased. We now look at approximate methods for circumventing this problem and at asymptotic analysis as $Q\to\infty$, comparing both approaches with exact numerical procedures we shall describe.

## 2. HIGH Q SCATTERING FROM AN OSCILLATOR AND A SIMPLE LATTICE

### Asymptotic Analysis for the Lattice

Physically one starts from the view point that large energy transfers (frequencies, u) imply short times of interaction consistent with the exploration of small distances, that is, correspondingly large momenta. Mathematically one sees in integrals such as (1.9) a large parameter $\kappa^2$ suggesting expansion

## Asymptotic Analysis For The Isolated Harmonic Oscillator

Conventionally the asymptotic analysis of scattering laws arises from the short-time behaviour of the correlation functions of the motion of the scatterer (1.2). For a harmonic oscillator it would involve looking at times t  $1/\omega_0$ over which the restoring effect of the potential would not be felt. This leads to incorrect results. The first reason can be seen physically in that the system is periodic and there is an infinite sequence of equivalent origins in time, $t = n/\omega_0$. As for the lattice we observe that saddle points for the time integration (1.9) are not at the origin $t = 0$ but displaced in the complex plane. Moreover for the reason given above there is not one such point but an infinite series of equally important points, as had been noted in a total cross-section calculation by Wick[10]

If the simple procedure of small time expansion were to be followed through the result would inevitably be Gaussian (with corrections) in form. In fact the exact result (1.15) is a series of sharp lines further underlining the assertion above that, irrespective of the brevity of times and lengths over which the neutron may interact with its target, free-particle-like behaviour is only observed when the final state is part of a continuum. In the appendix we resolve the apparent difficulty that asymptotic analysis gives Gaussian-like forms whereas the exact result is distinctly non-Gaussian. There we also conclude that the overall envelope of the sharp lines is Gaussian at high momentum transfers and temperature. With coarse energy resolution it is just this that one would see, that is, free particle behaviour albeit with an effective temperature given in terms of the binding potential.

Although the asymptotic analysis cannot be carried out explicitly for a general potential, a few observations can be made. In general there will be some bound states and a continuum of unbound ones; the former will lead to some discrete energy transfers, with associated delta functions in energy in the expansion for the cross section. Again these will not come out of a single saddle point approximation. However we would expect that the unbound states would be more robust in terms of being treated by a single saddle point approximation.

There is a similarity between the adequacy of the single saddle point approximation for the harmonic lattice and its expected validity for the unbound states for a single particle potential: they both deal with continua of states, which in general imply that correlation functions of the type $\langle e^{iQr(t)} e^{-iQr(0)} \rangle$ will tend to zero as $t \to \infty$. Thus one might expect that the saddle point nearest $t = o$ should be the dominant one when calculating $S(Q,\omega)$.

## Approximate Analysis for the Lattice

Early work in this analysis was done by Egelstaff and Schofield[7]. The Gaussian approximation to the scattering law was derived by Sjölander[4] and Schofield and Hassitt[8] from a central limit theorem. Lovesey[6] presents a derivation of Sjölander's results. The approach is to expand the function $G_L(t)$ for short times about t=0 to get terms linear and quadratic in t, the former yielding the recoil shift, the latter the underlying Gaussian form. The result has the additional appeal that the zero (elastic) and one phonon processes are taken out explicitly so that for small Q, where they dominate, the expression is exact. The result is[6]:

$$S(Q,u) \;=\; e^{-2W(Q)}\left\{\delta(u) + \kappa^2 \frac{Z(u)}{\theta} n(u) + \frac{1}{\Delta} e^{-u/\Delta^2\gamma} F(x,y)\right\} \qquad (2.5)$$

where the first term in the brackets is the elastic line and the second the one phonon term. Also 2W(Q), the exponent of the Debye-Waller factor derives from the time independent part of (1.6) and (1.11). thus also defining $\gamma$:

$$2W(Q) \equiv \kappa^2\gamma \equiv \kappa^2 \int_0^\infty du \frac{Z(u)}{u} \coth(u/2T^*)$$

$$x = u/\Delta$$

$$y = 2W(Q)\exp\{-1/2(\Delta\gamma)^2\} \qquad (2.6)$$

$$\varepsilon = 3/4 \int_0^\infty du\, Z(u)\, u \coth(u/2T^*)$$

$$\Delta^2 = 4\varepsilon/3\gamma - 1/\gamma^2$$

The function F(x,y), which represents the higher-order phonon processes, term by term, is given by

$$F(x,y) = \sum_{p=2}^{\infty} \frac{1}{p!\sqrt{2\pi p}} y^p e^{-x^2/2p} \qquad (2.7)$$

We shall later check numerically whether or not the Sjölander result (2.5) misses any significant structural details in the incoherent scattering.

## Numerical Analysis for the Lattice

The difficulty with an analytic development of $S_L(Q,u)$ for the lattice is that at higher and higher Q the energy transfer is dominated by many-quanta (phonon) processes. The reader can see this by taking (1.11) in (1.9) and expanding the exponential. The first two terms are explicitly present in the Sjölander approximation (2.5) being the zero-phonon (elastic) and one-phonon terms. They are taken so as to represent low energy transfer events and are easily obtained by integrating over times t and then over frequencies u'. At higher $Q^2$ the plethora of higher terms dominate and it is wiser to avoid expansion altogether and to perform the integral over u' in the exponent and then the Fourier transform with respect to time t directly. Results indicate that the true picture often lies in between the one-phonon term on the one hand (which mirrors the density of states, Z(u)) and the smooth Gaussian-like result expected at the other extreme of $Q^2$. Indeed there are discernible features from successively higher-phonon processes before blurring occurs. A measure of this is seen in the departures from the approximate scheme of Sjölander.

The first requirement to follow through the suggested scheme of two direct Fourier transformations is to have a density of states. We have taken over, in one component form, a simple model of a lattice[9], namely a purely harmonic system characterised by a frequency $\omega_\ell$ (the frequency used to reduce all energies in the problem) and a ratio $\eta$ of the squares of the transverse to longitudinal speeds of sound. The cut off in the density of states occurs at $\omega_{max} = \omega_\ell\sqrt{(2\eta + 1)}$ or $u_{max} = \sqrt{(2\eta + 1)}$.

We present below the results of the exact numerical treatment along with the approximation (2.5) for comparison. In all cases we have taken $\eta = 1$ for simplicity. The reduced temperature, $T^*$, is defined as in (1.8), that is, on a scale set by $\omega_\ell$. We set it equal to 1 to see the effect of finite temperature, most noticeably in (a) intensity for $u < 0$ (neutron energy gain) given by the detailed balance term, (b) a smaller Debye-Waller factor and (c) a finite intercept at u = 0 given for small Q vectors, that is when 1-phonon processes dominate, by $\exp\{-2W(Q)\}T^*\kappa^2[Z(u)/u^2]_{u=0}$.

For $T^* = 0$ the one-phonon part of $S_L(Q,u)$ is simply the density of states (weighted as indicated above). A qualitative picture of Z(u) can therefore be gleaned by inspection of figure 1 and allowing for finite $T^*$.

The Debye-Waller factor $\exp\{-2W(Q)\}$ essentially gives the weight of the elastic line $\delta(u)$ which is not plotted. The intensity in the inelastic part plotted is $(1-\exp\{-2W(Q)\})$ whence the quoted value of $\exp\{-2W(Q)\}$ will give the relative importance

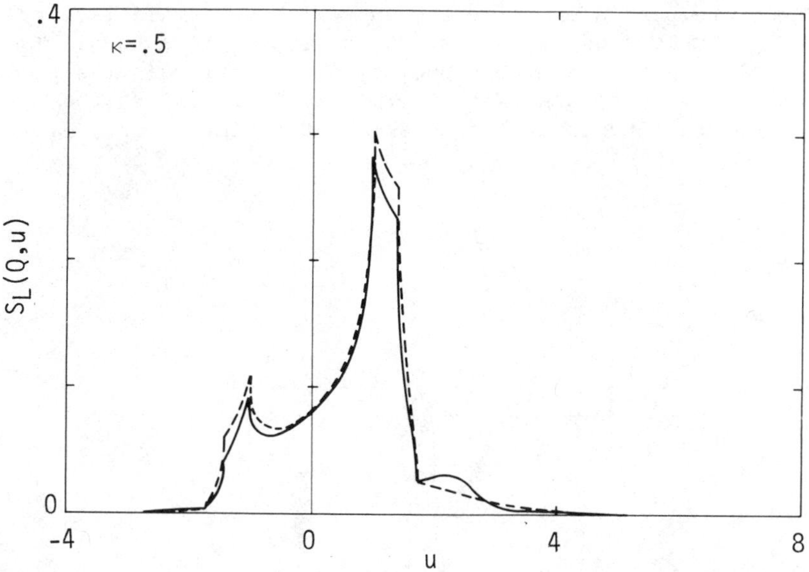


Figure 1: The scattering cross-section $S_L(Q,u)$ for neutron scattering from a lattice within the incoherent approximation. The reduced momentum transfer, κ, and energy transfer u are reduced by the characteristic momentum and energy associated with the lattice. Consult (1.7) and the energy reduction after (1.4). Here, and in subsequent figures, the exact result of the text is shown by a full line. An approximation due to Sjölander which extracts one-phonon processes exactly and treats the rest in a Gaussian manner is shown as a broken line. The elastic line in each case is not shown. The Debye-Waller factor gives its intensity relative to the total intensity of 1. For this case where κ = 0.5, the Debye-Waller factor is 0.604. The component of the overall scattering at this Q residing in the Gaussian approximate part will be denoted by "the Gaussian weight". Here it is 0.092. Since one-phonon processes dominate the approximation is seen to be very good. Two-phonon events in the exact result are visible. In figures (1) - (4) the reduced temperature, T*, is 1.

of the two contributions. We quote also the "Gaussian weight" in the case of the approximate form of $S_L(Q,u)$, (2.12). This refers to the total intensity of the term F(x,y), that is, the total intensity not in the elastic and one-phonon processes.

Figure 1 shows the exact (full line) and approximate (broken line) forms of $S_L(Q,u)$. The reduced momentum transfer, $\kappa$, is 0.5. Because the one-phonon processes dominate the inelastic part of $S_L$ and these are put in exactly, the approximate formula is very good. The only point missed is the two-phonon energy loss hump. The Debye-Waller factor is 0.604 (60% of the intensity in the elastic line) and the Gaussian weight is 0.092 implying 25% of the inelastic intensity is in the Gaussian approximate part, F.

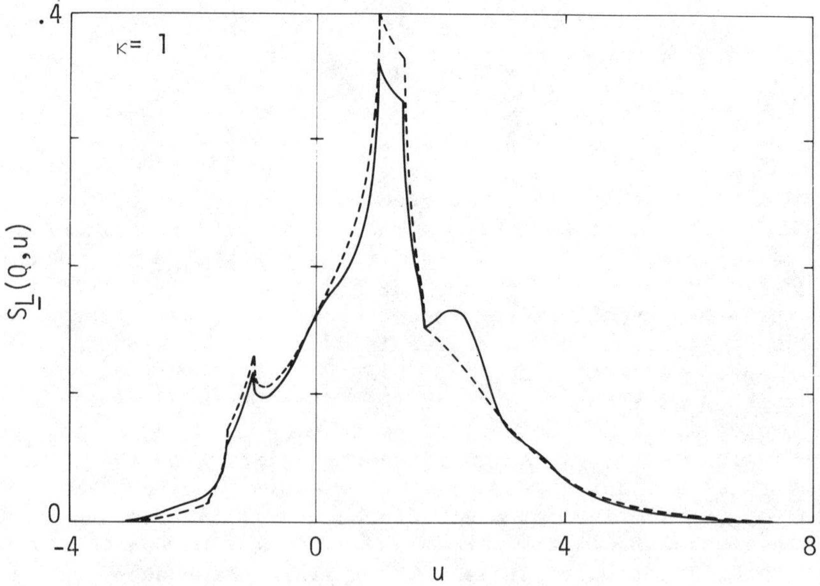


Figure 2: Neutron scattering cross-section $S_L(Q,u)$ of energy transfer, u. Reduced momentum transfer is now $\kappa = 1$. The Debye-Waller factor is 0.133 implying a dominance of inelastic events. The Gaussian weight 0.599 shows further that multi-phonon processes are at least equally important as one-phonon processes. There is thus structure missed by the Gaussian approximation.

In figure 2 the reduced momentum transfer has been doubled to $\kappa = 1$. The Debye-Waller factor of 0.133 implies most intensity is now inelastic. The Gaussian weight of 0.599 shows that multi-phonon processes are beginning to dominate and accordingly there is structure, two and three- phonons not reproduced by the Gaussian approximation. The gross features are however, accurately reproduced.

When Q is further increased to give $\kappa = 1.5$ (figure 3) we see a possibly important defect in the approximate results. Inelastic processes now totally dominate exp $-2W(Q) = 0.011$) and they are mostly multi-phonon events (Gaussian weight = 0.94). The now conspicuous two-phonon peak would not be expected if one placed

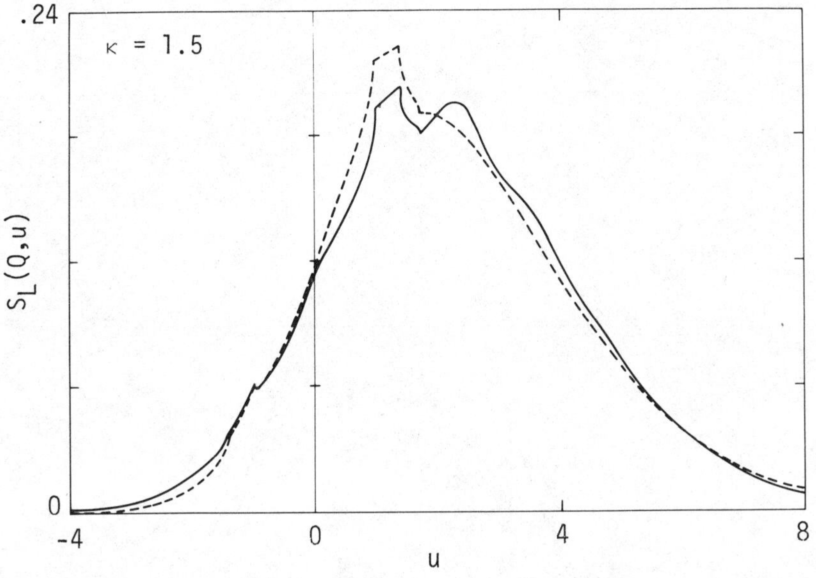


Figure 3: Neutron scattering cross-section $S_L(Q,u)$ as a function of energy transfer, u. Momentum transfer is now $\kappa = 1.5$. Multi-phonon inelastic events dominate since the Debye-Waller factor is 0.011 and the Gaussian weight is 0.94. If one were to rely on the approximate form (broken line) one would interpret the double peaks as implying split peaks in the density of states. In reality (solid line) the splitting is due to the important role of two-phonon processes.

reliance on the approximate form. One could instead incorrectly interpret the full line as exhibiting a split peak. Higher multi-phonon processes become increasingly blurred, an observation that takes us naturally to figure 4.

Now $\kappa = 2$ and the Debye-Waller factor and Gaussian weights are 0 and 1 respectively. No individual orders of multi-phonon processes are discernible and the Gaussian approximation becomes an excellent representation of S(Q,u).

It remains only to remark how the overall magnitude of energy transfer increases rapidly with Q (the scale is smaller in figure 4). In figure 4, and also for even higher values of Q not shown, the maximum in S occurs at lower u in the approximate form than in the exact result. Both results still have a noticeable difference from the asymptotic result $u_{max} = \kappa^2$.

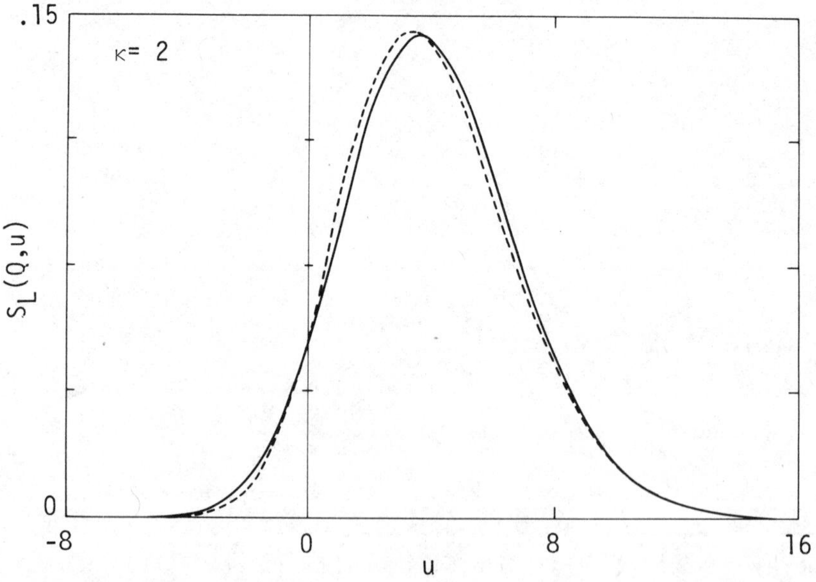


Figure 4: Neutron scattering cross-section $S_L(Q,u)$ as a function of energy transfer, u. Reduced momentum transfer is $\kappa = 2.0$. The Debye-Waller factor and the Gaussian weight are essentially 0 and 1 respectively. Multi-phonon events blur together and the Gaussian approximation is excellent.

## 3. THE EFFECT OF CRYSTAL ENVIRONMENT ON THE OBSERVATION OF INTERNAL MODES OF MOLECULES (9)

We turn now to the study of a binary system with masses and force constants that we vary to explore a range of different regimes. This is a simple model of a situation frequently encountered in chemical spectroscopy, where the light scatterer, typically a proton, is bound in a molecule which is itself bound in a lattice. The vibrational density of states consists of two distinctly different components (when the masses and force constants are sufficiently dissimilar). At low frequencies there is a broad spectrum that is proportional to $\omega^2$ for $\omega \to 0$ and terminates at $\omega_{max}$, which arises primarily from the vibrations of the heavy lattice particles. The second component arises from the vibrations of the light particles; it is centred at a frequency that is large compared with $\omega_{max}$, and it has a width which is small compared with $\omega_{max}$. We shall often refer to these two components of the vibrational density of states as the lattice and oscillator, or low and high frequency, bands.

Two effects of the coupling of the molecules in the lattice can arise. When the internal mode and characteristic lattice frequencies are well separated the effect is purely kinematic, that is, the phonons cause an additional Doppler shift to the molecular vibrations. When the frequencies are not well spaced then in addition one must consider the mixing or hybridisation of the lattice and internal modes, that is, the motions lose their underlying character. It turns out that both effects can lead to detailed structure in the response function especially if Q is high. New neutron sources with high flux at the energies required for chemical spectroscopy, that is neutron frequencies capable of probing the dynamical structure of molecules, will revolutionise this subject but because of the attendant high Q's care will have to be taken not to interpret structured lines. We show that with harmonic systems structure arises that could otherwise be attributed to more interesting effects such as anharmonicity, site splittings, etc.

Before considering the full calculation which includes the mixing of the two types of motion we consider the case when they are well separated in their timescales and ask how lattice vibrations effect the oscillator lineshape purely through kinematic or Doppler effects[11]. Let the position of the molecule be $\underline{R}(t)$ and the position of the oscillator bound to it be $\underline{r}(t)$ within the molecular frame. Then $\underline{r} + \underline{R}$ is the appropriate displacement in the expression (1.1) governing incoherent neutron scattering. When the lattice and oscillator bands are well separated, we expect minimal correlation between the two motions hence a separable expression for the correlation function

$$F(Q,t) \doteq \langle e^{-i\underline{Q}.\underline{R}(o)} e^{i\underline{Q}.\underline{R}(t)} \rangle_R \langle e^{-i\underline{Q}.\underline{r}(o)} e^{i\underline{Q}.\underline{r}(t)} \rangle_r \tag{3.1}$$

$$= F_L(Q,t)\, F_o(Q,t)$$

This result can be expressed as a convolution of the two parts labelled L (lattice) and o (oscillator) respectively,

$$S(Q,\omega) = \int_{-\infty}^{\infty} d\omega'\, S_L(Q,\omega-\omega')\, S_o(Q,\omega') \tag{3.2}$$

Recalling the analysis of section 2 for a single harmonic oscillator and taking the n quanta term in the expansion (1.15) one can immediately perform the convolution in (4.2) to give, no longer reducing frequencies:

$$S^{(n)}(Q,\omega) = S_L(Q,\omega-n\omega_o)e^{-2W_o(Q)} I_n(z)e^{n\omega_o/2T} \tag{3.3}$$

with z and $2W_o$ defined by (1.16) and (1.17). If in addition we assume that $\omega_o \gg T$ we obtain, by expanding the Bessel function as in (4.20)

$$S^{(n)}(Q,\omega) = \frac{1}{n!} S_L(Q,\omega-n\omega_o)e^{-2W_o(Q)} (Q^2/2m\omega_o)^n \tag{3.4}$$

The line shape is evidently now determined by the lattice function $S_L$, (1.9) and (1.11), and it is no longer the delta function otherwise implied by the time-scale separation. Egelstaff and Schofield[7] suggest this approach of convolution of two separated effects directly from a consideration of a density of states which includes a band plus one or more split off high frequency modes. The correlation function $F_L(Q,t)$ corresponding to this $S_L(Q,\omega)$ is now written,

$$F_L(Q,t) = e^{-2W_L(Q)}\exp \int \frac{Q^2}{2M\omega_\ell} \frac{d\omega'}{\omega_\ell} Z_L(\omega')\coth(\omega'/2T)\cos\omega't+i\sin\omega't\} \tag{3.5}$$

where $Z_L(\omega)$ is the density of states in the lattice scaled by $\omega_\ell$ and M is the mass of the lattice particles.

A naive analysis of (3.5) would be to ignore the time dependence, equivalent to taking only the elastic part n=0 of an expansion of $S_L(Q,\omega)$. One would then conclude that the contribution of $S_L$ to S is simply exp $(-2W_L(Q))$. This lattice

Debye-Waller factor would be vanishingly small at the Q vectors typically employed in a neutron scattering experiment[12,13,14] to investigate the high frequency oscillator behaviour around $\omega = \omega_o$. Despite the appearance of the lattice Debye-Waller factor in the expression for the oscillator line, the experiments referred to above give finite oscillator intensities. This apparent conflict has motivated earlier investigations[9,11,12]. Intuitively one can picture the motion of the lattice, ie. the molecular contribution, to be frozen over the very short times the neutron interacts with the rapidly moving oscillator. There is consequently no reduction in intensity due to molecular motions. Mathematically if an expansion in phonon number of the type (1.5) corresponding to (3.5) is taken to all orders then the effect of the $e^{-2W_L}$ term is nullified. Placzek[15] recognised that the expansion in phonon number is heavily compensating. Alternatively we can avoid such an expansion and evaluate the exponent in (4.5) for short times $t \leqslant 1/\omega_{max}$, whereupon we get roughly the Gaussian asymptotic form for $S_L(Q,\omega)$ that we discuss in section 2 at (2.2). This is not of vanishingly small amplitude unless Q is extremely large, in which limit it is extremely large. In any case inspection of the underlying cross-section expressions (1.1) and (1.2) for incoherent scattering from a lattice point shows that the integrated intensity under the curve, $\int d\omega\, S_L(Q,\omega) \equiv F_L(q,t=0) = 1$, whence there is no reduction in total intensity in scattering from a proton component of a molecule due to the latter's binding in a crystal. Rather the crystal causes intensity to be transferred from the sharp line expected at $\omega=n\omega_o$ to a line, broadened and shifted increasingly as Q increases, determined by $S_L(Q,\omega)$. The final result in the asymptotic limit for $S_L(Q,\omega)$ is then

$$S^n(Q,\omega) = \frac{1}{n!}\left(\frac{Q^2}{2m\omega_o}\right)^n \exp\{-(\omega-n\omega_o-Q^2/2M)^2/2\Gamma^2 - 2W_o(Q)\} \qquad (3.6)$$

where the small argument expression for $I_n(z)$ has been employed. This is the result obtained by Griffin and Jobic[11].

Numerical Results for the Molecular Crystal

For the molecular crystal the same model density of states was taken[9] for the underlying lattice of molecules as for the harmonic system[3] reported on in section 2. With a density of states the lattice contribution to the oscillator line shape and intensity (3.3) can be calculated. As Q increases the n quanta line at $\omega=n\omega_o$ gets broadened by a series of curves similar to the actual phonon curves, figures (1) - (4).

The reader is referred to a systematic development of line shape[9] resultant from kinematics. To make contact with experiment we estimate the effect of the lattice in the case of hexamethylenetetramine (HMT), examined by Dolling and Powell[16]. There are normal modes of motion involving protons which, because of their low associated mass compared with the molecular mass, are well separated in frequency from the lattice frequency $\omega_\ell$. We deduce[16] that $\eta=.5$ and $\omega_{max} = 2.4\times10^{12}$ by whereupon quantities $\kappa$, $T^*$ and u reduced by $\omega_\ell$ have the correspondence to experimental values according to

$$\begin{aligned} Q &= 1.822\kappa \quad (k \text{ in } Å^{-1}) \\ T &= 85T^* \quad (T \text{ in } {}^oK) \\ E &= 7.0\omega \quad (E, \text{ neutron energy transfer, in meV}) \end{aligned} \tag{3.7}$$

We now examine the effect of increasing Q. Figure (5) shows

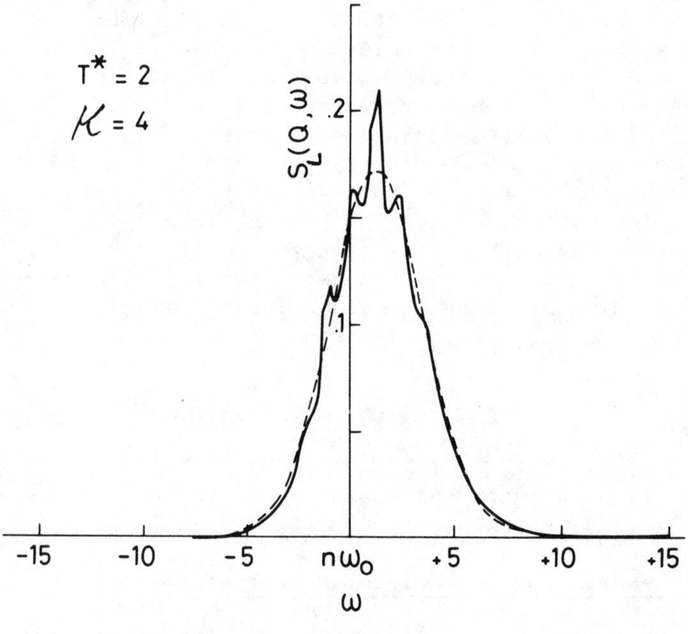


Figure 5: $S_L(Q,\omega)$ against energy transfer $\omega$. This function gives the line profile, due to the effect of phonons, for the oscillator-like transition otherwise sharp at $n\omega$. The temperature $T^* = 2$ and momentum transfer $\kappa = 4$ are both reduced in terms of the frequency $\omega_\ell$ and is seen to be in a wide range about $n\omega_o$. This is due to many-phonon events. The broken line is the Gaussian approximation to the lattice response function and misses a lot of structure in the line shape. The elastic (unshifted component) to $S_L$ is not indicated but has a relative weight of only 0.012.

$S_L(Q,u)$ for $\kappa=4$. Multiphonon effects now dominate the line shape but much structure is still evident, both in the region $u = n \pm u_{max}$ mirroring the density of states, and in the region beyond this where successive multiphonon humps in $S_L(Q,u)$ are evident. The Debye-Waller factor is now .012 indicating that essentially all the weight resides in the figure as shown and little in the unshifted, unbroadened component of the line not plotted. With currently available resolutions (often a small fraction of typical values of $\omega_{max}$) it is clear that even at such $\kappa$ values (equivalent to Q=7.3Å$^{-1}$ in the example of HMT) a structure consisting of four peaks and perhaps shoulders could be seen. We emphasise that this result has been achieved within an harmonic approximation and assuming no intrinsic structure to the oscillator

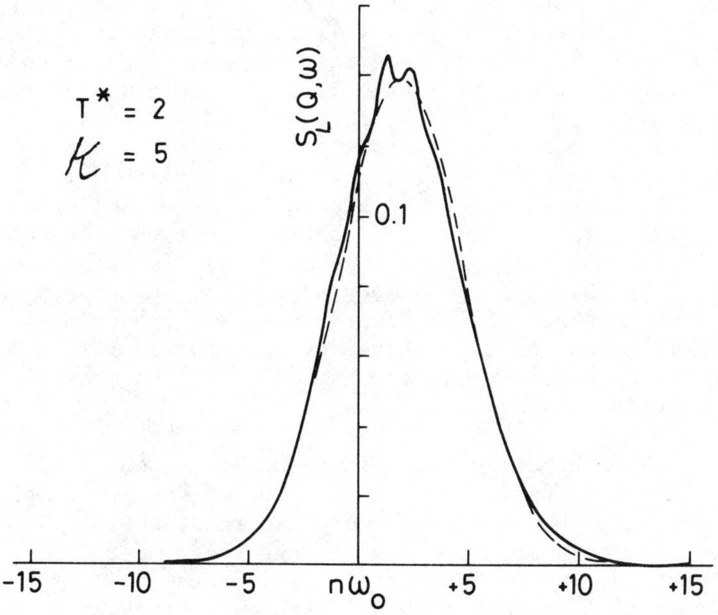


Figure 6: $S_L(Q,\omega)$ against energy transfer $\omega$. Reduced temperature is still T* = 2 but the reduced momentum transfer is increased to $\kappa = 5$ with the effect that multi-phonon processes are further stimulated, blurring out features and making the Gaussian approximation (broken line) still better.

function. A cautionary note must therefore be sounded about a too rapid recourse to more complicated possible explanations of observations of apparently split peaks or fine structure when inelastically scattering from internal modes of molecules or from other oscillators in a lattice.

Proceeding to a higher value of $\kappa$ ($\kappa$=5)with $T^*=2$ we see in figure (6) the smoothing out of phonon effects to give an envelope that is evidently tending to a Gaussian shape. The Debye-Waller factor $\exp(-2W_L(Q))$ for the lattice is now vanishingly small and essentially all of the $S_L$ contribution to $S^{(n)}$ in (3.4) resides in the part of $S_L$ broadened and shifted, and plotted in the figure.

## Mode Hybridisation

When the frequency of the internally bound proton, $\omega_0$, approaches $\omega_\ell$ characterising the lattice modes the modes no longer retain their identity but mix. Since this is a totally harmonic system an exact solution of the eigen problem is possible, the result being a distorted density of states for the lower, formerly phonon, band and a broadening of the formerly isolated oscillator spike. The oscillator band is now wide and consequently the scattering law for neutron scattering from the internal modes, involving proton motion, will have both kinematic and hybridisation components to their width. In reference[9] a model system, mimicking the interplay between slow internal modes and the motion within the molecular crystal, is diagonalised. The results at a fixed temperature $T^*=2$ show increasing broadening as $\kappa$ is increased from $\kappa$=1 to 2,figures (7) and (8). The spikes represent alternately and in turn the fundamentals and harmonics of the "lattice" and "oscillator" bands. A large number of many-quanta events rapidly enter $S(Q,\omega)$. Neglect of the lower band and its harmonics would lead to their erroneous identification with the oscillator states and hence the conclusion that the harmonics of the oscillator level were too closely spaced.

To conclude the discussion of the mixed harmonic solid as a model for chemical spectroscopy in the solid state we emphasise that the advent of high neutron energy sources, while revolutionising such science, also brings the need to eliminate the effect of the lattice in interpretation. At the low Q found in electromagnetic techniques such as infrared spectroscopy, lattice modes are unimportant (the intensity being in the unshifted component of $S_L$). With neutrons one can see that highly structured spectra can result from even such a trivial, harmonic model system.

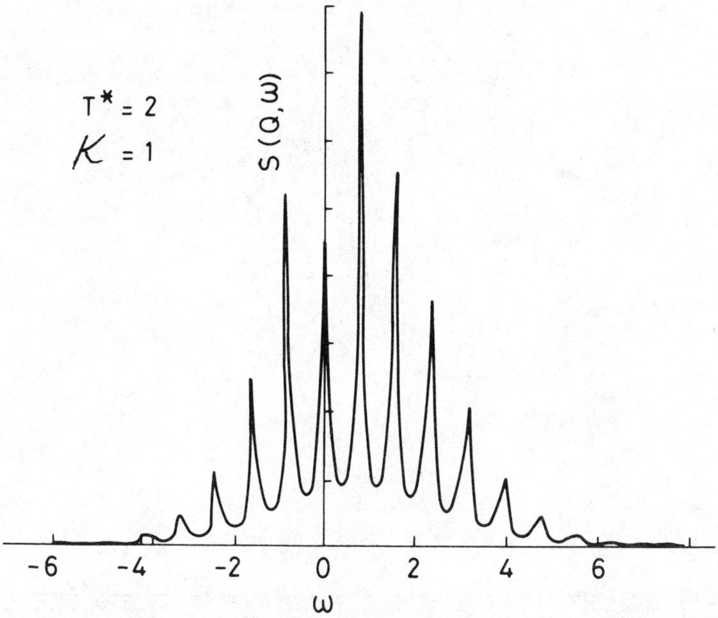


Figure 7: S(Q,ω) against energy transfer ω, in units of the energy $\omega_\ell$ characterizing the lattice band. This is for the case of an oscillator band made widely by the close proximity of the intrinsic molecular level to the phonon band of the molecular crystal. As in previous figures the elastic component is not plotted but at the reduced momentum transfer and temperature, $\kappa = 1, T^* = 2$ chosen it is in any case essentially zero. Peaks away from zero frequency are alternately of "lattice" or "oscillator" character, three orders of multi-quanta events being discernable.

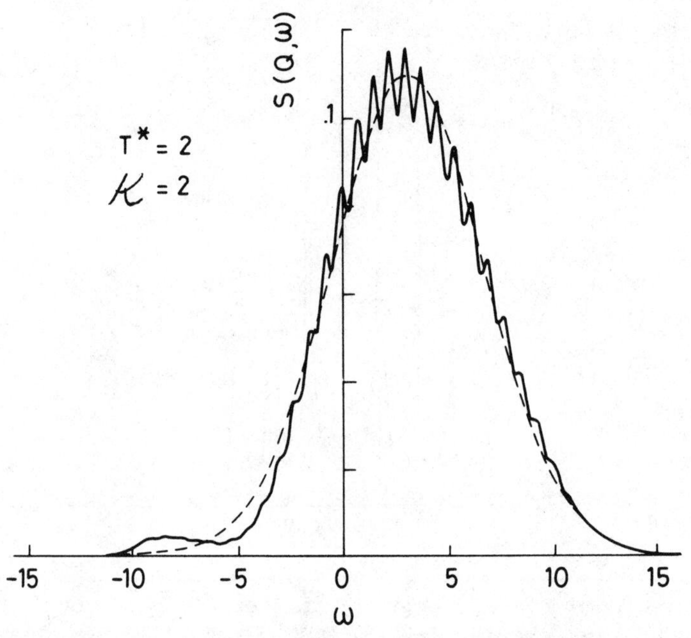


Figure 8: S(Q,ω) against frequency ω (in units of $w_\ell$). Reduced momentum and temperature are κ = 2, T* = 2. Band mixing is the same as in the previous two figures. Now the envelope of scattering processes comes close to the Gaussian approximation (broken line) and individual features reflecting the band structure (at various orders) blurred out.

## Conclusions

Finally we comment on a physical interpretation of the $Q \to \infty$ limit of $S(Q,\omega)$, that appears frequently in the literature. Firstly the leading term of an expansion, exploiting $1/Q$ being small, is often called the ''Impulse Approximation'', and its validity is associated with a ''short interaction time''. We think that the term ''Impulse Approximation'' is misleading for the following reasons. Classically, an impulsive force is one that transfers a finite amount of momentum to a body in an infinitesmal time; this implies that the force is very large, so during the time of its application, other forces acting on the body can be neglected. Hence the body acts like a free particle as it is being struck by the force. Now to transplant this notion to scattering, consider the neutron to be the body that applies a force to the scattering centre (we could consider it the opposite way round), then imagine the neutron to be initially in a state which is a wave packet (we need a wave packet to make contact with classical notions) with mean wave vector $k_i$. This wave packet is then scattered by the scattering centre into a far more diffused state comprising the original wave packet proceeding straight on and a scattered ''disturbance'' moving out from the scattering centre. If we assume that we are working within the Born approximation, and so exclude resonant scatterers, then an ''interaction time'' of the neutron with the scattering centre is determined by the passage time of the incident wave packet (which becomes the ''unscattered part'' of the state which describes the neutron after the collision), as the constraint of working within the Born approximation implies that the scattered part of the wave function is not rescattered and so cannot linger at the scattering centre. Given that we are using the Fermi Pseudopotential, which has zero range, then the interaction time is roughly the time the incident wave packet takes to pass the scattering centre, namely $(hk/m)^{-1}(\Delta x)$ where $\Delta x$ is the width of the incident wave packet, and we neglect dispersion causing this to change with time. Implicit in this is the assumption that the width of the incident wave packet is large compared to the spread in the initial state of the scattering centre - if it's not then the above expression becomes $(hk/m)^{-1}$ $(\Delta X)$ where $\Delta X$ is the spreading the scattering centre wave function. Note that the Fermi pseudopotential makes these estimates easy to make, as a finite range R would necessitate several expressions for the interaction time, depending which particular inequality involving $\Delta x$, $\Delta X$ and R was being considered.

Thus we arrive at the conclusion that the natural definition of ''interaction time'' in a quantum mechanical context implies that for the scattering to be considered impulsive, it is the incident momentum which should be large, not the momentum transfer, $hQ$, which does not enter the expression for ''interaction time''.

In conclusion we find that the limit of large momentum transfer does yield free particle like behaviour when scattering from a lattice, and for single particles may yield this, apart from a granularity in the allowed energy transfers. However the usual intuitive interpretation associating a ''short interaction time'' with $Q \rightarrow \infty$ does not seem to be correct, and we have no intuitive picture with which to replace it.

## 4. POWDER AVERAGING FOR SPECTROSCOPY AT HIGH Q

So far we have examined the consequences of high momentum transfer on two types of problems, that of an oscillator or a lattice where multi-quanta processes are stimulated by high Q and an approach to quasi-free behaviour is seen, and that of where the environment increasingly modifies the perception of internal degrees of freedom of molecular oscillators. High Q appears to be a disadvantage that neutrons at these relatively high energies have over low Q techniques. In reality neutrons have many advantages to offer: apart from being sensitive to proton motion and to motions not optically active, they also allow a fuller description of motion because of the finite size of Q compared with a reciprocal dimension of the amplitude of motion. We continue considering harmonic motion. It is characterized by two quantities, either a mass and a spring constant or, equivalently, a frequency and an amplitude of motion. Hitherto spectroscopy has yielded only a frequency, $\omega_0$. New neutron sources will yield epithermal neutron fluxes sufficient to follow line intensities over a range of Q values enabling a mean square displacement and hence also an amplitude of motion to be determined. The two pieces of information will allow the character of the normal mode to be assigned, even in a complicated system such as a molecule, via a knowledge of the effective mass.

One further problem subsequently arises (we shall henceforth put the kinematical problems of the lattice to one side). If the sample is a powder and the molecular modes anisotropic then cross-sections must be averaged over all angles. Conventionally line intensities involving such quantities as $\exp(-2W(\underline{Q}))$, now a function of the vector $\underline{Q}$, have been averaged by expanding the exponential and averaging term-by-term[17,18]. For the reasons given above, that is because $W(\underline{Q})$ involves terms like $Q^2\langle r^2\rangle$, a mean square displacement and momentum transfer and these by construction are comparable inverses of each other, the region of interesting Q variation is just where the exponential many <u>not</u> be expanded. The expansion can be avoided by direct numerical integration for the powder average[19]. We wish to outline an alternative analytical attack that can reduce the problem to that of simple tabulated functions, for instance, error functions[5].

Returning to (1.2) for $F(\underline{Q},t)$ the vector character of $\underline{Q}$ and $\underline{r}$ was glossed over. In general $\underline{Q}$ and $\underline{r}$ will not be parallel and, since $\underline{r}$ relates to a frame embedded in the molecule, different orientations of a molecule as encountered in a powder average will cause $F(\underline{Q},t)$ to be dependent on the directions of $\underline{Q}$. More than one normal mode will cause there to be several contributions to the displacement $\underline{r}$ which can be independently averaged, analogous to the case in (3.1) for independent lattice and internal normal modes. Each of these too will have a different relation to $\underline{Q}$. Our task is to average the final result over directions of the molecule or, equivalently, over directions of $\underline{Q}$. It can be simply shown (see the appendix of reference 5) that in the presence of several normal modes labelled $\nu,\lambda,\ldots.$ the result (1.15) for the response function of an oscillator can be generalized by recognising that independent modes give a product of the correlation functions $F(\underline{Q},t)$

$$S(\underline{Q},\omega) = \exp\{-2W(\underline{Q})\} \left[\prod_{\nu\neq\lambda} I_o(z_\nu)\right] I_{n_\lambda}(z_\lambda)e^{n_\lambda\omega_\lambda/2T} \delta(\omega-n_\lambda\omega_\lambda) \quad (4.1)$$

where we have considered the scattering to involve $n_\lambda$ quanta of the $\lambda^{th}$ mode with fundamental frequency $\omega_\lambda$. If the mode polarisation vector is $\underline{e}^\lambda$, a unit vector, and the associated mean square displacement is $1/2m_\lambda\omega_\lambda\sinh(\omega_\lambda/2T)$ then a mean square displacement tensor $\underline{\underline{B}}^\lambda$ can be constructed from the unit vectors and magnetude, thus.

$$\underline{\underline{B}}^\lambda = \underline{e}^\lambda\underline{e}^\lambda/2M_\lambda\omega_\lambda\sinh(\omega_\lambda/2T) \quad (4.2)$$

whence the scalars in the exponent of (1.2b) are generalised to $\underline{\underline{B}}^\lambda:\underline{Q}\,\underline{Q}$ which defines $z_\lambda$ for each mode $\lambda$

$$z_\lambda = \underline{\underline{B}}^\lambda:\underline{Q}\,\underline{Q} \quad (4.3)$$

Likewise $2W_\lambda(\underline{Q})$ is $\underline{\underline{B}}^\lambda:\underline{Q}\,\underline{Q}\cosh(\omega_\lambda/2T)$, generalising (1.10). To get the complete Debye-Waller factor we add the W factors

$$2W(\underline{Q}) = \sum_\lambda 2W_\lambda(\underline{Q}) = \{\sum_\lambda \underline{\underline{B}}^\lambda\cosh(\omega_\lambda/2T)\}: \underline{Q}\,\underline{Q} \equiv \underline{\underline{A}}: \underline{Q}\,\underline{Q} \quad (4.4)$$

thus also defining $\underline{\underline{A}}$. Double contraction is in the Dyadic notation, : .

Line intensities, J, are then of the form

$$J_{n_\lambda} \propto I_{n_\lambda}(z_\lambda)\exp\{-\underline{\underline{A}}: \underline{Q}\,\underline{Q}\} \quad (4.5)$$

It remains to powder-average these expressions over all directions. The conventional method of powder averaging involves expanding[17,18] the $\exp\{-\underline{\underline{A}} : \underline{Q}\,\underline{Q}\}$ term and expressing all tensors in a cartesian[17-19] representation. We suggest that both paths should be avoided :

(i) The quantity $\underline{\underline{A}} : \underline{Q}\,\underline{Q}$ is of order $Q^2/(2\bar{m}\omega)$ where $\bar{m}$ and $\omega$ are a characteristic mass and frequency for the system. At thehigh energy transfers required to observe these intra-molecular transition one often has a very small final energy for the neutron whence $Q^2/2m \sim n\omega$ with n being the number of quanta transferred to the system. This means that $\underline{\underline{A}} : \underline{Q}\,\underline{Q}$ is of order n, and hence $\exp\{-\underline{\underline{A}} : \underline{Q}\,\underline{Q}\}$ is not amenable to expansion.

(ii) In a system with final spherical symmetry a spherical representation of tensors is suggested. Recognition of the transformation properties of tensors yields immediate simplifications in scattering problems with such symmetry[20].

Under rotations the elements of a tensor $\underline{\underline{A}}$ can be analysed[21] into invariant subgroups, characterized by an angular momentum label $\ell$, within which they transform. Combinations of such group elements, characterized by the magnetic quantum number m, can be taken and these transform as the eigen functions $Y_m^{(\ell)}$ of the spherical operator $\hat{L}^2$. The matrix elements of the rotation operator $\underline{L}$, the Wigner matrices $D_{mm'}^{(\ell)}(\underline{\Theta})$, give the transformation[22] of such elements $A_m^{(\ell)}$ of A, ie.

$$A_m^{(\ell)} = \sum_{m'} D_{mm'}^{(\ell)}(\Theta) A_{m'}^{(\ell)} \qquad (4.6)$$

where $\underline{\Theta}$ indicates the three Euler angles specifying the rotation. Relevant examples of the invariant subgroup elements in the cylindrically symmetric case are

$$A_o^{(o)} = \mathrm{Tr}(\underline{\underline{A}})/\sqrt{3} \qquad (4.7)$$

(the trivial group of a rotationally invariant element) and

$$A_o^{(2)} = 2\{A_{zz} - A_\perp\}/\sqrt{6} \qquad (4.8)$$

where $A_\perp$ refers to the mean value in the perpendicular plane

$$A_\perp = \{A_{xx} + A_{yy}\}/2 \qquad (4.9)$$

Denote by lower case symbols tensors in the principal frames of the crystallites. We then have

$$A_m^{(\ell)} = \sum_{m'} D_{mm'}^{(\ell)}(\underline{\theta}) a_{m'}^{(\ell)} \qquad (4.10)$$

where $\underline{\Theta}$ represents the angles required to rotate a crystallite into the laboratory (LAB) frame. The average is now over $\underline{\Theta}$. Let us choose the LAB z axis to be along the scattering vector $\underline{Q}$. Relations (4.3) and (4.4) respectively reduce to

$$z_\nu = Q^2 B^\nu_{zz} \tag{4.11}$$

$$2W(\underline{Q}) = Q^2 A_{zz} \tag{4.12}$$

with the polar expressions[21] for the z-z components of the tensor being

$$A_{zz} = \frac{1}{\sqrt{3}} \{A^{(o)}_o + \sqrt{2} A^{(2)}_o\} \tag{4.13}$$

and

$$B^\nu_{zz} = \frac{1}{\sqrt{3}} \; B^{(o)^\nu}_o + \sqrt{2}\; B^{(2)^\nu}_o \tag{4.14}$$

The required expressions in terms of the orientation $\underline{\Theta}$ are:

$$A_{zz} = \frac{1}{\sqrt{3}} a^{(o)}_o + \sqrt{2}\, D^{(2)}_{oo}(\underline{\theta})\, a^{(2)}_o + \sqrt{2} \sum_{m' \neq o} D^{(2)}_{om}(\underline{\theta})\; a^{(2)}_{m'} \tag{4.15}$$

with[22] $D^{(2)}_{oo}(\underline{\Theta}) \equiv P_2(\cos\theta)$, $\theta$ being the polar angle of $\underline{\Theta}$ and $P_2$ the second Legendre polynomial. When $\underline{a}$ is cylindrical there is only one $\ell$=2 element, namely that with m=0, $a^{(2)}_o$. Likewise $B^\nu_{zz}$ is

$$B^\nu_{zz} = \frac{1}{\sqrt{3}} b^{(o)^\nu}_o + \sqrt{2} \sum_m D^{(2)}_{om'}(\underline{\theta})\, b^{(2)^\nu}_{m'} \tag{4.16}$$

where in general there are no requirements such that $b^{\nu(2)}_{m \neq o} = 0$. The tensors $\underline{b}^\nu$ are related to the mean square displacements of the modes $\nu$ in the principal molecular frame.

We now specialize to the case of cylindrical symmetry appropriate to $HCCo_3(CO)_9$. This is a molecule with a singly degenerate stretching mode and a doubly degenerate bending mode in the plane perpendicular to the first mode. It illustrates the use of intensity analysis of powder averaged lines. The tensors $\underline{b}^\nu$, are[23] such that a, their sum, has cylindrical symmetry.

They are

$$\underline{\underline{b}}^{\nu=1} = \begin{pmatrix} b_\parallel & & \\ & 0 & \\ & & 0 \end{pmatrix} \;;\; \underline{\underline{b}}^{\nu=2} = \begin{pmatrix} 0 & & \\ & b_\perp & \\ & & 0 \end{pmatrix} \;;\; \underline{\underline{b}}^{\nu=3} = \begin{pmatrix} 0 & & \\ & 0 & \\ & & b_\perp \end{pmatrix}$$

corresponding to stretch and degenerate bend respectively. The elements of $\underline{\underline{a}}$ are $a_{zz}=b_\parallel$ and $a_{xx}=a_{yy}=b_\perp$ whereupon (4.15) reduces to

$$A_{zz} = b_\perp + \Delta b\cos^2\theta \tag{4.17}$$

with $\Delta b=b_\parallel-b_\perp$. Likewise in (4.16) we have $b^{(2)}_{m\neq 0}=0$ for $\nu=1$ only, yielding

$$B^{\nu=1}_{zz} = b_\parallel \cos^2\theta \tag{4.18}$$

and for $\nu=2$ or 3 we get

$$B^{\nu=2,3}_{zz} = \frac{1}{\sqrt{3}}\left\{\frac{b_\perp}{\sqrt{3}} - P_2(\cos b\theta)\frac{b_\perp}{\sqrt{3}} \pm \frac{b_\perp}{\sqrt{2}}\left[D^{(2)}_{02}(\underline{\Theta}) + D^{(2)}_{0-2}(\underline{\Theta})\right]\right\} \tag{4.19a}$$

$$\equiv \frac{b_\perp}{2}\{1-\cos^2\theta\} \pm \frac{b_\perp}{2}\cos 2\phi\sin^2\theta$$

The quantity $B_{zz}$ will be used in (18) for $y_\nu$. One can prove[5] the intuitively obvious step that for degenerate modes the axes defining $\nu=2$ and 3 can be freely chosen, the natural ones being parallel and perpendicular to the component of $\underline{Q}$ in their plane, ie. $\phi=0$ and $\pi/2$.

This yields

$$B^{\nu=2}_{zz} = b_\perp\,(1-\cos^2\theta) \tag{4.19b}$$

and $B^{\nu=3}_{zz}=0$. This mode can subsequently be discarded since the corresponding z is zero and $I_n(z)$ is unity (n=o) or vanishes (n≠o).

We shall also limit ourselves to small values of z, the argument of the Bessel function $I_n$. This amounts to the limit of $T \ll \omega_\nu$ which is nearly always the case, whereupon the $[\sinh(\omega_\nu/2T)]^{-1}$ in (4.2) becomes exponentially small. We then expand the Bessel function for small values of its argument:

$$I_n(z) \sim (z/2)^n/n! \qquad (4.20)$$

We redefine b by extracting the $(\sinh)^{-1}$ from it so that is proportional to $1/(2m_\nu\omega_\nu)$, which is the form of a mean square displacement. This is consistent with the use of b in A since, in the same limit, $\coth(\omega_\nu/2T) \to 1$. The exponential remaining from the $\sinh^{-1}$ cancels with the detailed balance term in (4.1) leaving a factor of $2^n$. The resulting expressions for the line intensity upon averaging over the polar angle θ are then

$$J^{\nu=1}_{n=1} = Q^2 b_\parallel \exp\{-Q^2 b_\perp\} \int_o^1 dx\ x^2 \exp\{-Q^2 \Delta b x^2\} \qquad (4.21)$$

$$J^{\nu=2}_{n=1} = Q^2 b_\perp \exp\{-Q^2 b_\perp\} \int_o^1 dx(1-x^2)\exp\{-Q^2 \Delta b x^2\} \qquad (4.22)$$

where the variable change $x=\cos\theta$ has been made.

In reference 5 detailed formulae for the powder averaged intensities of fundamental, harmonic and combination modes can be found. In particular the variation of the ratios of their intensities with Q is useful since absolute measurements are not required. Here we simply examine the two fundamental intensities a little further to illustrate the method.

The results for J are is obviously in terms of tabulated functions; the error function for $\Delta b \geq 0$ or the Dawson integral for $\Delta b \leq 0$. Let us denote $\int_o^1 dx \exp(-Cx^2)$ by F(C), the relationship of which to the error function or Dawson integral is given in reference 5. We then have (with $C=Q^2\Delta b$):

$$J^{\nu=1}_{n=1} = Q^2 b_\parallel \exp(-Q^2 b_\perp)\ \{F(C) - \exp(-C)\}/2C \qquad (4.23)$$

and similarly

$$J^{\nu=2}_{n=1} = \tfrac{1}{2}\left(\frac{b_\perp}{\Delta b}\right) \exp\left(-Q^2 b_\perp\right)\ \{(2C-1)F(C)+\exp(-C)\} \qquad (4.24)$$

These formulae can be checked in the limit of isotropy, ie. $\Delta b=0$, $b_\parallel=b_\perp=b$. They now each yield

$$J_{n=1} = bQ^2/3 \exp(-Q^2 b) \qquad (4.25)$$

The additional factor of 1/3 comes from the average of the $\cos^2\theta$ term from the expansion of $I_1(y)$. The { } expressions are proportional to $Q^2$ as $Q \to 0$ by the same analysis.

The limit of large $Q^2$, ie. large $Q^2\Delta b$ can also be easily investigated by examining $F(Q^2\Delta b)$ in (4.23) and (4.24).

For large positive $Q^2\Delta b$ the line intensities are:

$$J_{n=1}^{\nu=1} = \frac{\sqrt{\pi}}{4}\left(\frac{b_\parallel}{\Delta b}\right) \exp\{-Q^2 b_\perp\}/Q\sqrt{\Delta b} \qquad (4.26)$$

$$J_{n=1}^{\nu=2} = \frac{\sqrt{\pi}}{2}\,\frac{b_\perp}{\Delta b} \exp\{-Q^2 b_\perp\}/Q\sqrt{\Delta b} \qquad (4.27)$$

The intensity of the mode $\nu=2$ is greater by an amount $2(Q^2\Delta b)b_\perp/b_\parallel$ than the intensity of the mode $\nu=1$.

For large negative $C=Q^2\Delta b$ we exploit the asymptotic behaviour of the Dawson integral. The resultant line intensities are

$$J_{n=1}^{\nu=1} = \tfrac{1}{2}\,\frac{b_\parallel}{|\Delta b|} \exp\{-Q^2 b_\parallel\} \qquad (4.28)$$

$$J_{n=1}^{\nu=2} = \tfrac{1}{2}\,\frac{b_\perp}{|\Delta b|} \exp\{-Q^2 b_\parallel\}/Q^2|\Delta b| \qquad (4.29)$$

Now the $\nu=2$ line is less intense by a factor $b_\perp/(b_\parallel Q^2\Delta b)$.

## The Apparent Debye-Waller Factor

However we have argued above that $Q^2 b$ is of order n by experimental construction and the above two limits are difficult to obtain except perhaps for the large Q limit obtaining at large n. On the contrary, it is observed that the different lines and their harmonics appear to have different Debye-Waller factors, not simply a common $\exp\{-Q^2 b_\perp\}$ or $\exp\{-Q^2 b_\parallel\}$ as the above limits suggest. The terms denoted by { } in (4.23) and (4.24) play a vital rôle.

The Debye-Waller factor is extracted[24] by casting the intensity in the form

$$J_n^\nu \sim (b^\nu Q^2)^n \exp\{-Q^2\alpha_n^\nu\} \qquad (4.30)$$

By inspection of (4.23) expanding out the { } for $Q^2\Delta b$ not too large one gets, taking the $\nu=1$ mode as an example :

$$\alpha_1^1 = (b_\perp + 3/5\Delta b) - \frac{(Q^2\Delta b)}{7}\Delta b + \ldots \qquad (4.31)$$

The first term contributes to the usual definition of the exponent of the Debye-Waller factor of the fundamental.

For the first harmonic one gets, using a result[5] equivalent to (4.23).

$$\alpha_2^1 = (b_\perp + 5/7\Delta b) + \ldots \qquad (4.32)$$

The first term differs from the first term of (4.31). This corresponds to experience that Debye-Waller factors apparently change from mode to mode. However in the light of the previous comments that the values of $Q^2b$ are unsuited to expansion, the reabsorption of pre-exponential terms into the exponent as logarithms and their subsequent expansion within the exponent may be unwise.

## CONCLUSIONS

We have given exact expressions for the mode intensities resulting from neutron scattering from a powder containing anisotropic oscillators. Such expressions are vital to understanding highly inelastic experiments since one cannot in general expand the exponential expressions for line intensities involving mean square displacement tensors and scattering vectors. The systematic attack on the problem is made by the use of spherical tensors.

APPENDIX : Asymptotic Analysis

(a) The Harmonic lattice

One proceeds with evaluating the integral over time in (1.9) with (1.11) by observing that the point of expansion is about a saddle point[7], not necessarily at t=0. Taking the time-dependent part of the exponent $G_L(t)$ to be f(t) we have to perform the integration

$$\int dt\ \exp(\kappa^2 f(t)) \tag{A.1}$$

The saddle point condition df/dt=0 yields from its real and imaginary parts the two conditions

$$\int du\ Z(u)\ n(u)\ \sin\ (ut_R)\ e^{-ut_I} = 0 \tag{A.2}$$

$$\int du\ Z(u)\ n(u)\ \cos\ (ut_R)\ e^{-ut_I} + u/\kappa^2 = 0 \tag{A.3}$$

where $t_R$ and $t_I$ represent the real and imaginary parts of t. A saddle point is clearly at $t_R = 0$ and $y_I = t_I^*(u)$, the solution of

$$\int du' Z(u')n(u')e^{-u't_I} = -u/\kappa^2 \tag{A.4}$$

We have not been able to find any other saddle points and will assume henceforth that there are no other significant contributions. The line $t_I = t_I^*(u)$ is locally the path of steepest descent at $t = 0 + it_I^*$ whereupon we deform the contour in A.1 to get

$$\exp\{Q^2 f(it_I^*(u)\}\int_{-\infty}^{\infty} dt_R\ \exp\{-\frac{\kappa^2}{2} A\ t_R^{\ 2}\} \tag{A.5}$$

where for S(Q,u) we get

$$S(Q,u) = (2/\pi\kappa^2 A)^{\frac{1}{2}}\ \exp\{ut_I^* + \kappa^2\int \frac{du'}{u'} Z(u')n(u')e^{t_I^*(u)u'}\}e^{-2W(Q)} \tag{A.6}$$

with $A \equiv d^2f/dt_R^{\ 2}$ being

$$A(u) = \int du'u'Z(u')n(u')e^{t_I^*(u)u'} \tag{A.7}$$

This is not a very useful form since the frequency dependence u is not only explicit but also implicit via $t_I^*(u)$ and $A(u)$. One can however, examine $S(Q,u)$ for its maximum. Recognising that, in its essential u dependence, (A.6) can be written:

$$S(Q,u) \sim \frac{1}{\sqrt{A}} \exp\{\kappa^2 g(u)\} \tag{A.8}$$

thus defining $g(u)$. Denoting d/du by primes, the maximum of S occurs at

$$A\kappa^2 g' - A'/2 = 0 \tag{A.9}$$

In the limit $\kappa^2 \to \infty$, where we neglect the A' term, explicit differentiation of $g(u)$ yields

$$g'(u) \equiv - t_I^{*'} \; t_I^* \; A = 0 \tag{A.10}$$

This can only be satisfied for either $t_I^* = 0$ or $t_I^{*'} = 0$ since A is positive definite. By differentiating equation (A.4) for $t_I^*$ an expression for $t_I^{*'}$ is obtained, namely

$$t_I^{*'} = - \kappa^2/A \neq 0 \tag{A.11}$$

whence there only remains the possibility that $t_I^* = 0$ for $u = u_{max}$. Inspection of (A.4) shows in turn that $t_I^* = 0$ implies a frequency at the maximum, $u_{max}$, given by

$$u_{max} = \kappa^2 \int_{-\infty}^{\infty} du'Z(u')n(u') = Q^2 \tag{A.12}$$

(using the identity on the Bose factor $n(-u') = -(1 + n(u'))$. This is the free particle result $\omega_{max} = Q^2/2m$ for the maximum. The scattering law (A.6) can be turned into a Gaussian form consistent with this equality between the first moment and the maximum by expanding the exponent about $u_{max}$. The u dependent part is $\frac{1}{2}(u-u_{max})^2/\kappa^2 A$. The width of the Gaussian, $\Gamma$, in contrast to the free particle result is then (2.3) obtained by setting $t_I = 0$ in (A.7) for A.

To summarise, the asymptotic form of the scattering law can be derived by a steepest descent analysis. The frequency of the maximum corresponds to a saddle point at $t = 0$ and a Gaussian approximate form can be constructed about this point as its maximum. In general though the result is not quite Gaussian as a result of the saddle point moving away from the origin of t-space as one moves in frequency away from the maximum. These results are in sharp contrast to the results of a saddle point analysis for the harmonic oscillator that we now treat.

(b) The Harmonic Oscillator

The time integrations (1.9) and 1.10) can be re-expressed as:

$$S(Q,u) = \frac{1}{\omega_o} \exp\{-2W(Q)\} \int_{-\infty}^{\infty} dt \exp\{zf(t)\} \tag{A.13}$$

with W(Q) and z as in (1.17) and (1.16) and

$$f(t) = iut/z + \tfrac{1}{2}(e^{it+\beta} + e^{-it-\beta}) \tag{A.14}$$

with $1/2T^* = \beta$. Proceeding as in the phonon case the saddle points are located by the simultaneous equations from $df/dt = 0$:

$$(X - 1/X) \cos t_R + u/z = 0 \tag{A.15}$$

$$(X + 1/X) \sin t_R = 0 \tag{A.16}$$

with $X = \exp(\beta - t_I)$ and $t_R$, $t_I$ the real and imaginary parts of t. Since $X \geqslant 0$ the roots are given by

$$t_R = n\pi \tag{A.17}$$

$$X^2 + (-1)^n (u/z) X - 1 = 0 \tag{A.18}$$

Thus there is not one but an infinity of saddle points, differing in character according to whether n is even or odd. It turns out[3] that the odd saddle points have exponentially small contributions. We treat only n even here.

The line of steepest descent is $t_I = \bar{t}_I$ whereupon a contribution to (A.13), a bar denoting evaluation at a saddle point, is

$$\exp(z\bar{f}) \quad \int dt_R \exp\{\tfrac{1}{2}(\frac{\partial^2\bar{f}_R}{\partial t_R^2})\, z(t_R - n\pi)^2\} \tag{A.19}$$

where

$$\begin{aligned}
\bar{f} &= \tfrac{1}{2}(X + 1/X) - \tfrac{1}{2}\, u\bar{t}_I + in\pi u/z \\
X &= \surd\{(u/z)^2 + 1\} - u/z \\
\bar{t}_I &= \beta - \ell n\, X \\
\partial^2\bar{f}_R/\partial t_R^2 &= -\tfrac{1}{2}(X + 1/X) < 0
\end{aligned} \tag{A.20}$$

Performing the Gaussian integral (A.19) about the saddle point we obtain:

$$\exp(in\pi u)(\pi/z(X + 1/X))^{\frac{1}{2}} \exp\{\tfrac{1}{2}(X + 1/X)z - u\bar{t}_I\} \tag{A.21}$$

The coefficients of z in the exponentials have the limiting values, remembering that u is also large:

$$\tfrac{1}{2}(X + 1/X) \sim u/z \text{ for } z \ll u \tag{A.22a}$$

$$\tfrac{1}{2}(X + 1/X) \sim 1 \text{ for } z \gg u \tag{A.22b}$$

Therefore in both limits of z (A.21) is exponentially large.

We now sum the contributions from all the even saddle points, which are on the line $t_I = \beta - \ell nX$. The result of summing all the phase factors from (A.21) is to recover sharp lines corresponding to the creation ($\ell > 0$) or annihilation ($\ell < 0$) of $\ell$ quanta:

$$\sum_{p=-\infty}^{+\infty} \exp(2\pi ipu) = \delta(u-\ell) \tag{A.23}$$

For S(z,u) we now have

$$S(z,u) = \exp(-2W(z)) \sum_{\ell} \left\{\frac{\pi}{z(X + 1/X)}\right\}^{\frac{1}{2}} \exp\{\tfrac{1}{2}(X + 1/X)z - u\bar{t}_I\}\, \delta(u-\ell) \tag{A.24}$$

In the limits $z \gg u$ and $z \ll u$ we can also make contact with the exact result (1.15):

$z >> u$ ($\equiv \ell$): using (A.22) we get

$$S(z,u) = e^{-2W(z)} \frac{e^z}{\sqrt{2\pi z}} e^{-u/2T^*} \delta(u-\ell) \tag{A.25}$$

We find this agrees with the exact result by noting that the asymptotic form of $I_n(z)$ is

$$I_n(z) \sim e^z/\sqrt{2\pi z} \quad \text{for } z >> n \tag{A.26}$$

<u>$z << u$:</u> Using (A.20) for X we get for S

$$S(z,u) = \exp(-2W(z)) \left(\frac{2\pi}{u}\right)^{\frac{1}{2}} \left(\frac{1}{2}\frac{z}{u}\right)^u \exp(u - u/2T^*) \delta(u-\ell) \tag{A.27}$$

Terms can be gathered in this expression to give (setting $u = \ell$ for the $\ell^{th}$ line)

$$(z/2)^\ell/\sqrt{\ell}\,(\ell/e)^\ell \sim (z/2)^\ell/\ell! \sim I_\ell(z) \tag{A.28}$$

employing Stirlings formula and the small z expansion for $I_n(z)$, valid for $z << \ell$ ($\equiv u$).

In summary we have reobtained the exact result of a series of sharp lines with Bessel function weights by an analysis involving an infinite series of saddle points appropriately deformed off the real axis. Equally it can be seen that at high temperatures and Q the envelope of these lines has a Gaussian (free particle) character.

REFERENCES


1. Howard J and Waddington T C: Advances in Infrared and Raman Spectroscopy, Chapter 3, Vol.7. Clark R J H and Hester R E (eds). Heyden and Sons, 1980.

2. Springer T: Hydrogen in Metals I, Chapter 4. Alefeld G and Völkl J (eds). Berlin: Springer-Verlag, 1978.

3. Gunn J M F and Warner M : Z Phys B (in the press)

4. Sjölander A, Art. für Fysik 14, 315 (1958).

5. Tomkinson J, Warner M and Taylor A D : Mol. Phys 51, 381 (1984)

6. Lovesey S W, "Theory of Neutron Scattering from Condensed Matter", Vol.I. Oxford: Clarendon Press, 1984.

7. Egelstaff P A and Schofield P, Nucl. Sci. Eng. 12, 260 (1962).

8. Schofield P and Hassitt A, Progress in Nuclear Energy, Series I, Vol 3, page 194; London: Pergamon Press, 1959.

9. Warner M, Lovesey S W and Smith J, Z.f.Phys. B51, 109-126 (1983).

10. Wick G C, Phys. Rev. 94, 1228 (1954).

11. Griffin A and Jobic H: J. Chem. Phys. 75, 5940 (1981).

12. Jobic H, Ghosh R E and Rencuprez A: J Chem. Phys. 75, 4025 (1981).

13. Vibrational spectroscopy on a pulsed neutron source employs relatively high momentum transfers, for example: Boland B C, Mildner D F R, Stirling G C, Bunce L J, Sinclair R N and Windsor C G:Nucl. Inst. Meth. 154, 349 (1978).

14. The limitation to working at relatively high momentum transfer Q( .5Å$^{-1}$) when performing neutron spectroscopy is analysed, for example: Brugger R M, Strong K A and Grant D M: p.323 Neutron Inelastic Scattering, Vol.II. Vienna: International Atomic Energy Agency, 1968.

15. Placzek G: Phys. Rev. 86 377 (1952), 93 895 (1954), 105 1240 (1957).

16. Dolling G and Powell B M: Proc. Roy. Soc. Lond. A319, 209 (1970).

17. Reynolds P.A., Kjems J.K. and White J.W. J. Chem. Phys 56, 2972 (1972).

18. Thomas M.W. and Ghosh R.E. Mol. Phy. 29 (1975).

19. Wright C.J J Chem. Soc. Faraday II, 73, 1497 (1976).

20. Warner M. Colloid and Polymer Science, 261, 508 (1983).

21. Berne B.J. and Pecora R. ''Dynamic Light Scattering,'' Chichester, Wiley & Sons (1976).

22. Davydov A.S. ''Quantum Mechanics,'' Oxford, Pergamon (1965).

23. Howard J., Tomkinson J., Eckert J., Goldstone J. and Taylor A.D. Rutherford Appleton Laboratory Report RL-83-032 and Mol. Phys. (submitted).

24. Howard J. and Tomkinson J. Chem. Phys. Letts. 98 239 (1983)

INDEX